REVERSE ENGINEERING OF THE DEEP-CYCLE AUTOMOTIVE BATTERY

REVERSE ENGINEERING OF THE DEEP-CYCLE AUTOMOTIVE BATTERY

UNDERSTANDING THE

DEEP-CYCLE BATTERY

IN THE TWENTY-FIRST CENTURY

FRANK EARL

REVERSE ENGINEERING OF THE DEEP-CYCLE AUTOMOTIVE BATTERY
UNDERSTANDING THE DEEP-CYCLE BATTERY IN THE TWENTY-FIRST CENTURY

iUniverse books may be ordered through booksellers or by contacting:

iUniverse
1663 Liberty Drive
Bloomington, IN 47403
www.iuniverse.com
844-349-9409

ISBN: 978-1-5320-4254-6 (sc)
ISBN: 978-1-5320-4255-3 (e)

Library of Congress Control Number: 2018901579

Print information available on the last page.

iUniverse rev. date: 05/20/2021

CONTENTS

PREFACE

THE ALL-ELECTRIC VEHICLE IS already in place to curve our ferocious appetite for gasoline from crude oil. However; those in the field of automotive battery technology say that, they haven't found a perfect chemical composition to a longer-lasting battery for it. The big question is that, why do they need one when we could replenish our batteries' chemical energy with electrical energy as well.

I found if we could add energy back to our batteries while taking it out, then we could increase their efficiency and overall capacity. Without finding that, perfect chemical composition with a longer chemical reaction for them. Thus; we could rely more heavily on, the electric motor than the gasoline engine to propel our basic form of transportation as well.

Most people think if it was possible to add energy back to our batteries while taking it out, then why those in the field of automotive and automotive battery technology haven't thought of it before. It's not because they haven't, they just have swept it under the rug as nothing more than wishful thinking. How could anyone dismiss this possibility if they knew anything about the mechanics behind our batteries?

Taking an in depth analysis of the mechanics behind, the charging cycle of a lead acid automotive starting battery. I found it's hard to believe it'll be nothing more than wishful thinking to add energy back to it while taking energy out. Given that; its plates in each cell, they're connected in series-parallel with one another by their straps, tab connectors and the electrolyte solution in each cell.

When the battery is charged, electrons could flow from a strap to a tab connector on one plate. And then, flow across that plate through the electrolyte solution to an adjacent plate. Without flowing to and from, the same tab connector on the same plate. Thus; one plate, it's simultaneously charged and discharged to another by way of the electrolyte solution in each cell.

Since each cell in the battery is connected in series with its terminal connections, then its plates in one cell. They're simultaneously charged and discharged to its plates in the next cell by way of an electrical bus. Thus; we don't have to, stop adding electrons to its plates in one cell in order to add them to its plates in the next cell; although, they're separated by a non-conducting material celled plastic as well.

Wherein; it's easy to see why, it would be more than wishful thinking to add energy back to the battery while taking it out. If energy doesn't have to go in, the same way it has to come out in order to charge the battery plates or for them to be discharged as well. Dismissing the idea of adding energy back to the battery while taking it out, it sounds like nothing more than a campaign of disinformation.

I've complied and compared data from books written by those in the field of automotive battery technology. To come to a logical conclusion that, it's scientifically possible to add energy back to the battery while taking it out. If we change its contemporary design, so, energy could enter it on one terminal and exit it on another terminal at the same time as well.

Looking at the mechanical makeup of the battery, it's nothing more than a collection of 2-volt batteries. That is connected in series with one another and housed together in a plastic case to make up the whole battery as we know of today. When it's charged as a whole, we're simultaneously adding and subtracting energy from one battery while simultaneously adding and subtracting energy from another as well.

It seems like there's a disconnection between how those in the field of automotive battery technology design the battery and how, some thinks its cycling processes will be carried out within the realm of physic as well. Are we focusing on, the right science when it comes to the energy input and output process of the battery? I found the answer is no because we should be focusing on mechanics and not physics

The mechanical make up of the battery and mechanics behind its charging cycle. It points toward the laws of mechanics and not the laws of physics when it comes to its energy input and output process. It makes me wonder why most battery experts turn their focus toward the laws of physics when it comes to a conversation about the energy input and output process of the battery as well.

While earning my Associate Degree in Science in the field of heating, cooling and refrigeration, I learned how electrical energy will flow to our load sources from a power source. Based on, the connections made between them; such as, a series, a parallel or a series-parallel connection. These types of connections will determine how, energy will flow between a battery plates and cells as well.

The energy input and output process of a battery isn't determined by the laws of physics; but, it's determined by the laws of motion and the effect of forces on bodies. If anyone thinks otherwise, it's nothing more than an illusion created by energy having to go in the battery. The same way it has to come out in order to charge the battery plates or for them to be discharged to a load source as well.

Those in the field of automotive battery technology say, we've to stop the discharging cycle of the battery in order to start its charging cycle because of physic. This is nothing more than a campaign of disinformation as well; given that, we could charge its plates in each cell without having to stop charging them in one cell in order to charge them in the next cell; although, they're separated by plastic as well.

Then charging the battery plates while discharging them to a load source, it shouldn't be any different than charging the battery plates in one cell while discharging them to its plates in the next cell as well. How difficult it would be for those in the field of automotive battery technology to design the battery; so, its plates could be charged while being discharged to the load source as well.

Based on, the mechanics behind the charging cycle of the battery. The idea of adding energy back to it while taking energy out, it becomes more than wishful thinking if energy doesn't have to go in the same way it has to come out. Then the possibility of increasing, the efficiency and overall capacity of the battery by adding energy back to it while taking energy out becomes possible as well.

Basically; it's not that those in the field of automotive battery technology, they haven't thought of adding energy back to the battery while taking energy out. As another option to increase its efficiency and overall capacity without finding that, perfect chemical composition with a longer-lasting chemical reaction for it. They'll turn around and use the laws of physics to discredit the process as well.

I found using the laws of physics to discredit the process of adding energy back to the battery while taking it out. It's nothing more than a stall tactic to delay the electric vehicle from becoming our primary source of transportation since gasoline from crude oil has become a vital part of our global economy for more than a century as well.

Some folks believe switching from relying heavily on, the gasoline engine to relying more heavily on the electric motor to propel our vehicles could have an adverse effect on our global economy as well. Maybe; this is the reason and not the lack of battery technology, why the electric vehicle hasn't become our primary source of transportation for more than a century as well.

ACKNOWLEDGEMENTS

INNOVATION COMES FROM THOSE who dare to think outside the box and dream beyond the stars to improve the existence of humankind. Contrary to popular belief, scientists don't know everything; thus, most innovations or the improvement of existing innovations come from the ideas of ordinary people. So; don't think, you've to be a scientist or a genius because everyday people do have ingenious solutions as well.

My research on the lead acid automotive starting battery was prompted by Mr. Paul Lutz, whom I met on a business trip. He told me, "what if we could charge our batteries while discharging them." "Then we wouldn't need to find a chemical composition with a longer-lasting chemical reaction for them." It made sense because we could replenish their chemical energy with electrical energy as well.

I thought if we could add energy back to our batteries while taking it out. Then we could increase the overall travel ranges of our vehicles on battery power. Without find that, perfect chemical composition with a longer-lasting chemical reaction for their batteries as well. My efforts was to find away to add energy back to our batteries while taking it out without being impeded by their designs.

In my first book "Miracle Auto Battery", I tried to explain why it would be possible to add energy back to our batteries while taking it out; but, it didn't resonate with my readers. After re-evaluating my first book, I decided to write another while taking a different approach. There aren't many books written about the mechanical structure of a battery and how it'll play a role in its cycling processes as well.

I decided to write this book in layman terms to help the general public to understand. The mechanical structure of a battery is just as important as its chemical structure when it comes to new battery technology. I like to give thanks to those who supported me; my wife (Felita), Derrick Watkins, a master electrician: Prof. Gary Hayward and Jon Eddy; my HVACR technician instructors as well.

I also I like to acknowledge those battery experts in Australia because their ideas gave me courage to write this book as well. Given that; in America, battery experts have no interest in adding energy back to our batteries while taking it out and it seems like they don't see the benefits in it as well. On the other hand; some battery experts in Australia, they already see the benefits in it.

I've heard many different opinions why it'll be impossible to add energy back to our batteries while taking it out and it'll be no benefits in it because of the laws of physics as well. The laws of physics, they're vast and complex. And sometimes, they're misinterpreted or miss-applied by some; so, I thought it would be helpful to incorporate some of those opinions into my book because they inspired me to write it as well.

SUMMARY OF CONTENTS

Chapter 1; Observation of an Auto Battery, in general what does the average person knows about the cycling processes of an automotive starting battery. They probably know it could be charged or discharged and we've to stop its discharging cycle one in order to start its charging cycle. It has to be charged on the same terminal connection it was discharged on. The average person doesn't know too much.

Chapter 2; Research on an Auto Battery, Allesandro Volta made the first primary cell battery or the first non-rechargeable battery in 1796. A little more than sixty years later, then Gaston Plante' made the first secondary cell battery or the first lead acid automotive battery in 1859; otherwise, known as a rechargeable battery. However; crude oil, it was discovered around that same period of time. And then, people started experimenting with the electric car in the late 1890s.

Chapter 3; Mechanical Aspects of an Auto Battery, to ascertain whether it's due to the laws of physics or it's due to the laws of mechanics why a secondary cell battery can't be charged while being discharged. Wherein; it seems suspect that, it's due to the laws of physics because we could charge or discharging the same battery. However; we've to charge it on the same terminal connection that, it was discharged on; so, we've to explore its mechanical structure and not the laws of physics.

Chapter 4; Chemical Aspects of an Auto Battery, if we understand the paradox of the oxymoron behind the discharging cycle of a secondary cell battery. Then we might realize a chemical solution, it might not be the best solution to increase its efficiency and overall capacity as well. Maybe; sending, an electrical charge back to it while using it might be the best solution since adding electrical energy back to it restores its chemical energy as well.

Chapter 5; Electrical Aspects of an Auto Battery, to qualify as an electrical circuit, there must be a power source, a conducting source to carry the current and an electrical device to use it. Thus; a lead acid automotive starting battery, it has all three attributes of an electrical circuit during its discharging cycle. Since its plates, are the power source and its terminals, straps, tab connectors and the electrolyte solution in each, are the conducting sources to carry the current to a load source.

Chapter 6; Lack of Information about an Auto Battery, I didn't know voltage could flow through one terminal of a battery and out the other; but, current had to enter or exit the same terminal connection of the battery in order to cycle it. Basically; I knew, what the average person knows about an automotive starting battery when it comes to installing it in their vehicle. I had to connect a red cable to its positive terminal and a black cable to its negative terminal connection to properly install it in my vehicle.

Chapter 7; Misconceptions about an Auto Battery, since energy has to go in a battery the same way that, energy has to come out in order to cycle it. Then it leads to misconceptions about what we could or couldn't do with it because of the laws of physics. Then our perceptions why it can't be charged while being discharged, they're nothing more than an illusion created by the contemporary design of the battery.

Chapter 8; Focusing on the Wrong Science, based on the mechanical make up of a lead acid automotive starting battery as we know of today. It seems like we're focusing on the wrong science when it comes to its energy input and output process. It seems like there's a disconnection between how we think its cycling processes will be carried out and how, they're actually carried out within the realm of physic as well.

Chapter 9; The Laws of Physics; if we observe, how the mechanical structure of a lead acid automotive starting battery will play a role in its cycling processes. Then we'll find it's ridiculous to assume it can't be charged while being discharged because of the laws of entropy, inertia or the laws of conservation of energy since energy as well. Energy is simultaneously added and subtracted from the battery plates in each cell and yet, they're still charged; although, energy is lost to heat during the process.

Chapter 10; The Wrong Mechanics in Play, why a lead acid automotive starting battery can't be charged while being discharged, it's due to mechanics. Since energy has to go in it the same way energy has to come out as well; therefore, the wrong mechanics are in play to charge it while discharging it. Given that; it has to be charged on the same terminal connection, it was discharged on. Wherein; like charges, they can't simultaneously enter and exit the same opening while flowing in the opposite direction of one another.

Chapter 11; Erroneous Beliefs about an Auto Battery, it was non-rechargeable batteries before there were rechargeable batteries. When the first rechargeable battery was made in 1859, people were saying the same thing back then as people are saying today. About charging it while discharging it; wherein, it can't be done because of the laws of physics as well. It's nothing more than an erroneous belief created by the lack of information about the process.

Chapter 12; The Mechanics of the Current, we must understand a parallel circuit connection made between a battery and a load source. It doesn't work the same as a parallel circuit connection made between two load sources because of the different circumstances involved. We've to understand those different circumstances if we're going to comprehend why, a battery can't be charged while being discharged. Since current, it flows to a load source different than it does a battery.

Chapter 13; Focusing on the Right Science, if we're focusing on, the laws of entropy, inertia or the laws of conservation of energy to determine. If it's possible or not to add energy back to a secondary cell battery while taking it out, then we're focusing on the wrong science to ascertain that information. Given that; those laws of physics, they don't determine how energy is added or subtracted from the battery; but, the laws of motion and the effect of forces on bodies does.

Chapter 14; The Laws of Mechanics, how energy is added or subtracted from a battery plates in each cell and to the plates in the next cell. It's done by mechanics because they determine how its charging cycle will be carried out. If the battery plates in each cell, they're connected in series-parallel with one another. And each cell, it's connected in series with its terminal connections as well. Then we don't have to stop charging its plates in one cell in order to charge them in the next cell as well.

Chapter 15; The Right Mechanics in Play, in order to charge a battery plates in each cell without having to stop charging them in one cell in to charge them in the next cell, then the right mechanics has to be in place. Since the battery plates in each cell, they've only one tab connector each and each cell will be separated by a non-conducting material. Then energy has to flow in one direction among its plates in each cell by connecting them together series-parallel and flow in one direction to its plates in the next cell by connecting them together in series.

Chapter 16; A Run Capacitor, it could be charged or discharged on either terminal connection because it has separate terminals and plates for charging while discharging. Its plates are separated by a non-conducting material called dielectric. Using the mechanics of the alternating current, a run capacitor could be simultaneously charged and discharged within an Ac circuit. Its innovated design gives insight into why it'll possible to simultaneously charge and discharge a secondary cell battery as well.

Chapter 17; A Power Transformer, it has a primary and a secondary winding that is interconnected by an iron core. The iron core acts like a medium between the primary and the secondary winding. Wherein; the iron core, it's simultaneously charged and discharged to the secondary winding. The innovated design of a power transformer, it shows why. It'll be possible to simultaneously charge and discharge a secondary cell battery if it had separate terminals for charging and discharging.

Chapter 18; Restricted by Its Design, how a run capacitor and a power transformer are constructed, it shows we're not bound or restricted by the laws of physics when it comes to the energy input and output process of a secondary cell battery. We're only restricted by the design of the battery when it comes to a simultaneous cycling process of it; in which, its contemporary design is the reason. It can't be charged while being discharged because energy has to go in it the same way energy has to come out as well.

Chapter 19; A prototype of an Auto Battery, to prove it's due to the wrong mechanics in play to simultaneously charge and discharge a secondary cell battery. I had to venture outside the status quo to prove my theory. Given that; all of our secondary cell batteries, they've only one opening for energy to enter or to exit them. So; I had to figure out, how to build a prototype of a lead acid automotive starting battery; so, it could have two openings for energy to enter and to exit the battery at the same time as well.

Chapter 20; Analysis of a Prototype, adding an additional terminal connection to a cell with a terminal connection already in it, the terminal had to have its own strap and tab connectors leading to the same plates as the other terminal connection in that cell as well. I found voltage and current would flow from the strap connecting the plates together in parallel with the charging terminal. It'll be in reverse of voltage and current will flow to the strap connecting the plates to together in parallel with the discharging terminal.

Chapter 21; Facts versus Perceptions, those in the field of automotive battery technology, they've been designing our secondary cell batteries with only one opening for energy to enter or to exit them for more than a century. Conversely; they would turn around and say that, it's impossible to add energy back to our batteries while taking it out due to the laws of physics; although, energy has to go in them the same way it has to come out in order to cycle them as well.

Chapter 22; Reverse Engineering of an Auto Battery, we could use reverse engineering on the charging cycle of a secondary cell battery. If it had separate terminals for charging and discharging in the same cell and those terminals, they had their straps and tab connectors leading to the same plates in that cell as well. A surplus of electrons would be created in the cell with the charge and the discharging terminal in it. To reverse, the current flow back to a load source and the other cells in the battery as well.

Chapter 23; The Laws of Series-Parallel Circuitry, I found if a battery had separate terminals for charging and discharging in the same cell. And those terminals, they had their straps and tab connectors leading to the same plates in that cell as well. When there's a charge and a load source on the battery at the same time. It'll create a series-parallel circuit connection between them; in which, it'll allow the battery to be charged while being discharged to the load source as well.

Chapter 24; Is it Conceivable or Farfetched, after publishing my first book I learned that, there are others who believe it's possible to charge our secondary cell batteries while discharging them. If we change their contemporary designs, so, electrons could enter and exit them at the same time. It verified my experiments I carried out on a lead acid automotive starting battery; thus, I came up with the idea of charging a battery while discharging it with only three terminals instead of four as well-.

Chapter 25; Energy Input for an Auto Battery, we've some electric vehicles could get over three hundred miles on a single battery charge. The question is that, where their energy input going to come from when it's time to charge their batteries on the go. Find a charging station that may or may not exist in most cities and if so, they're few and far in between. Then the only alternative isn't to travel far from home; in which, it defeats the purpose of buying an all-electric vehicle to save on gas when travelling far from home.

Chapter 26; Energy Output from an Auto Battery, I discovered there're many benefits in a simultaneous cycling process of our secondary cell batteries. One benefit in particular is that, when the charging voltage and current has to flow into one cell in a battery before flowing to a load source. It'll increase the energy output of the cell current has to enter in first with the help of the charging voltage; thus, increasing the total amp capacity of that cell to help power the load source as well.

Chapter 27; Reluctant to Change the Status Quo, for those in the oil, automotive and the automotive battery industries. It seems like there's no incentive for the all-electric vehicle to become our primary source of transportation. Since half of every barrel of oil that is pumped out of the earth, it'll be used for gasoline. We'll no longer need parts for the gasoline engine or automotive starting batteries; therefore, a large portion of sales and jobs related to the gasoline engine will be gone; thus, not a big incentive to change the status quo.

Chapter 28; Technology Already on the Shelf, we've lots of technologies already on the shelf that could decrease or eliminate altogether our gasoline consumption; but, it's hidden from the general public because of profits to be made from crude oil. There're many different web sites that shows different types of energy sources that could be used to power our basic form of transportation. One in particular is www.panacea-bocaf.org; also, go to Bob@RGEneryg.com as well.

INTRODUCTION

Reverse Engineering of an **Auto Battery**; think about an electronic alternator regulator control system under the hood of a gasoline powered automobile. It'll allow an alternator charging system to create a surplus of electrons at the vehicle's voltage and current regulatory system in order to reverse the current flow back into its battery to charge it.

I found if we allow the charging current to flow into one cell in the battery before flowing to the vehicle's voltage and current regulatory system. Then we could use reverse engineering on the charging cycle of the battery because the surplus of electrons will be created in the cell the charging current has to enter in first.

Then the charging current, it'll be reversed back through the other cells in the battery and toward the vehicle's voltage and current regulatory system as well. Thus; charging, the battery while discharging it to the voltage and current regulatory system; unlike, the traditional charging cycle of the battery when there's a charge and a load source on it at the same time.

Using reverse engineering on the charging cycle of our batteries, then we could increase their efficiency and overall capacity while increasing the overall travel ranges of our vehicles on battery power as well. Imagine we could travel as far as we want with our vehicles on battery power. Without having to stop, the discharging cycle of their batteries in order to start their charging cycle as well.

It'll be like having two sets of batteries to power our vehicles; given that, we could switch back and forth between the charge and discharging cycle of their batteries without missing a beat. Using a lead acid automotive battery because it's the first rechargeable battery made, I'm going to describe in detail why it'll be possible to use reverse engineering on its charging cycle.

I found if the battery has separate terminals for charging and discharging in the same cell. And those terminals, they'll have their own straps and tab connectors leading to the same plates in that cell. Then we don't have to stop its discharging cycle in order to start its charging cycle because we could use the same technique that is used to charge its plates in each cell.

The technique I'm talking about, it's a series-parallel circuit connection. It allows us to charge the battery plates in each cell without having to stop charging one plate in order to charge another; although, they've one tab connector each and they're separated by a non-conducting porous material as well.

Using a series-parallel circuit connection between a charge and a load source in their relationship with a battery, then we could charge the battery while discharging it to the load source without missing a beat. I'm going to focus on three aspects of a lead acid automotive starting battery; its mechanical, chemical and electrical aspects as well.

Focusing on those three aspects of the battery, a layperson could understand why it'll be possible to switch back and forth between its charge and discharging cycle without missing a beat when it comes to its cycling processes. While solving an age old mystery, why we could charge or discharge the same battery; but, we can't charge it while discharging it because of physics.

I'm not going to use the laws of physics or sophisticated terms or equations to explain why we could switch back and forth between the charge and discharging cycle of the battery without missing a beat; but, I'm going to use the laws of motion and the effect of forces on bodies to explain why it would be possible to switch back and forth without missing a beat.

I'm going to use the mechanical structure of the battery to explain why. It's not necessary to stop its discharging cycle in order to start its charging cycle. Using line up on line and percept up on percept to explain why, it functions the way that it does. If I didn't, you probably wouldn't believe why it could be charged while being discharged as well.

Have you ever thought about why, we've to stop the discharging cycle of a lead acid automotive starting battery in order to start its charging cycle? Since we don't have to stop charging, its plates in one cell in order to charge them in the next. Although; the battery plates in each cell, they're separated by a non-conducting material called plastic as well.

I've never thought about it until I started exploring the mechanics behind the charging cycle of the battery and then, it didn't make any sense. Why most in the field of automotive battery technology, they believe we've to stop its discharging cycle in order to start its charging cycle because of the laws of physics. It doesn't make any sense if you understand the mechanics behind the battery's charging cycle.

I had to get familiar with the mechanical structure of the battery in order to understand, why we don't have to stop charging its plates in one cell in order to charge them in the next. Given that; it's barely any material written about its mechanical structure and how, it'll play a role in its cycling processes; in which, the material mainly focus on its chemical structure.

I had to take the battery apart piece by piece to see, how its mechanical structure will play a role in its cycling processes. By cutting off, the top of its plastic casing while leaving, its terminals, plates and cells intact. So; I could understand, how electrons will flow through it during its cycling processes by using ammeters and voltmeters while carrying out my experiments.

I found the battery plates behaved like load sources during their charging cycle. I was able to determine this because of the knowledge, I had acquired in the field of heating, cooling and refrigeration while earning my Associate Degree in Science. It dawn on me one day, energy will flow to and from a battery plates the same way it'll flow to and from our load sources as well.

After realizing that, it was clear the laws of physics will have little to do with why. We've to stop the discharging cycle of the battery in order to start its charging cycle; but, it had more to do with the laws of motion and the effect of forces on bodies why we had to stop its discharging cycle. In order to, start its charging cycle or let's say it has more to do with its design.

While venturing into the field of heating, cooling and refrigeration, I learned how to use capacitors and power transformers in order to help power our electrical devices; such as, heating, cooling and refrigeration systems. I found electrons had to simultaneously enter and exit a run capacitor or a power transformer in order to carry out their functions.

A run capacitor, it has multi-terminal connections and they've their own plates and straps leading to them. In which; its plates, they're separated by a non-conductive material called dielectric. Thus; the run capacitor, it could be simultaneously charged and discharged back into an Ac circuit on either terminal connection in order to carry out its functions.

On the other hand; I found that, a power transformer is an electrical device will produce electrical current in a second circuit through by electromagnetic induction; however, it has a primary and a secondary winding that is interconnected by an iron core. As a result; the iron core, it's simultaneously charged and discharged to carry out the functions of the power transformer as wells.

After my observation of a run capacitor and a power transformer, I realized it's not due to the laws of physics why a lead acid automotive starting battery. It can't be simultaneously charged and discharged to a load source, like a run capacitor or a power transformer. However; it's due to, the design of the battery why it can't be charged while being discharged to a load source.

I know this may sound crazy to most in the field of automotive battery technology because it goes against. Their conventional wisdom about the cycling processes of the battery; however, I'm not talking about a conventional cycling process of it; but, I'm talking about an unconventional cycling process of it if it has separate terminals for charging and discharging like a run capacitor.

Wherein; the battery terminals for charging and discharging, they'll be interconnected by the same plates. Like a primary and a secondary winding, are interconnected by an iron core inside a power transformer as well. With this unconventional design of the battery, it would allow us to simultaneously charge and discharge it to a load source as well.

Most people, they don't know what a simultaneous cycling process of a battery means, not alone why it'll be beneficial. I found it's the same reason we charge it when it gets low on charge; that is, to replenish its chemical energy. A simultaneous cycling of a lead acid automotive starting battery isn't unprecedented. It goes through one each time, it's charged.

We haven't realized it or recognized it yet; but, it's not out of the realm of physic. We just think it is because we haven't realized it or recognized it yet. Wherein; the process, it'll be nothing more than an extension of the battery charging cycle. If energy, it could enter and exit the battery at the same time, like it does a run capacitor or a power transformer as well.

New battery technology, it doesn't always come in form of a chemical composition; but, it could come in the form of changing the design of a battery as well. I've written this book in hope that, it'll change our perspective about our secondary cell batteries. Since their mechanical structures, are just as important as their chemical structures as well.

In chapters 3, 4 and 5, there're some known facts about the mechanical structure of a lead acid automotive starting battery and how it'll play a role in its energy input and output process. After reading those chapters, then you decide for yourself. Whether there's a mechanical rather than a chemical solution to increase the efficiency and overall capacity of the battery as well. I believe so if the right mechanics, are in play as well!

CHAPTER 1

OBSERVATION OF AN AUTO BATTERY

AFTER MR. PAUL LUTZ plant, the idea of charging an automotive battery while discharging it in my head. I went out and bought a few lead acid automotive starting batteries and a couple of deep cycle (lead acid) automotive batteries as well. To observe, how their designs will play a role in their cycling processes. In general, what does the average person knows about those things?

We probably know we could charge or discharge the same battery and we've to charge it on the same terminal connection that, it was discharged on. Other than that, we probably don't know too much about it. At first; all I knew was that, I had to connect the red cable to the positive terminal and the black cable to the negative terminal of a battery to properly install it in my car.

Basically; I knew, what an average person knows about an automotive starting battery when it comes to installing it in their vehicles. However; I didn't know that, current had to go in the battery the same way it had to come out. I thought current could flow through one terminal and out the other. Until; I started observing, the mechanics behind the cycling processes of the battery.

Upon my observations, it seems like stopping one cycling process in order to start the other is insufficient and time consuming. Likewise; observing, the paradox behind the chemical composition of the battery. It seems like there's no chemical composition out there with a longer-lasting chemical reaction for it; although, we keep searching for one; but, we should be searching for a better way to cycle it as well.

For some in the field of automotive battery technology, it seems like adding energy back to our secondary cell batteries while taking it out. It's not an option to increase their efficiencies and overall capacities; given that, it'll be n benefits in it due to the laws of physics. Although; adding, electrical energy back to our batteries restores their chemical energy as well.

Thinking there's no benefits in adding, energy back to our batteries while taking it out because of the laws of physics. Then this line of thinking, it limits our options down to finding a chemical composition with a longer chemical reaction than the ones we already use in our batteries today. What qualifies as a perfect chemical composition to a longer-lasting battery any ways?

It has been more than a century since the first secondary cell battery was made. And yet, we haven't found a perfect chemical composition with a longer chemical reaction for it. Wherein; it seems like there's no such thing as a perfect chemical composition to a longer-lasting battery out there other than the ones, we already have on the shelf.

Based on, my observations. It seems like the best solution to overcome the paradox behind our secondary cell batteries. Without finding that, perfect chemical composition for them. It's to add energy back to them while taking it out since adding electrical energy back to them. It diminishes the chemical residue left on their plates from their discharging cycles.

I find as long as we could replenish our batteries' chemical energy with electrical energy. Then their chemical compositions, they're perfect for the job of increasing the overall travel ranges of our vehicles on battery power; however, we haven't realized this for more than a century. Given that; we've been overlooking, the technology that is already on the shelf.

It seems like our batteries' potentials for the electric vehicle to become our primary source of transportation. It lies in their mechanical structures as well as their chemical structures as well. The reason we called them secondary cell batteries because their chemical structures, they're two directional compared to our primary cell batteries chemical structures.

We could only subtract electrons from our primary cell batteries because their chemical structures are one directional. We could add or subtract electrons from our secondary cell batteries' chemical structures; but, we can't charge them while discharging them; although, their chemical structures are two directional. It begs the question is it due to the laws of physics?

This question, it must be answered if we're going to increase the overall travel ranges of our vehicles on battery power. Without finding that, perfect chemical composition with a longer-lasting chemical reaction for their batteries. So far those in the field of automotive battery technology, they've failed to find that perfect chemical composition to a longer-lasting battery.

On the other hand; those in the field of automotive battery technology, they're overlooking the facts that. We could replenish our secondary cell batteries' chemical energy with electrical energy as well. There're many questions needed to be answered; especially, one in particular. Are those in the field of automotive battery technology sincere about making, the electric car our primary source of transportation?

During my observations on the cycling processes of a deep cycle (lead acid) automotive battery, I found if energy could enter and exit it at the same time. Then it seems possible to increase its efficiency and the overall capacity while increasing, the overall travel ranges of our vehicles on battery power as well. Given that; adding electrical energy back to the battery, it restores its chemical energy as well.

I've spent more than a decade observing, the mechanics behind the cycling processes of a lead acid automotive starting battery. Searching for the answer why, we could charge or discharge the same battery; but, we can't charge it while discharging it because of the laws of physics. As some in the field of automotive battery technology, they'll eloquent states.

Observing the inner workings of a lead acid automotive starting battery, I've found the answer why it can't be charged while being discharged. It's not due to the laws of physics; but, it's due to the contemporary design of the battery. Given that; we could charge, its plates in each cell without having to stop charging them in one cell in order to charge them in the next cell.

Although; the battery plates in each cell, they're separated by a non-conducting material called plastic. If we change its contemporary design, so, energy doesn't have to go in it the same way energy has to come out. In order to charge its plates or for them to be discharged to a load source, then it seems like we could charge the battery while discharging it to the load source as well.

Wherein; we could have a mechanical rather than a chemical solution to increase the battery efficiency and overall capacity as well. However; some or most in the field of automotive battery technology, they believe it's impossible to add energy back to our secondary cell batteries while taking it out due to the laws of conservation of energy as they claim.

I believe some or most in the field of automotive battery technology, they haven't observed the inner workings behind the charging cycle of a lead acid automotive starting battery as well. They'll find its plates in each cell, are connected in series-parallel with one another by their straps, tab connectors and the electrolyte solution in each cell. And each cell, it's connected in series with its terminal connections as well.

Given how the battery is constructed, we don't have to stop charging its plates in one cell in order to charge them in the next cell; however, we still can't add energy back to them while using them to power a load source as well. After observing, the mechanical structure of the battery. I still find it's not due to the laws of physics why it can't be charged while being discharged to a load source as well.

Sometimes we're given scientific explanations by those in the scientific community why, things work or don't work; but, it doesn't mean those explanations are infinite. What it does mean is that, their explanations are based on their understanding of how things work or don't work; thus, contrary to popular belief. Scientists don't know everything as we assume they do.

Less than seventy years ago, scientists thought that sea life couldn't exist in the deepest parts of the oceans because of the lack of sunlight and the amount of water pressure that existed. After mankind developed technology, that allowed us to explore the deepest parts of the oceans. We found sea life does exist in the deepest parts of the oceans as well.

Those in the scientific communities, they were wrong to assume sea life doesn't in the deepest parts of the oceans because their assumption was based on sea life as they knew of. What about the Wright brothers, scientists thought they were crazy when they tried to fly in their weird machines. Likewise; scientists thought, Thomas Edison was crazy when he said sound could be recorded as well.

Those in the scientific community, they give us many explanations why things work or don't work. Sometimes it makes sense and sometimes, it doesn't. If we've had accepted every explanation, given by those in the scientific communities why things work or don't work. We probably still will be living in the dark ages because scientists don't get everything right.

Remember just over five hundred years ago everybody thought that the earth was flat. Until; Christopher Columbus, he went out an investigated the matter for himself and he found that the earth was round. When things don't make sense, sometimes we've to go against the status quo and investigate the matter for ourselves because science isn't always exact.

This is why I decided to investigate, the reason why we could charge or discharge the same battery; but, we can't charge it while discharging it because of the laws of physics as most in the field of automotive battery technology believes. It doesn't make any sense based on my observation of a lead acid automotive starting battery because it doesn't line up with, the mechanics behind its charging cycle.

Given that; the battery plates in each cell, they're connected in series-parallel with one another by their straps, tab connectors and the electrolyte solution in each cell. And each cell, it's connected in series with its terminal connections as well. I found electrons could flow from a strap to a plate. And then, flow across that plate through the electrolyte solution to an adjacent plate.

Thus; electrons, they don't have to flow to and from the same tab connector on the same plate in order to flow to an adjacent plate in each cell. Therefore; the plates in each cell, they're simultaneously charged and discharged amongst themselves. Since each cell is connected in series by way of an electrical bus, then the plates in one cell are simultaneously charge and discharged to the plates in the next cell as well.

Although; the battery plates in each cell, they're separated by a non-conducting material called plastic. However; the plates in each, they're still charged because they're connected in series with one another by way of an electrical bus. Observing how energy is taken from the battery plates in one cell; so, it could be added to its plates in the next cell to charge all the plates in each cell.

It doesn't seem logical to blame it on the laws of physics why, we can't add energy back to the battery while taking it out to power a load source. It seems disingenuous because energy has to go in the battery the same way energy has to come out in order to charge its plates or for them to be discharged to a load source as well. It seems like it has nothing to do with the laws of physics.

Based on, my observations. It seems like it has more to do with the laws of mechanics why we can't add energy back to the battery while taking it out to power a load source. In other words; I found that, it has more to do with. How we go about adding and subtracting energy from the battery and not due to the laws of physics in and of themselves as some people may think.

Some in the field of automotive battery technology think, if we tempt to carry out both conversion processes of the battery at the same time. They'll cancel themselves out; therefore, no gain because energy lost to heat will occur in both conversion processes of the battery. After observing the mechanics behind its charging cycle, it seems disingenuous to make this kind of an assumption.

Although; energy, it'll be lost to heat in both conversion processes of the battery. It doesn't mean if we tempt to simultaneously carry out both conversion processes, they'll cancel themselves out. Based on, the mechanics behind the charging cycle of the battery. Given that; its plates in each cell, they could consume and store energy while it's supplied to its plates in the next cell as well.

It seems like any energy lost to heat during the charging cycle of the battery plates, it'll be made up by the charging source as well. Therefore; we could assume adding energy back to the battery while taking it out, it'll be made up by the charging source as well. Then it'll seem disingenuous to assume it'll be no benefits in adding energy back to the battery while taking it out as well.

Reading the literature of those in the field of automotive battery technology on, how the cycling processes of a lead acid automotive starting battery will be carried out. I compiled and compared that data to come to a scientific conclusion. It'll be possible to add energy back to the battery while taking it out; although, energy is lost to heat in both of its conversion processes.

I found this energy lost to heat in both conversion processes of the battery, it'll have no bearing on. Whether it'll beneficial or not to adding energy back to the battery while taking it out. How we think things will work or don't work, it'll be based on what we know or don't know about the mechanics behind the cycling processes of a lead acid automotive starting battery as well.

If we don't know or can't equate, what'll happen during the cycling processes of the battery. Most likely we'll not know, what'll happen if we're adding energy back to it while taking energy out. Thus; our assumptions, they'll be based on what we know or don't know about the cycling processes of the battery; in other words, our assumptions will be our best guess.

For instance; if we know that, each cell in a lead acid automotive starting battery is nothing more than a battery in and of itself. That is connected in series with one another and housed together in a plastic case to make up the whole battery as we know of today. We might assume when it's charged as a whole, we're simultaneously adding and subtracting energy from one battery to another.

Then it'll be hard to assume we can't add energy back to the battery while taking it out to power a load source as well. Based on, the mechanical make up of the battery and the mechanics behind its charging cycle. Since each cell in the battery, it's nothing more than a battery in and of itself that is connected in series with one another and housed in a plastic case to make up the whole battery.

Upon observing, the mechanical structure of a lead acid automotive starting battery and how it'll played a role its cycling processes. It' seems like its mechanical structure will allow energy to be simultaneously added and subtracted from its plates in one cell to its plates in the next cell as well. It has been a serious miss calculation about, what we could or couldn't do with it as well.

To assume, we can't add energy back to the battery while taking it out and it'll be no benefits in it because of the laws of physics. Then those assumptions in and of themselves, they contradict the mechanics behind the charging cycle of the battery. Likewise; we can't use the laws of entropy, inertia or the laws of conservation of energy to discredit the process as well.

Given that; those laws of physics, they'll not apply if we're adding energy back to the battery while taking it out. It'll not be a good depiction of those laws of physics because they depict energy not being added back to a system. Observing, the mechanics behind the charging cycle of the battery. I found any energy lost to heat during its charging cycle, it'll be made up by the charging source as well.

Therefore; we've to assume that, any energy lost to heat while adding energy back to the battery while taking it out. It'll be made up by the charging source because we're charging the battery as well. Then it'll be disingenuous to use the laws of entropy, inertia or the laws of conservation of energy to discredit adding energy back to the battery while taking it out as well.

If we're using the laws of entropy, inertia or the laws of conservation of energy to discredit adding energy back to the battery while taking it out. Then we're missing, the whole point about adding energy back to the battery while taking it out as well. Adding energy back to the battery while taking it out, it'll prohibit those laws of physics from coming into play; that is, the whole point.

If we can't comprehend, what'll happen during the normal charging cycle of the battery when it's just a charge on it? Then we'll not comprehend, what'll happen if we're adding energy back to battery while taking it out or how the laws of physics will play a role in it as well. Then how could, we get it right if we don't understand what'll happen during the normal charging cycle of the battery?

I find it has little to do with the laws of entropy, inertia or the laws of conservation of energy why we've can't add energy back to the battery while taking it out. I'm not saying those in the field of automotive battery technology, they're inept to come to this conclusion. What I'm saying is that, their claims don't add up with the mechanics behind the charging cycle of the battery.

It seems like some or most in the field of automotive battery technology today, they're using the same old myths that has been handed down for one generation to the next. Why energy, it can't be added back to our batteries while taking it out because of the laws of physics. In which; it doesn't make any sense if you observe, the mechanics behind the charging cycle of a lead acid automotive starting battery.

Sometimes, we've to carry out the experiment for ourselves in order to get it right because science isn't always exact. It might be based on assumptions and past events that might be relevant or irrelevant to the matter at hand. In this case, past events will be irrelevant because it'll be based on the contemporary design of the battery; therefore, the assumptions will be flawed as well.

I've heard many assumptions ranging from, it's impossible to charge a battery while discharging it because energy lost to heat will occur in both of its conversion processes. Or we can't charge it while discharging it; given that, its charge and discharging cycle. They're two different processes; therefore, they can't exist at the same time if we attempt to charge it while discharging it.

On the other hand; I heard that, it's impossible to charge a battery while discharging it because the polarity between its negative and positive electrode. They must be in direct opposite of the polarity between them during its discharging cycle; therefore, its charge and discharging cycle can't exist at the same time. In which; these assumptions, they're based on the contemporary design of our batteries.

However; based on, my observations of the cycling processes of a lead acid automotive starting battery. I find those assumptions will be flawed if energy could enter and exit our batteries at the same time. Wherein; those assumptions, they only hold true because energy has to go in our batteries. The same way it has to come out in order to charge or discharge them.

Here's why; if energy could enter and exit our batteries at the same time, then we could charge them while discharging them. If we're adding energy back to them while taking it out, then any energy lost to heat will be made up by the charging source as well. Thus; the assumptions why they can't be charged while being discharged because of the laws of physics, it's based on their contemporary designs.

For example; if energy, it could enter and exit a battery at the same time. If the charging voltage is higher than the normal battery voltage, then the charging voltage will prevail over the battery voltage. Thus; the charge and discharging cycle of the battery, they could exist at the same time in this case. This is how, each cell in a battery is charged in the first place because the charging voltage is higher than the normal battery voltage.

Once we observe the mechanics, behind the charging cycle of a lead acid automotive starting battery. Then those assumptions why, its charge and discharging cycle can't exist at the same time because of the laws of physics. They don't pan out if energy could enter the battery on one terminal and exit it on another terminal at the same time and then, those assumptions become flawed.

Based on, my observations adding energy back to a battery while taking it out. It has little to do with the laws of electrostatic or the laws of electromagnetic. Likewise; the assumption, it'll be no benefits in adding energy back to a battery while taking it out. It has little to do with the laws of entropy, inertia or the laws of conservation of energy as well

These old myths handed down for one generation to the next why, we can't charge our secondary cell batteries while discharging them because of the laws of physics. They only hold true because we design our batteries, so, energy has to go in them the same way it has to come out as well. The myths are true; but, they're based on the contemporary design of our batteries.

Here's the thing; the interior structure of a lead acid automotive starting battery, it allows energy to be simultaneously added and subtracted from its plates in one cell to its plates in the next cell. Without having to, stop charging them in one cell in order to charge them in the next cell; thus, charging all of the battery plates in each cell at once without stopping.

Here's why; because energy, it could flow in one direction from the battery plates in one cell to its plates in the next cell. Thus; the laws of electrochemistry, the laws of electromagnetic, entropy, inertia or the laws of conservation of energy will not be an issue. Since energy, it could flow in one direction from the battery plates in one cell to its plates in the next cell as well.

We can't add energy back to our secondary cell batteries while taking it out because the wrong mechanics are in place since energy has to go in them the same way it has to come out. If we don't change the contemporary design of our batteries, then we can't achieve the goal of adding energy back to them while taking it out to increase their efficiency and overall capacity as well.

It seems like it's a catch twenty-two situation with those in the field of automotive battery technology. They say it's no benefits in changing the contemporary design of our batteries; so, energy could enter and exit them at the same time. But; they'll turn around and say that, it'll be impossible to add energy back to our batteries while taking it out and it'll be no benefits in it as well.

Here's thing; it seems like those in the field of automotive battery technology, they're down playing the benefits in adding energy back to our batteries while taking it out. Thus; it'll be no need to, change their designs; so, energy could enter and exit them at the same time. Therefore; if we don't change their designs, then we can't achieve the goal of adding energy back to them while taking it out as well.

Wherein; we can't increase the overall travel ranges of our vehicles on battery power without finding that, perfect chemical composition to a longer-lasting battery as well. So; it seems like, it's a catch twenty-two situation because those in the field of automotive battery technology. They want to down play another option to increase, the overall travel ranges of our vehicles on battery power as well.

The mechanics behind the charging cycle of a lead acid automotive starting battery suggests. It proves there's another option to increase the efficiency and overall capacity of our batteries. Without finding that, perfect chemical composition with a longer-lasting chemical reaction for them. If energy, it could enter and exit them at the same time as well.

Here's why; electrical energy, it could enter and exit each cell in a lead acid automotive starting battery at the same time in order to replenish, the chemical energy of its plates in each cell. Without having to, stop charging them in one in order to charge them in the next cell. So; it seems like, there's a mechanical option to increase our batteries' efficiency and overall capacity as well.

Based on, my observations of the mechanics behind the charging cycle of a lead acid automotive starting battery. The laws of physics don't justification why we shouldn't change the contemporary design of our secondary cell batteries; so, energy could enter and exit them at the same time. As those in the field of automotive battery technology, they might articulate or speak about.

For too long, we've received disinformation about what we could or couldn't do with our secondary cell batteries because of the laws of physics. It has led us to focus too much on the chemical structure of our batteries and not enough on their mechanical structures. If our hope depends on us finding a perfect chemical composition with a longer-lasting battery, then we're screwed.

Given that; those in the field of automotive battery technology, they've been searching for more than a century. For a perfect chemical composition to a longer-lasting battery and yet, they haven't found one. I believe if they come to the realization as those who produce batteries for our electronic devices; such as, Laptops, Notebooks, iPod and cell phones.

Those in the field of automotive battery technology, they'll realize new battery technology doesn't always come in the form of a chemical composition; but, it could come in the form of changing the design of the battery as well. On the other hand; that perfect chemical composition with a long-lasting chemical reaction for our batteries, it might not exist as well.

Here's why; the paradox behind our secondary cell batteries, it has haunted those in the field of automotive battery technology for more than a century. And yet, they haven't found a chemical composition will allow them to pack an infinite number of electrons within our batteries without facing that dreadful paradox of taking electrons out as well.

Observing, how the chemical composition of a lead acid automotive starting battery works. I found it works like a double edge sword because it'll allow electrons to be added or subtracted from it. But; at the same time, it limits the amount of electrons that could be subtracted from it as well. This is true with, the other types of chemical compositions used in our batteries today as well.

So; our ability to add or subtract electrons from the same chemical compositions of our batteries, it creates a cause and affect scenario. Given that; as more and more electrons are released from our batteries, then less and less electrons could be released from them. Because of the chemical residue, that is left on their plates by their chemical compositions as well.

In which; the chemical residue left on our batteries' plates, it prevents them from releasing more electrons as well. Wherein; it seems like, it's the price to be paid for being able to add or subtract electrons from the same chemical composition of our batteries as well. After observing, how the chemical composition of a lead acid automotive starting battery will behave.

I found it was strange to hear some in the field of automotive battery technology to say. It'll be no benefits in adding electrons back to our batteries while taking electrons out. Given that; adding electrons back to our batteries, it diminishes the chemical residue left on their plates during their discharging cycles. Wherein; charging them, it'll allow more electrons to be added back to them as well.

For more than a century; those in the field of automotive battery technology, they've been suggesting there's no other option to increase the overall travel ranges of our vehicles on battery power; but, to find a chemical composition with a longer-lasting chemical reaction. In which; it sounds like, a paradox because adding electrons back to our batteries do replenish their chemical energy as well.

For the last two centuries battery technology, it has progressed from the basic voltaic-cell to the advanced lithium-ion battery. And yet; those in the field of automotive battery technology, they still haven't found a chemical composition will give back the same amount of electrons put in; thus, it seems like there's no such thing as a perfect chemical composition.

Unfortunately to say that; it seems like, the words "perfect chemical composition to a longer-lasting battery" is nothing more than a ruse to justify why the electric vehicle hasn't become our primary source of transportation for more than a century as well; in other words, it's nothing more than a campaign of disinformation for the general public.

Here's why; during my observations of a lead acid automotive starting battery, I found the assumption it'll be no benefits in changing its design. It's related to the assumption it'll be no benefits in adding energy back to it while taking energy out as well. It seems like these two assumptions are intertwined with one another because without one, the other don't exist.

Not only did I had to examine the mechanical structure of a lead acid automotive starting battery and how it played a role in its cycling processes; but, how the laws of physics played a role in it as well. I found each cell in the battery is nothing more than a battery in and of itself; that is, connected in series with one another and housed in a plastic case to make up the whole battery.

I found when the battery is charged as a whole, we're simultaneously adding and subtracting energy from one battery while simultaneously adding and subtracting energy from another battery. Because of how, they're interconnected with one another. In essence; we're taking energy from one battery while adding, it to another by way of an electrical bus.

Given that; the batteries, they're separated with a non-conducting material called plastic; wherein, it seems like it'll be no difference than adding energy back to a battery while taking it out to power a load source by way of an electrical bus as well. Thus; it brings into question, are we getting disinformation from those in the field of automotive battery technology.

Since some or most in the field of automotive battery technology claim that, it'll be impossible to add energy back to a lead acid automotive starting battery while taking energy out and it'll be no benefits in it because of the laws of physics as well. Given how the battery is charged as a whole in the first place; that is, taking energy from one cell while adding it to another.

Not only do the claims made by some or most in the field of automotive battery technology. It's disingenuous and contradicts, the mechanics behind the charging cycle of the battery; but, they contradict the laws of physics as well. Given that; we could charge the battery plates in each cell without having to, stop charging them in one cell in order to charge them in the next cell as well.

Observing the mechanics behind the charging cycle of the battery, it raises one question. Are those in the field of automotive battery technology focusing on the wrong science, when it comes to the energy input and output of the battery? Given that; they think, it's impossible to add energy back to it while taking energy out and it'll be no benefits in it because of the laws of physics as well.

It seems like those in the field of automotive battery technology, they view the battery; as though, it's a closed system; in which, it's true because of its contemporary design. Thus; it'll be no surprise, they'll reject my observations because their answers is based on energy going in the battery the same way it has to come out. Not energy going in the battery one way and coming out another.

After several years of observing, the mechanics behind the charging cycle of a lead acid automotive starting battery. I found electrons don't have to go in the battery the same way they've to come out in order cycle it. It's just a misconception created by the contemporary design of the battery; given that, electrons have to go in it the same way they've to come out.

Wherein; I found the battery plates in each cell, they're connected in series-parallel with one another by their straps, tab connectors and the electrolyte solution in each cell. And each cell, it's connected in series with its terminal connections as well. Thus; electrons, they could flow from a strap to a tab connector on one plate and then flow across that plate through the electrolyte solution to another plate.

Therefore; electrons, they don't have to flow to and from the same tab connector on the same plate in order to flow to and from the same plate. Understanding, how electrons are added and subtracted from the battery plates in each cell during its charging cycle. I found it's not necessary to stop its discharging cycle in order to start its charging cycle based on its interior design.

However; based on, the exterior design of the battery. I found we've to stop its discharging cycle in order to start its charging cycle because it's charged on the same terminal connection that, it's discharged on. In view of my findings, it'll be possible to charge it while discharging it if we change its contemporary design; so, energy could enter and exit it at the same time.

I found we could use the same technique; that is, used to simultaneously charge and discharge its plates in one cell to its plates in the next cell. We could use it to simultaneously charge and discharge it to load source. Without having to, stop the discharging cycle of the battery in order to start its charging cycle; thus, bridging the gap between them as well.

Observing, what happens on the inside of the battery during its charging cycle. I found it has little to do with the laws of physics why it can't be charged while being discharged; but, it has more to do with the laws of mechanics. If we think it's due to the laws of physics why the battery can't be charged while being discharged, then we're focusing on the wrong science as well.

If we're focusing energy lost to heat in both conversion processes of the battery why, it can't be charged while being discharged. Then we've stray away from the concept of adding energy back to the battery while taking it out. Given that; we'll be focusing on, the value of energy after it has been lost to heat within a closed system.

However; if we're adding energy back to the battery while taking it out, then we've to account for the value of the energy being added back to the battery as well. It seems like those in the field of automotive battery technology, they're overlooking this aspect of the concept of adding energy back to the battery while taking it out; in which, they've stray away from the concept as well.

After I gained knowledge about the mechanical makeup of a lead acid automotive starting battery and how it played a role in its cycling processes, it was hard to believe. It can't be charged while being discharged and it'll be no benefits in it as well. In the next chapter; let's briefly research, the origin of the battery because it was the first rechargeable battery made in 1859 as well.

CHAPTER 2

RESEARCH ON AN AUTO BATTERY

AFTER MY OBSERVATIONS OF a lead acid automotive starting battery, I decided to do a little research on the battery. I found Alexandra Volta made the first primary cell battery or a non-rechargeable battery in 1796. Then in 1859, Gaston Plante' made the first lead acid automotive battery or first the secondary cell battery; but, crude oil was discovered around that same period of time as well.

In the late 1890s about thirty years after the first rechargeable battery was made, then we started experimenting with the electric car. However; in the early 1900s about eleven years after the first electric car was made, then came the gasoline powered automobile. Thus; the idea of the electric car becoming our primary source of transportation, it was swept under the rug.

Since it was, less inconvenient to add gasoline back to our empty gas tanks than it was to add electrons back to our highly discharged batteries. At the advent of crude oil more than a century ago, we've limited the lead acid automotive battery potential. To make, the electric car our primary source of transportation by designing the battery with only one opening for electrons to enter or to exit it.

It has been more than a century since the first electric car was made and yet, we're still using gasoline from crude oil to power our basic form of transportation as well. Wherein; our gasoline powered automobiles, they release carbon monoxide into our atmosphere on a daily basis and they may cause global warming as well. But; on top of all of that, a growing demand for crude oil over the pass forty years.

This growing demands for crude oil, it has superseded laws put in place by US Congress to protect some of our natural environments from oil drilling. Our growing demands for foreign oil, it has become a threat to our national security; in which, it affects how we negotiate our foreign policies as well. It has been more than a century since the first lead acid automotive battery was made.

And yet, we haven't realized the cycling processes of the battery are a product of its design. During my observation of the battery, I found adding energy back to the battery while taking it out. It'll not be out of the realm of physic as some believe it to be; but, it'll be nothing more than an extension of the battery charging cycle if we're charging it while discharging it.

I found how we add or subtract electrons from the battery. It'll be essential in decreasing or eliminating our gasoline consumption from crude oil. If we design the battery so we don't have to charge it on the same terminal connection that, it was discharged on. Then we could intermittently add electrons back to it while taking electrons out as well.

It'll be a mechanical rather than a chemical solution to increase the efficiency and overall capacity of the battery. Without finding that, perfect chemical composition with a longer-lasting chemical reaction for it as well. During my research, I found that. The battery cycling processes, they're carried out by the laws of mechanics based on its design.

In which; how, the battery plates and cells are interconnected with its terminal connections. It'll determine how its cycling processes will be carried out; that is, one process at a time or both processes simultaneously. We must think outside the box when it comes to the cycling processes of the battery because it could be simultaneously charged and discharged.

Wherein; this process, it'll be a mechanical rather than chemical solution to increase the battery efficiency and overall capacity. Since adding, electrical energy back to it restores its chemical energy as well. Thus; a mechanical solution, it'll be a more readily available solution to increase the overall travel ranges of our vehicles on battery power than a chemical solution will.

I found no reason why we can't add energy back to our secondary cell batteries while taking it out, other than energy has to go in them the same way it has to come out. In view of my research on the lead acid automotive starting battery, the idea of a mechanical solution to increase its efficiency and overall capacity isn't inconceivable or farfetched.

While carrying out my research, I found current could only enter or exit the same terminal connection of a lead acid automotive starting battery in order to charge it or discharge it. Then I find that, the only difference between its charge and discharging cycle. It's the direction that the current is flowing between its negative and positive electrode during its cycling processes.

In other words; current, is either flowing to or from the same electrode in order to enter or exit the battery. Wherein; it makes, it impossible to add current back to the battery while taking it out. We've to change the design of the battery in order to add current back to the battery while taking it out. Since the battery, it has only one opening for current to enter or to exit it.

On the other hand; if we continue to believe, there's no other way to carry out the cycling processes of battery; but, to stop one cycling process in order to start the other because of the laws of physics. Then we'll continue to design it with only one electrode for current to enter or to exit it. In which; I find that, it'll limit the battery potential more so than its chemical composition does.

What I found during my research is that, a chemical solution might not be the best solution to a longer-lasting battery due to the paradox of our secondary cell batteries. Their chemical compositions, they limit the amount of electrons that could be released from them. Compared to, the amount of electrons that was stored in them during their charging cycles.

However; I found that, a mechanical solution might be the best solution to combat the paradox of our secondary cell batteries. Given that; adding electrical energy back to them, it diminishes their internal resistance and restores their chemical energy as well. Wherein; it seems like, we don't need to find a perfect chemical composition to a longer-lasting battery after all.

Also; I found that, it's more to it than the lack of battery technology why the electric vehicle hasn't become our primary source of transportation for more than a century. Maybe; special-interest groups like big oil and those who are in cahoots with them. They've been controlling the outcome of our automotive and automotive battery technology for more than a century.

To keep us depend, more heavily on the gasoline engine than the electric motor to propel our basic form of transportation. History has proven that we're not lacking in the area of automotive technology. Then the big question is that, why we're still depending on gasoline from crude oil to power our basic form of transportation more than a hundred years later.

In the mid-1990s; General Motors, they started experimenting with the EV-1. An all-electric vehicle; but, it practically disappeared overnight. The question is still asked to this day, who killed the electric car. Maybe; those in the field of automotive and automotive battery technology, they're in cahoots with big oil to delay the electric vehicle from becoming our basic form of transportation.

Given that; it seems like those in the field of automotive and automotive battery technology, they're all interconnected in some form or fashion with big oil when it comes to the gasoline powered automobile. In spite of the negative impact of oil drilling and oil burning has on our environment, we're still willing to depend heavily on gasoline from crude oil as well.

Maybe; we're behaving, like the four little monkeys that always complained about their situation; but, they never did anything about it. You remember the story about the four little monkeys that played in the tresses all day long in the sun shine. Until; the sun stop shying and it started to rain and then, they started to complain about the cold rain while they huddled together beneath a tree. Promising to build a house; as soon as, the rain stops and the sun came out again.

When the rain stopped and the sun came out again, the four little monkeys forget all about their promises and started playing in the trees again. Until; it started to rain again and then, they started making their promises all over again. To build a house as soon as, it stop raining and the sunshine came out again. In essence, are we like the four little monkeys?

Here's why; we only think about, the need for a clean inexpensive alternative source of energy to power our basic form transportation. When there's an oil disaster or when gasoline prices, they become so unbearable during the week. But; when it ends, we stop complaining about the need for a clean inexpensive alternative source of energy to power our vehicles.

On the other hand; the only time, we complain about the need for a clean inexpensive alternative source of energy to power our basic form of transportation. It's when the news media, they start showing all those negative effects of oil drilling and oil burning has on our environment. When they stop showing all those negative effects, then we stop complaining as well.

The truth of the matter is that, as long as, we're drilling for oil. Then oil disasters will happen and if we don't have a clean inexpensive alternative source of energy to power our vehicles. Then we'll continue to spew poisonous gases into our atmosphere on a daily basis as well. Those who control the price of a barrel of oil, they could raise the price to whatever they want.

It's nothing we could do about price gouging at the pump as individuals; but, complain. However; those businesses, not associated with big oil in some form or fashion. They could bring together their economic resources. To develop, a clean inexpensive alternative source of energy to power our basic form of transportation to end price gouging at the pump as well.

Then we could dictate the price of a barrel of oil by driving down its demand; as a result, minimizing the effect of oil disasters and oil burning has on our environment as well. If we think big oil and those who are in cahoots with them, they're going to give up billions of dollars a year in gasoline sales because of our environmental concerns, we better think again because it's not going to happen.

We need to think about our own future and our children future as well because big oil and those who are in cahoots with them, they're not because it'll be taking money out of their pockets if they give up gasoline from crude oil. However; the oil deposits in the earth, they're not going to last forever. So; will, we have a clean inexpensive alternative source of energy when that day comes?

Is it going to be like the movie "Road Warrior" with Mel Gibson as Mad Max? In the movie, people were scavenging and killing for gasoline to power their automobiles because gasoline was scarce and money couldn't buy it. We might want to take this movie as a warning because it might become our reality, if we don't have an alternative source of energy to power our vehicles.

During my research, I found if we're able to add energy back to our secondary cell batteries while taking it out. It'll open the door to many opportunities to increase the overall travel ranges of our basic form of transportation on battery power. Without finding that, perfect chemical composition with a longer-lasting chemical reaction for our batteries.

For instance; the CSX trains, they're one good example on how to decrease our fuel consumption. Have you seen the commercials about the CSX trains? They could move one ton of freight almost 500 hundred miles on a single gallon of fuel. Since they only rely on their electric motors to propel them; given that, they only use their diesel engines to power their generator systems.

Wherein; the CSX trains, they use their generator systems to charge their batteries; so, they could power their electric motors to propel them. If we could add energy back to our automotive batteries while taking it out, then we could deplore some similar to the CSX trains to minimize our fuel consumption because we could rely more heavily on the electric motor than the gasoline engine.

Big oil and those who are in cahoots with them, they don't want the electric vehicle to become our primary source of transportation. Given half of every barrel of oil that is pumped out of the earth is used for gasoline to power our basic form of transportation. Thus; if the electric vehicle, it becomes our primary source of transportation and then big oil loses.

They'll lose half of their annual sales and thousands of jobs that are related to the production of gasoline as well. On the other hand; those in the field of automotive technology, they might be in cahoots with big oil to delay the electric vehicle from becoming our primary source of transportation as well.

Here's the thing; not only will the electric vehicle decrease our fuel consumption; but, it'll affect other aspects of the automotive industry as well. We'll no longer need gas tanks, fuel pumps, fuel injection systems, gasoline engines or aftermarket parts for our vehicles. Thus; the automotive industries, they'll lose a large portion of their annual sales and tens of thousands of jobs as well.

Likewise; it'll affect, other types of businesses that is directly or indirectly related to the automotive industry as well. If we do the math, then hundreds of thousands of jobs will be lost. If the electric vehicle, it becomes our primary source of transportation. Thus; the lack of battery technology, it might not have anything to do with why the vehicle hasn't become our primary source of transportation as well.

This might explain why adding, energy back to our secondary batteries while taking it out isn't an option for those in the field of automotive battery technology. To increase, the overall travel ranges of our vehicles on battery power. They might be in cahoots with big oil to delay the electric vehicle from becoming our primary source of transportation as well.

Here's the thing; although, the all electric vehicle will increase sales for the deep cycle automotive battery; but, it'll eliminate sales for the automotive starting battery as well. However;

the aftermarket sales for the deep-cycle automotive batteries, it'll not increase the annual income for those in the field of automotive battery technology as well.

Given that; the deep cycle automotive batteries, they've slow turnover rates because they could be cycled hundreds of times. Economically; speaking, it might not be enough incentive for those in the field of automotive battery technology for the electric vehicle to become our primary source of transportation as well.

So; some or most in the field of automotive battery technology, they could be in cahoots with big oil. Given that; they've been claiming for more than a century, they need to find that perfect chemical composition to a longer-lasting battery. Before the electric vehicle, it could become our primary source of transportation. And yet, they haven't found one as well.

During my research, I found there's a reluctance to change the status quo from relying heavily on the gasoline engine to relying more heavily on the electric motor to propel our basic form of transportation. Also; it seems like those in the field of automotive battery technology, they're reluctant to change the contemporary design of our deep cycle automotive batteries as well.

You would think those in the field of automotive battery technology, they'll know subtracting electrons from a deep cycle automotive battery. It'll increase its internal and electrical resistance; wherein, it'll decrease its efficiency and overall capacity as well. Thus; it could decrease, the overall travel range of a vehicle on battery power as well.

On the other hand; those in the field of automotive battery technology, they should know adding electrons back to a deep cycle automotive battery. It'll decrease its internal and electrical resistance while increasing its efficiency and overall capacity as well. Thus; it could increase, the overall travel range of a vehicle on battery power as well.

So; it seems like, if electrons could enter and exit our deep cycle automotive batteries at the same time. Then it'll be an option to increase their efficiency and overall capacity. Thus; increasing, the overall travel ranges of our vehicles on battery power. Without finding that, perfect chemical composition to a longer-lasting battery as well.

Then the question is why, we've to wait for those in the field of automotive battery technology to find. A perfect chemical composition to a longer-lasting battery before the electric vehicle, it could become our primary source of transportation. Given that; we could replenish, the chemical energy of our deep cycle automotive batteries with electrical energy as well.

My observation of the cycling processes of a lead acid automotive starting battery revealed that. The quickest way for the electric vehicle to become our primary source of transportation, it's to add energy back to its batteries while using them. It raised serious questions about the effort put forth by those in the field of automotive battery technology as well.

Wherein; the statements made, by some in the field of automotive battery technology. It'll be impossible to add energy back to our batteries while using them and it'll be no benefits in it because of the laws of physics as well. It seems like those statements are stall tactics because they've aligned themselves with big oil and those in the automotive industry as well.

On the other hand; do any of them have a contingency plan, other than using gasoline from crude oil to power our basic form of transportation? It's no doubt that the planet is going to run out of crude oil; but, the question is when. If we don't have alternative sources of energy similar to gasoline, then certain jobs related to the production of gasoline will be gone any ways.

Nevertheless; whatever, rationale might be used to delay the electric vehicle from becoming our primary source of transportation. It'll be a price to be paid for our fossil fuel consumption because it's no such thing as a free lunch. We've to decide what price we're willing to pay for burning fossil fuel to power our basic form of transportation.

I found it'll have a short term effect on our globe economy if we don't burn fossil fuel to power our basic form of transportation. Or it'll have a long term effect on our environment if we keep on burning fossil fuel to power our automobiles. Nevertheless; whichever direction we chose, it'll be a price to be paid. But; the question is, what price we're willing to pay.

We don't need be an expert in the field of meteorology to figure out what's going on with our weather patterns; unless, our heads are in the sand. The average person could see our weather patterns are changing from season to season and could discern globe warming is real. However; some people, they deny climate change for economic gains as well.

We can't assume nothing will happen if we keep spewing cubit tons of poisonous gases into our atmosphere on a daily basis. Such as; carbon monoxide spewing from our automobiles, coal burning plants and not to mention other types of toxic gases released into our atmosphere on a daily basis as well. If you don't believe climate change is real, then it's okay.

However; you can't assume that, it'll be no consequences for releasing poisonous gases into our atmosphere on a daily basis as well. We've to ask ourselves where the rest of the carbon monoxide goes if it's not consumed by trees or algae. On the other hand; where do the other types of poisonous gases go if they're not consumed by trees or algae as well?

Most likely those poisonous gases, they'll get trapped into our atmosphere because it's no evidence showing they'll float into outer space. Knowing this, we still continue to spew them into our atmosphere on a daily bases. Eventually; it'll become so saturated with them, we'll have to breathe more and more of them in; until, we can't breathe anymore.

We use hundreds of millions of barrels of crude oil every year to meet our ever growing demand for it and it's getting worse. Crude oil isn't only used for gasoline to power our basic form of transportation; but, it's used for hundreds of other applications as well; such as, explosives, synthetic materials, lubricants, and so forth.

However; the biggest consumer of crude oil by far, is the gasoline powered automobile. Since half of every barrel of oil, that is pumped out of the earth is used for gasoline to power our gasoline powered automobiles. If we could come up with, a clean inexpensive alternative source of energy to power our basic form of transportation.

We could reserve our crude oil for other applications while decrease our demand for it and saving our planet in the process. The least expensive and more readily available alternative source of clean energy to power a vehicle is our deep cycle automotive batteries; however; they could only power the vehicle for a short period of time before we've to add electrons back to them.

What compound the problem for our deep cycle automotive batteries to power our vehicles is that, we've to stop using the vehicles in their electric mode in order to add electrons back to their batteries. In which; this process, it causes us to rely more heavily on the gasoline engine than the electric motor to propel our basic form of transportation as well.

More than a century has passed since the first lead acid automotive battery was made and yet, its potential for us to rely more heavily on the electric motor than the gasoline engine is uncertain. We've sent people to the moon, space probes to other planets and yet, we still haven't produced a battery that could power our vehicles sufficiently to meet our transportation needs.

Likewise; those in the field of automotive battery technology, they've tried many different types of chemical compositions. To increase, the efficiency and overall capacity of our deep cycle automotive batteries; but, haven't succeeded. What eludes those in the field of automotive battery technology about our secondary cell batteries, in which, they failed to realize.

If they've to add electrons to our secondary cell batteries when they're first assembled, then at some point the user has to add electrons back to them in order to keep using them. Given that; their chemical compositions, they don't generate their own electrons. Then the next question is that, why spend decades searching for a perfect chemical composition to a longer-lasting battery.

When we've to add, electrons back to our secondary cell batteries anyways if we're going to keep using them. Scientists say that, the only endless source of energy is nuclear fusion; but, we could be centuries away from developing such a process to be used in our automobiles. Then our next alternative is to add electrons back to our batteries while taking electrons out.

Since we could replenish our batteries' chemical energy with electrical energy, then the next step is to design them; so, energy could enter and exit them at the same time. It'll be a more readily available solution than looking for a perfect chemical composition to increase their efficiency and overall capacity; in which, it seems like there's no such thing.

After all, battery technology has been around for two centuries. In all that time, it has progressed from the basic voltaic-cell battery to the advanced lithium-ion battery. And yet, we haven't found a chemical composition will increase the overall travel ranges of our vehicles on battery power. In order for us to, rely more heavily on the electric motor than the gasoline engine.

What I didn't understand during my research on, the lead acid automotive starting battery is that. Those in the field of automotive battery technology, they know during its discharging cycle. Its internal resistance increases and its voltage level drops as we continue to use it; but, during its charging cycle. Its internal resistance decreases and its voltage level rises as we continue to charge it.

Therefore; it seems like adding, energy back to our secondary cell batteries while taking it out will be an obvious solution to increase their efficiency and overall capacity as well. On the other hand; some or most in the field of automotive battery technology, they claim it'll be no benefits in adding energy back to our batteries while taking it out because of the laws of physics.

During my research, I find their claims don't add up with the mechanics behind charging cycle of a lead acid automotive starting battery. Since its plates in each cell, they're separated by a non-conducting material called plastic. Thus; energy, it's taken from the battery plates in one cell by way of an electrical bus; so, it could be added to its plates in the next cell as well.

It seems like it'll be no different than adding energy back to the battery plates while taking energy from them by way of an electrical bus to power a load source as well. Since each cell in the battery, it's separated by a non-conducting material called plastic. What's the difference between the two processes that, the laws of physics will allow one and not the other?

I found the difference is mechanics because energy, it could flow in one direction from the battery plates in one cell to its plates in the next cell in order to charge them in each cell because of the contemporary design of the battery. However; it's the same reason, we can't charge it while discharging it to a load source because of its contemporary design as well.

Given that; energy, it has to go in the battery the same way it has to come out in order to charge the battery plates or for them to be discharged to a load source as well. In my research, it seems certain that the mechanical make up of the lead acid automotive starting battery is the reason it can't be charged while being discharged to a load source as well.

Here's why; the only difference between simultaneously charging and discharging the battery plates in one cell to its plates in the next cell, compare to simultaneously charging and discharging them to a load source is mechanics. Because energy, it could flow in one direction from the battery plates in one cell to its plates in the next cell; but; energy can't flow to a load source this way.

Here's another thing that, I noticed. A lead acid automotive starting battery plates behave like load sources during their charging cycle. Then it's a matter of how, the battery plates and a load source will be interconnected with one another in their relationship with a charging source. If we're trying to, charge the battery while discharging it to the load source.

It seems like those in the field of automotive battery technology, they're giving us the run around when it comes to the conversation about adding energy back to our secondary cell batteries while taking it out to power a load source. In which; some in the field of automotive battery technology, they're using the laws of physics to discredit this type of process.

During my research, I found the idea it'll be no benefits in adding energy back to our secondary cell batteries while taking it out due to the laws of physics. It's nothing more than a ruse orchestrated by some or most in the field of automotive battery; so, we could continue to rely more heavily on the gasoline engine than the electric motor to propel our vehicles.

I found as long as we could replenish our secondary cell batteries' chemical energy with electrical energy. Then they're suited for the job for us to rely more heavily on the electric motor than the gasoline engine to propel our vehicles. Not only we should be searching for a prefect chemical composition; but, we should be searching for a better way to cycle our batteries as well.

Maybe; those in the field of automotive technology, they rather spend their time searching for a perfect chemical composition to a longer-lasting battery instead of finding a batter way to cycle our batteries. Have anyone stop to think that designing, our deep cycle automotive batteries with only one negative and one positive terminal connection wasn't a good idea?

It probably was a good idea for the lead acid automotive starting batteries in the early 1900s. They were designed with one purpose in mind; that is, to start the engines of our gasoline powered automobiles after that, the batteries sit there waiting to be charged. Thus; designing them with only one negative and one positive terminal connection, it doesn't pose a problem for the user.

Wherein; lead acid automotive starting batteries, they're not a constant power source for a vehicle; thus, they could be charged at any time. However; deep-cycle automotive batteries, they were designed to be a constant power source for a vehicle. Then designing them with only one negative and one positive terminal connection, it posed a problem for the user.

Given that; the user, has to stop using the batteries in order to charge them. Then the design concept of the batteries, they contributed to the downfall of the electric vehicle because they didn't facilitate the usage that they were needed for. For example; you wouldn't use, a primary cell battery to start the engine of a gasoline powered automobile on a daily basis.

Since you can't charge the battery after using it, so, it'll not be suited for the job to start the vehicle on a constant or a continuous basis. Likewise; designing, our deep-cycle automotive battery. So; they've to be charged on the same terminal connection that, they were discharged on. If you really think about it, they're not good idea batteries for the electric vehicle.

Since you'll have to stop using the vehicle in order to charge its batteries, then they'll not facilitate the usage that they're needed for when it comes to the electric vehicle on a daily basis.

Think about the gas tanks on our gasoline powered automobiles, they're portable storage devices for gasoline; so, we could carry it with us when we travel as well.

On the other hand; deep-cycle automotive batteries, they're portable storage devices for electrical energy; so, we could carry it with us when we travel as well. A gasoline powered automobile, it could only travel a certain number of miles on a single tank of gasoline before we've to add gasoline back to its tank to continue our journey.

Likewise; an electric vehicle, it could only travel a certain number of miles on a single battery charge before we've to add electrons back to its batteries to continue our journey as well. Thus; our deep cycle batteries, they're no different than our gas tanks if we're going to use more than their stored capacities before we reach our destination as well.

Then the real difference between our gas tanks and our deep-cycle automotive batteries is that. It'll take longer to add electrons back to our highly discharged batteries than it would take to add gasoline back to our empty gas tanks. You wouldn't allow your vehicle to run out of gas before you attempt to add gasoline back to its gas tank; given that, it might cause you major inconvenience.

Likewise; we shouldn't allow our deep cycle automotive batteries to become highly discharged before, we attempt to add electrons back to them. But; we do because of the inconvenience of having to stop using, our electric vehicles in order to charge their batteries. We usually travel as far as possible on a single battery charge before, we attempt to charge them.

Here's the difference, our gas tanks have two openings for gasoline to enter and to exit them. Thus; we could add, gasoline back to them without having to turn off the vehicle's engine. Not saying that, we should because it's illegal in most states; but, just to say it's possible. However; our deep cycle batteries, they only have one opening for electrons to enter or to exit them.

Imagine if our vehicles' gas tanks, they had only one opening for gasoline to enter or to exit them. When we pulled up to a gas pump, we've to turn off their engines and disconnect their fuel lines from their gas tanks in order to add gasoline back to them. This would cause, major inconveniencies for us; thus, we probably wouldn't buy vehicles like that.

Wherein; buying, an all-electric vehicle will cause major inconveniencies to add electrons back to its batteries. Because you'll have to, stop using it in order to charge its batteries; given that, they only have one opening for electrons to enter or to exit them. On the other hand; it wouldn't make any sense to, delay the vehicle from becoming our primary source of transportation as well.

While spending decades searching for, a perfect chemical composition to longer-lasting battery; given that, we could replenish our batteries' chemical with electrical energy as well. Wherein; there's a more readily, available solution to the problem; that is, to design our batteries; so, energy could enter and exit them at the same time as well.

Then we wouldn't have to stop using our vehicles in order to charge their batteries because we could charge them while using them. Nevertheless; some or most in the field of automotive battery technology, they can't see beyond finding a perfect chemical composition to a longer-lasting battery because they've myopia; given that, they're too close to problem.

I found if we changed the contemporary design of our batteries, so, energy could enter and exit them at the same time. Then we could charge them while using them. In the next chapter; let's briefly explore, the mechanical aspects of a lead acid automotive starting battery. To see if, it could be charged while being used without being impeded by the laws of physics.

Remember a lead acid automotive battery or a secondary cell battery was the first rechargeable battery made in 1859 by Gaston Plante'. If we want to know anything about our automotive batteries today, we've to start from the beginning to understand. What we could or couldn't do with our batteries today because we need to understand the basic.

CHAPTER 3

MECHANICAL ASPECTS OF AN AUTO BATTERY

AFTER OBSERVING AND CARRYING out a brief search, on a lead acid automotive starting battery. I was still unconvinced by some in the field of automotive battery technology. It was impossible to add electrons back to the battery while taking them out and it'll be no benefits in it as well. It weighed heavily on my mind because we could charge or discharge the same battery.

During my research, the only difference I found between adding electrons back to a lead acid automotive starting battery while taking them out. Electrons had to go in the battery the same way they've to come out in order to charge its plates or for them to be discharged to a load source as well. I was convinced we could charge the battery while using it if we changed its contemporary design.

To verify my assumptions, I had to explore the mechanical structure of the battery since it could be charged or discharged; but, it couldn't be charged while being discharged. So; I suspected that, it had something to do with the design of the battery. Because we've to charge it on the same terminal connection that, it was discharged on as well.

Although; my assumptions, they were based on my observation and research on the battery. I found it wasn't enough to determine it's not due to the laws of physics; but, it's due to the design of the battery why we can't add electrons back to it while taking them out. Since science isn't always exact because most of the time, it's based on assumptions as well.

Thus; sometimes, scientists don't always get it right the first time when they use assumptions because of unknown variables might exist. Therefore; scientists, they've to actually carry out the experiment for themselves to understand the unknown variables that might exist in order to get it right the second time or third time or how many trials it takes.

If I were going to find out, was it due to the laws of physics or the design of the battery why it couldn't be charged while being discharged. The best way to find out is to take the battery apart piece by piece and see why it functions the way that it does. Maybe; I could figure out, why electrons couldn't be added back to the battery while they were taken out as well.

After taken a lead acid automotive starting battery apart piece by piece, I learned that each cell in it had a certain number of negative and positive plates in it. The negative and the positive plates in each cell, they were set-up in sequence; such as, a negative and then a positive plate in this sequence; until, each cell had an equal number of negative and positive plates in it.

The negative and positive plates in each cell, they were separated by porous rubber separators; so, the plates wouldn't touch one another in each cell. Also; the porous rubber separators, they would allow electrons to flow between the plates in each cell by way of the electrolyte solution in each cell as well.

Figure 1
Traditional 6-volt Deep-cycle (lead acid) Battery

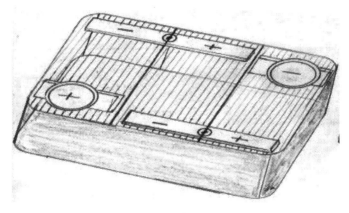

(A series connected battery)

All the battery negative plates in each cell, they're interconnected with one another by a single strap. And likewise; all of its positive plates in each cell, they're interconnected with one another by a single strap as well. Thus; the negative plates in each cell, they're connected in parallel with one another and likewise, the positive plates in each cell are connected in parallel with one another as well.

However; the negative and the positive plates in each cell, they're connected in series with one another by way of the electrolyte solution in each cell. And each cell, it's connected in series with one another by way of an electrical bus. So; the battery plates in each cell, they're connected in series-parallel with one another and each cell is connected in series with its terminal connections as well.

Since the battery plates in each cell, they're connected in series-parallel with one another by way of their straps, tab connectors and the electrolyte solution in each cell. I found electrons could flow from a strap to a tab connector on one plate. And then, flow across that plate through the electrolyte solution to an adjacent plate without flowing to and from the same tab connector on the same plate.

Given that; the plates in each cell, they're connected in series-parallel with one another. Then one plate could consume and store electrons while they're supplied to an adjacent plate by way of the electrolyte solution in each cell. So; each plate, it could be charged during its charging cycle without having to stop charging it in order to charge another plate.

Since each cell, it's connected in series with one another. Then the plates in one cell, they could consume and store electrons while they're supplied to the plates in the next cell by way of an electrical bus. So; all the plates in each cell, they could be charged during their charging cycle without having to stop charging them in one cell in order to charge them in the next.

Given my training in the field of heating, cooling and refrigeration, I realized that a series-parallel circuit connection was connected in series with another series-parallel circuit connection. In order to, charge the battery plates in each cell without having to stop charging them in one cell in order to charge them in the next.

I found it makes each cell in the battery nothing more than a battery in and of itself. That is connected in series with one another and housed together in a plastic case to make up the whole battery as we know of today. Thus; when it's charged as a whole, we're actually simultaneously charging and discharging one battery while simultaneously charging and discharging another.

Wherein; I found that, the battery plates in each cell. They're constructed in parallel with one another like a battery with its plates and cells connected in parallel with its terminal connections. In which; a single strap, it'll connect all of its negative plates to its negative terminal. And a single strap, it'll connect all of its positive plates to its positive terminal as well. See figure 2.

However; a series constructed battery, like a lead acid automotive starting battery. In which; its plates and cells, they're connected in series with its terminal connections because a single strap. It'll connect all of its negative plates together in series with its positive plates in each cell in this sequence; until, all of its plates in each cell are connected in series with its terminals.

In the field of heating, cooling and refrigeration, we call electrical devices loads or load sources. Such as; a light bulb, an electric motor, a ceiling fan or a safety switch that could consume energy from an electrical power source. Thus; connecting, our load sources together in series. Then electrons have to pass through one load source in order to flow to another to power them all.

Since a lead acid automotive starting battery plates, they behave like load sources consuming energy during their charging cycle. Then it makes it possible to charge its plates in each cell without having to stop charging them in one cell in order to charge them in the next; given that, they're connected in series with one another and they behave like load source as well.

I found electrons could flow across one plate through the electrolyte solution to an adjacent plate because they're connected in series with one another by way of the electrolyte solution in each cell. Also; electrons, they could flow from the plates in one cell to the plates in the next cell because each cell is connected in series with one another as well.

It's similar to electrons flowing through one load source in order to flow to another; such as, light bulbs, electric motors, ceiling fans and safety switches. Since the battery plates, they consume electrons during their charging cycle. Like our electrical devices; such as, light bulbs, electric motors, ceiling fans and safety switches as well.

In figure 1, it shows a top inside view of a 6-volt deep-cycle (lead acid) automotive battery; wherein, its cells are connected in series with its terminal connections. Not only does connecting the battery cells in series with its terminal connections. It allows its plates in one cell to consume and store electrons while supplying them to the plates in the next cell as well.

However; there's another benefit in connecting, the battery plates and cells together in series with its terminal connections as well. It'll allow us to increase the voltage potential of its plates in one cell. For instance; the voltage potential of a lead acid automotive starting battery plates in each cell, they could range from 2.25 to 2.75 volts each.

If we needed 6 or 12-volts to carry out a particular application, then we could connect the battery cells together in series with its terminal connections. In order to, increase its voltage potential if we need more than 2 volts. This is one more benefit in connecting, the battery cells together in series; but, it only increases its voltage potential of its plates in one cell.

During my research, I found there are other benefits in connecting a battery cells together in series with its terminal connections as well. In which; I'll explain later on. However; I found that, it's a down side in connecting a battery cells together in series with its terminal connections as well. Given that; its cells, they're like links in a chain because it's only as strong as the weakest link in it.

Thus; a battery, it's only as strong as the weakest cell in it if its cell are connected in series with one another in their relationship with its terminal connections. For example; if one cell in a battery, it has an amp potential of 60 amps. And the other cells in it, they've an amp potential of 100 amps each. Then we could only draw 60 amps from the battery.

Given that; once, the 60 amps have been used up from the lowest within the battery. Then the battery is said to be dead because we can't draw anymore amps from it. But; there's another down side in connecting, its cells together in series as well. Regardless of the number of cells it has, its total amp draw will only equal to the plates in one cell in it during its discharging cycle.

Likewise; connecting a group of batteries together in series to increase voltage potential for particular application, it has a down side to it as well. Given that; a group of batteries, they'll only be as strong as the weakest battery in the group. For example; if one battery in the group is only a 60 amp capacity battery and the other batteries, they're 100 amps capacity batteries.

Then we could only draw 60 amps from the group of batteries; given that, once the amp capacity of the lowest battery has been used up. And then, the batteries are said to be dead because we can't draw anymore amps from the group of batteries. Unless; we switch out the battery with lowest amp capacity, then we could draw more amps from the group batteries.

Likewise; if a group of batteries, they're connected in series with one another. And their cells, they're connected in series with their terminal connections as well. Then the total amp draw from the group of batteries, it'll only equal to the plates in one cell in one battery in the group of batteries. Similar to one battery cells connected in series with its terminals as well.

For instance; if we've six batteries and each battery has three cells each, then we've a total of eighteen cells. Regardless of the number of cells in the group of batteries, their total amp draw will only equal to the plates in one cell in one battery in the group of batteries. This is the down side, for connecting a group batteries together in series and their cells are connected in series with their terminal as well.

The up side for connecting a battery cells together in series with its terminal connections. It gives us more voltage potential. However; the down side is that, it limits its amp potential to its plates in one cell. Likewise; if a group of batteries, they're connected in series with one another and their cells are connected in series with their terminal connections as well.

It'll give us more voltage potential; but, the down side is that. It'll limit the amp potential to the plates in one cell in one battery in the group of batteries. Here's another thing if we connect a group of batteries together in parallel from the outside. For example; all of the negative terminal connections of the batteries, they're interconnected by a single cable.

However; all of the positive terminal connections of the batteries, they're interconnected by a single cable as well. But; on the inside of the batteries, their cells are connected in series with their terminal connections. Then the total voltage potential from the group of batteries, it'll only equal to the voltage potential of those plates in one cell in each battery.

Given that; on the outside, all of the batteries. They're connected in parallel with one another. Thus; we'll only receive, the voltage potential of a cell within each battery with a terminal connection in it. Since the other cells in the batteries, they'll be connected in series with a cell with the terminal connection in it to increase its voltage potential.

However; the total amp draw from the group of batteries, it'll only equal to the total amp capacity of those plates in one cell in each battery. Given that; on the outside, all of the batteries are connected in parallel with one another as well. Thus; we'll only receive, the amp potential of the plates in one cell in each battery with a terminal connection in it.

On the other hand; if a battery cells, they're connected in parallel with its terminal connections instead of series. Then the opposite will happen. For instance; the total amp draw from the battery, it'll equal to the sum of all of its plates in each cell. But; its voltage potential, it'll only equal to the sum of its plates in one cell.

In figure 2, it shows a top inside view of a deep cycle (lead acid) battery with its cells connected in parallel with its terminal connections.

Figure 2

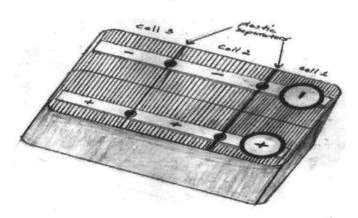

(A parallel connected battery)

In general; battery manufactures, they don't normally build deep-cycle marine or automotive batteries with their cells connected in parallel with their terminal connections; unless, it's for a particular application that requires 2 volts or less. How a parallel battery is constructed; a single strap connects, all of its positive plates together in parallel with its positive terminal.

Likewise; a single strap, it connects all of its negative plates together in parallel with its negative terminal connection. Thus; how, a parallel battery is constructed. It's similar to the construction of the plates in each cell in a lead acid automotive starting battery. Since all the positive plates in one cell, they're interconnected by a single strap connecting them together in parallel.

All negative plates in one cell, they're interconnected by a single strap connecting them together in parallel as well. However; the plates in one cell, they're connected in series with the plates in the next cell. By connecting, the strap that is connecting the plates together in parallel in one cell to the strap that is connecting the plates together in parallel in the next cell as well.

Thus; one set of positive plates, they end up connected in parallel with the positive terminal and one set of negative plates end up connected in parallel with the negative terminal connection. See figure 1. If we use the conventional current flow concept; that is, current could only enter or exit the positive terminal connection of a battery during its cycling processes.

If the battery cells, they're connected in series with its terminal connections. Then we'll only have access to the amp potential of the plates in the cell with the positive terminal connection in it. Given that; the other cells in the battery, they'll be connected in series with the cell with the positive terminal connection in it to increase its voltage potential.

Therefore; we could receive, the voltage potential of all of the battery plates in each cell and receive only the current potential of its plates in one cell during its discharging cycle. But; if a battery cells, they're connected in <u>parallel</u> with its terminal connections. We'll have access to the amp potential of all of its plates in each cell and receive only the voltage potential of its plates in one cell during its discharging cycle.

On the other hand; if a battery cells, they're connected in series with its terminal connections. Then during its charging cycle, the charging current will be the same throughout each cell in it. But; the charging voltage, it'll be divided evenly amongst its plates in each cell. However; if a battery cells, they're connected in parallel with its terminal connections.

Then during its charging cycle, the charging voltage will be the same throughout each cell in it. But; the charging current, it'll be divided amongst the battery plates in each cell based on their individual resistance. It seems like how a battery cells are interconnected with its terminal connections. It'll determine how voltage and current will enter or exit it.

So; if a battery, it has only three cells and they're only two volts each with a current potential of 125-amps. Then each cell has an amp capacity of 250-amps; that is, 2-volts time the current potential of one cell. Thus; a 6-volt battery with its cells connected in series with its terminal connections, it'll have a total amp capacity of 750-amps; that is, 6-volts time's the current potential of one cell.

If the battery cells are connected in parallel with its terminal connections and each cell is still 125 amps, then it'll still have a total amp capacity of 750-amps. Because it'll be the current potential of all of its plates in each cell, times the voltage potential of its plates in one cell. That is 125 amps from the plates in each cell equal 375 amps, time's 2-volts from its plates in one cell.

After examining, the mechanical structure of a series or a parallel constructed battery. I found how a battery plates and cells are interconnected with its terminal connections. It'll determine its total amp capacity. Let me explain this in a different way for those electric vehicles enthusiasts. Given that; it's the same way, they connect their batteries together on the outside when one isn't enough.

For example; if the voltage potential of one battery isn't enough for a particular application. Then the electric vehicle enthusiasts, they'll connect one battery positive terminal to another battery negative terminal in order to increase voltage potential; thus, increasing the voltage and current potential for the plates in one cell in one battery in the group of batteries.

Therefore; the batteries, they're connected in series with one another. On the other hand; in order to increase, voltage potential for one battery. Those in the field of automotive battery technology, they'll connect the negative plates in one cell to the positive plates in the next cell or vice versa; until, all the cells in the battery is connected in this sequence.

Thus; the battery cells, they'll be connected in series with one another in their relationship with its terminal connections. Remember each cell in a lead acid automotive starting battery is nothing more than a 2-volt battery in and of itself. So; it's no different, then connecting a group of batteries together in series to increase voltage potential for a particular application as well.

Now in order to, increase current potential for a particular application when one battery isn't enough. The electric vehicle enthusiasts, they must connect one battery negative terminal to another battery negative terminal connection. And then, connect their positive terminal connections together. In order to, increase current potential for a particular application when one battery isn't enough.

Thus; the batteries, they'll be connected in parallel with one another. On the other hand; in order to increase, current potential for one battery. Those in the field of automotive battery technology, they'll connect all of the battery negative plates to its negative terminal with a single strap. And then, connect all of its positive plates to its positive terminal with a single strap as well.

Thus; the battery cells, they'll be connected in parallel with one another. However; if the electric vehicle enthusiasts, they wanted to increase current and voltage potential for a particular application and one battery isn't enough. They must connect a certain number of batteries together in series-parallel with one another; until, they reach their desired goal.

For instance; if the electric vehicle enthusiasts, they needed 250 amps and 12-volts for a particular application; but, they only had four-six volt batteries and they were only 125 amps each. Then two of the four batteries must be connected in parallel with one another. And then, the other two batteries must be connected in series with the other two batteries.

In other words; the electric vehicle enthusiasts, they've to connect two of the batteries together in parallel to increase current potential to 250 amps. And then, they must to connect the other two batteries in series with the other two batteries; that is, connected in parallel with one another to increase voltage potential to 12-volts in order to achieve their goal.

In short; the electric vehicle enthusiasts, they'll connect two of the batteries' negative terminals together. And then, they'll connect the other two batteries' negative terminals to the positive terminal connections of the other two batteries; that is, connected in parallel with one another by their negative terminal connections in order to connect the batteries in series-parallel.

In essence; when electric vehicle enthusiasts, they connect their batteries together on the outside in order to increase voltage and current potential for their electric vehicles. It's no different than those in the field of automotive battery technology connecting a battery cells together on the inside in order to increase its voltage or current potential as well.

Upon exploring, the mechanical structure of a lead acid automotive starting battery. It's safe to assume its design is the deciding factor in determining how voltage and current will enter or exit it, not the laws of physics as some may allude to. This is why I thought, it would be wise to take the battery apart piece by piece to see why it functions the way that it does.

Given all the rumors about why a lead acid automotive starting battery can't be charged while being discharged and it'll be no benefits in it because of the laws of physics as well. After exploring, the mechanical structure of the battery. It seems like some in the field of automotive battery technology, they don't know or they were disingenuous about their comments.

Wherein; those in the field of automotive battery technology, they determine how energy will enter or exit a battery based on how it's designed. Observing a lead acid automotive starting battery during its discharging cycle, it seems like those in the field of automotive battery technology designed it to power a particular type of load source based on its energy requirements.

For instance; if a load source, it requires 6-volts and 20 amps to operate it. Then those in the field of automotive battery technology, they'll determine the number of plates and the size of the plates will be needed in each cell in the battery. In order to, reach the desired voltage and current potential of the battery in order for it to power the load source as well.

Wherein; we find those in the field of automotive battery technology, they determine how a battery plates and cells will be interconnected with its terminal connections in order to meet the requirements needed to power a particular load source as well. On the other hand; observing the charging cycle of a lead acid automotive starting battery, it tells the same story.

It seems like those in the field of automotive battery technology, they design the battery plates similar to load sources; so, they could consume and store energy from a particular type of power source as well. Based on, the size of the plates and the number of them in each cell within the battery. And how, they'll be interconnected with its terminal connections as well.

Since those in the field of automotive battery technology, they're the architect of the design of a lead acid automotive starting battery. Then they'll determine how energy will enter or exit it, not the laws of physics. So; there's a disconnect between, how those in the field of automotive battery technology design the battery and how some thinks its cycling processes will be carried out as well.

Using, the laws of physics to discredit adding energy back to our secondary cell batteries while using them to power our vehicles. It seems disingenuous for some in the field of automotive battery technology to make such a claim; unless, it's due to the lack of information. In other words; some, they don't have sufficient information about the cycling processes of the battery.

Remember if a battery cells, they're connected in series with its terminal connections. Then we could only receive the current potential of its plates in one cell and the voltage potential of all of its plates in each cell during its discharging cycle. On the other hand; during its charging cycle, the charging current will be the same throughout each cell in it.

However; the charging voltage, it'll be divided evenly amongst the battery plates in each cell. Now; if a battery cells, they're connected in parallel with its terminal connections. Then we could receive the current potential of all of its plates in each cell and only the voltage potential of its plates in one cell during its discharging cycle; but; on the other hand during its charging cycle.

The charging voltage will be the same throughout each cell in the battery. However; the charging current, it'll be divided amongst its plates in each cell based on their individual resistance. Upon my observation and research on the lead acid automotive starting battery, I find some or most in the field of automotive battery technology are focusing on the wrong science.

When it comes to the general public, it's no doubt we could charge or discharge the same battery. Especially; if we've cell phones, we probably charge their batteries more than fifty times a week. However; if someone asks, is it possible to charge a cell phone battery while using the phone? The answer probably be no because of the oxymoron behind its cycling processes.

Wherein; this oxymoron, it challenges our perceptions about the cycling processes of the battery as well. If you've notice when you're using, your cell phone while charging its battery. Sometimes, you're able to charge it while using it as well. What I've learned is that, it depends on the amount of electrons that the cell phone will consume during the charging cycle of its battery.

In which; I found that, it'll determine how difficult it'll be to charge the cell phone battery while using the phone. What we don't realize is that, it's not just one thing; but, it's a host of things coming together to prevent the battery from being charged while being used as well. What's amazing is that, we think it's due to the laws of physics; but, we're wrong as well.

However; the last thing come to mind is that, it's due to the design of the battery. What I found is that, the mechanical structure of a lead acid automotive starting battery. It creates an oxymoron when it comes to its cycling processes. Since electrons, they've to go in it the same way they've to come out. Thus; we think, its charge and discharging cycle can't exist at the same time.

Since we've to charge the battery on the same terminal connection that, it was discharged on. Then its charge and discharging cycle, they're the opposite of one another; in which, it makes them an oxymoron. Although; we could, charge or discharge the same battery; but, we can't charge it while discharging it; in which, it's not due to the laws of physics as we think.

It's due to the oxymoron behind the cycling processes of the battery. Wherein; the mystery, behind the paradox of its cycling processes. It lies with its design because electrons have to go in it. The same way they've to come out in order to charge its plates or for them to be discharged to a load source; thus, we can't charge the battery while discharging it to the load source as well.

At first; I didn't understand, the paradox of the oxymoron behind the cycling processes of the battery; until, I analyzed its mechanical structure and how it'll play a role in its cycling processes. And then, I begin to understand it. In which; it shed light on, why it's not due to the laws of physics why the battery can't be charge while being discharged to a load source as well.

However; we've overlooked, the oxymoron behind the cycling processes of our secondary cell batteries for more than a century. Given that; energy has to go <u>in</u> them, the same way energy has to come <u>out</u> as well. During my research on the mechanical structure of a lead acid automotive starting battery, I found an oxymoron existed with its chemical structure as well.

Not only taking electrons from the battery limits the amount of electrons that, we could take from it because we're taking electrons from it. However; its chemical composition, it'll limit the amount of electrons that could be released from it because it leaves a chemical residue on, the battery plates during its discharging cycle; so, there's a host of things coming into play.

The oxymoron behind the cycling processes of a lead acid automotive starting battery. It shows why it's impossible to add electrons back to it while taking them out. Because we've to charge it on the same terminal that, it was discharge on. We haven't found a perfect chemical composition because they leave chemical residue on the batteries plates as well.

In the next chapter; let's briefly explore, the chemical aspects of a lead acid automotive starting battery because it might help explain, the mystery behind the oxymoron of its discharging cycle as well.

CHAPTER 4

CHEMICAL ASPECTS OF AN AUTO BATTERY

UNDERSTANDING WHY, AN OXYMORON existed behind the discharging cycle of a lead acid automotive starting battery. We might spend less time searching for a perfect chemical composition as well. Since it seems like, a chemical solution might not be the best solution to a longer-lasting battery. If we truly understood, the oxymoron behind its discharging cycle as well.

Let's talk about the chemical structure of the least expensive and the most widely used automotive battery in the world. The lead acid automotive starting battery because it'll help us understand, the origin of our secondary cell batteries. We use lead-peroxide or lead dioxide for the positive plates (anodes) and sponge lead for the negative plates (cathodes) of the battery.

In which; the battery plates, they're pressed into a lead antimony grid for the hardness. We use distill water and sulfuric acid for its electrolyte solution which makes up its chemical composition. A twenty-five year old French physicist named, Gaston Plante' made the first lead-acid automotive battery or the first secondary cell battery in 1859.

He discovered using certain chemical elements for the battery chemical composition. It would allow him to add or subtract electrons from it; thus, its chemical composition is two directional. It's a down side for its chemical composition being two directional as well. Given that; it limits, the amount of electrons that could be subtracted from it as well.

Let's look at the chemical discharging equations of the anode and the cathode for a lead acid battery; that is, shown in figure 3A.

Figure 3A

DISCHARGE

Anode: $PbO_2 + 4H^+ + SO_4^= + 2e^- \rightarrow PbSO_4 + 2H_2O$

Cathode: $Pb + SO_4^= - 2e^- \rightarrow PbSO_4$

(The discharging chemical equations)

The factors affecting the efficiency and overall capacity of a lead acid automotive starting battery, they're caused by a chemical reaction allowing electrons to be released from its lead plates. This chemical reaction, it releases two electrons for every sulfate-ion radical that is bonded to the battery lead plates during their discharging cycle.

Thus; the lead and the sulfate-ions, they bond together to create lead-sulfate; in which, it's referred to as the internal resistance of the battery. The bonding between the lead and the sulfate-ions, it weakens the battery electromotive force as more and more electrons are released from its plates. Then more and more of the lead-sulfate will accumulate on its plates as well.

The accumulation of the lead-sulfate, it reduces the active area of the lead acid automotive starting battery plates to release more electrons. Thus; it decreases, its plates' efficiency and overall capacity as well. During my research, it seems like there's no such thing as a perfect chemical composition to a longer-lasting chemical reaction for it.

I found the same chemical composition that allows electrons to be subtracted from the battery plates. It'll limit the amount of electrons that could be subtracted from its plates as well. In which; this might explain, why the oxymoron exists behind the discharging cycle of the battery. Given that; as more and more electrons, are released from its plates than less and less are released.

For more than a century those in the field of automotive battery technology, they've been searching for a perfect chemical composition to a longer-lasting battery. And yet, they haven't found one will not limit the amount of electrons could be released from it. All the chemical compositions used in our batteries today, they limit the amount of electrons released from them as well.

However; the process of replenishing, a lead acid automotive starting battery chemical energy with electrical energy is threefold. First; the charging voltage, it must be higher than the normal battery voltage to reverse the current flow back into the battery to charge it. Secondly; the charging electrons, they must break down the lead-sulfate build up on the battery plates.

Thus; the sulfate-ions, they'll be added back to the battery electrolyte solution when the bonds between the lead and the sulfate-ions has been broken. Then the electrolyte solution of the battery will be restored back to its original state. Those electrons that weren't used up during the electrolysis process, they'll be stored on the battery plates. It's the third and final step.

However; some battery experts say that, the catalyst in the chemical reaction of the battery. It comes from the sulfate-ions; a mixture of water and sulfuric acid that, makes up its electrolyte solution. Here's the thing; the definition of a catalyst, it's a substance that increases the rate of a chemical reaction without itself undergoing any changes.

So; when the battery is discharged, then the reaction between the lead and the sulfate-ions will create lead-sulfate. In which; it's the substance that coats, the battery plates and decreases their active areas for releasing more electrons. Thus; during the discharging cycle of the battery, it removes the sulfate-ions from its electrolyte solution; thus, it weakens it.

On the other hand; during the charging cycle of the battery, the chemical reaction separates the lead from the sulfate-ions. Wherein; releasing, the sulfate-ions back into the electrolyte solution while increasing the active area of the battery plates. For more electrons to be restored on them or to be released from them, I'll talk about releasing electrons from them later on.

The correlation between electrons entering or exiting the battery, it points toward the electrons being the catalyst and not the sulfate-ions. In figure 4A, it shows the charging equations of the anode and the cathode of a lead-acid automotive starting battery plates.

Figure 4 A

CHARGING

Anode: $PbSO_4 + 2H_2O - 2e^- \rightarrow PbO_2 + 4H^+ + SO_4^=$

Cathode: $PbSO_4 + 2e^- \rightarrow Pb + SO_4^=$

(The charging chemical equations)

If you notice in the discharge and the charging equation of the battery in figures 3A and 4A (reading from the left to the right of the arrows), you'll see electrons (2e-) must be the catalyst that determines the state of the battery. For instance; when electrons, they're released from the battery plates. Then Pb+SO4 become PbSO4, in which, it becomes lead-sulfate when they're bonded together.

However; when electrons are added back to the battery, then PbSO4 are separated back into Pb+SO4. In which; the bonds between, the lead and sulfate-ions are broken. Thus; it seems like electrons, they increase the rate of the chemical reaction within the battery without themselves undergoing any changes. Not the sulfate-ions when electrons enter or exit the battery.

Wherein; it appears the sulfate-ions, they're not the catalyst that increases the rate of the chemical reaction within the battery. In short; the lead and the sulfate-ions, they become lead-sulfate. The electrolyte solution becomes mostly water when electrons are subtracted from the battery; but; when electrons are added back to the battery.

Then it separates the lead and the sulfate-ion while diminishing the lead-sulfate on the battery plates. The mostly water electrolyte solution is restored back to its original state; a mixture of 67% water and 33% sulfuric acid. For these reasons alone, it seems like the sulfate-ions aren't the catalyst that increases the rate of the chemical reaction without undergoing any changes.

Wherein; it seems like, the sulfate-ions. They only create an electron highway for electrons to travel within the battery from a more negative to a more positive point within it. Thus; electrons, they create the rate of change without themselves undergoing any changes when they're added or subtracted from the battery during its cycling processes.

Here's why; as more and more electrons, are released from the battery. Then more and more of the lead-sulfate, it'll build up on the battery plates and its voltage level will drop. But; the opposite, it'll happen as more and more electrons are added back to the battery plates. Then more and more of the lead-sulfate will be diminished from the battery plates and its voltage level will rise.

Understanding the chemical reactions; that are, taking place within the lead acid battery during its cycling processes. It seems like its efficiency and overall capacity is based on, how many electrons are put in versus how many electrons are taken out. Then adding electrons back to it while taking them out, it seems like it'll eliminate the oxymoron behind its discharging cycle as well.

Then it doesn't make any sense to assume, it'll be no benefits in adding electrons back to the battery while taking them out. While spending decades searching for, a perfect chemical composition with a longer chemical reaction for it as well. Since adding electrons back to the battery, it diminishes its internal resistance and restores its chemical energy as well.

Wherein; it seems like, there's a mechanical rather than a chemical solution to increase its efficiency and overall capacity as well. Here's why; when we're taking electrons out, it reduces the amount of electrons left in the battery; given that, we're taking them out. However; because we're taking them out, it'll increase the amount of lead-sulfate accumulating on the battery plates.

In which; this accumulation of lead-sulfate, it'll reduce the amount of electrons that could be released from the battery plates. Then less and less electrons could be released from its plates because of the ever increasing lead-sulfate accumulating on them. So; the battery chemical composition, it works like a double edged sword because we can't get back the same amount of electrons put in.

All because of the chemical residue that is accumulating on the battery plates because we're taking electrons out. Wherein; its chemical compositions, it becomes its own worst enemies when it comes to its efficiency and overall capacity as well. Not only releasing electrons from its plates, it'll reduce their capacity; but, it'll increase their resistance as well.

If we're going to take electrons out and not put any back right away, then the residual effects of taking electrons out. It's going to reduce the amount of electrons that could be taken out; given that, we're taking electrons out as well. It really doesn't matter if a chemical composition could hold an infinite number of electrons because it'll not give back the same amount of electrons put in.

The oxymoron behind the discharging cycle of our secondary cell batteries, it has driven those in the field of automotive battery technology. To search for, a perfect chemical composition to a longer-lasting battery for more than a century while overlooking a mechanical solution to the problem; that is, adding electrons back to our batteries while taking electrons out.

In the past decade or so those in the field of automotive battery technology thought that, the lithium-ion crystals were the chemical composition needed to increase the overall travel range of our vehicles on battery power. And yet; the lithium-ion crystals, they haven't begun to make the electric vehicle our primary source of transportation as well.

Here's why; the lithium-ion crystals, they undergo the same oxymoron as our other chemical compositions used in our secondary cell batteries today. Given that; as more and more electrons, are released from the lithium-ion crystals. Then less and less electrons could be released from the crystals, given the ever increasing chemical residue left by them on the battery plates as well.

The lithium-ion crystals, they don't behave any different than the other chemical compositions already used in our other secondary cell batteries today. However; the lithium-ion crystals endothermic chemical reactions, they don't act like most chemical compositions endothermic chemical reactions; especially, during their discharging cycles.

Most chemical compositions endothermic chemical reactions, they absorb heat during their discharging cycles. Wherein; the lithium-ion crystals, they get hot during both of their conversion processes. For instance; unlike, a lead acid automotive starting battery because it releases heat during its charging cycle and it absorbs heat during its discharging cycle.

Although; the lithium-ion crystals, they give more voltage per cell than the lead acid automotive battery. Wherein; the lithium-ion crystals, they're praised for their high voltage per cell. And yet, we're no closer in making the electric vehicle our primary source transportation then we were more than a century ago as well.

Some battery experts say that, the lithium-ion batteries. They're not safe because they don't absorb heat during their discharging cycles. In which; they could pose a problem, when they're stored or in use; given that, they could cause a fire or exploded. The question is that, is it worth the risk of having a little more voltage per cell? And yet, it doesn't bring us any closer in revolutionizing the electric vehicle as well.

In spite of the oxymoron behind the discharging cycles of our secondary cell batteries, it's a mechanical rather than a chemical solution to increase their efficiency and overall capacity. Without finding that, perfect chemical composition for them as well. All we've to do is design them; so, energy could be added back to them while it's taken out as well.

Then the oxymoron behind our secondary cell batteries discharging cycles, they become irrelevant to the overall travel range of our vehicles on battery power as well. Some people say that, I'm obsessed with the idea of charging our batteries while discharging them. In which; they say, it doesn't make any sense for a person to be so obsessed with an idea.

The reason I'm so obsessed with charging our batteries while discharging them because it's the key to increase the overall travel ranges of our vehicles on battery power. Without finding that, perfect chemical composition to a longer-lasting battery for them. Wherein; it seems like those in the field of automotive battery technology, they haven't found one yet.

After more than a century searching for a perfect chemical composition to a longer-lasting battery and haven't found one yet, it's time to change course. After taken, an in depth analysis of the mechanics behind the charging cycle of a lead acid automotive starting battery. It seems like it'll be a benefit in adding energy back to the battery while taking it out as well.

For more than a century, we've overlooked a simple solution to increase the efficiency and overall capacity of our secondary cell batteries. Without finding that, perfect chemical composition to a longer chemical reaction for them. Given that; we're focusing on the wrong science, when it comes to their energy input and output process as well.

On the other hand; some in the field of automotive battery technology, they claim even if energy could enter and exit our secondary cell batteries at the same time. We still couldn't add energy back to them fast enough to charge them while using them; given that, it's a lethargic process of adding energy back to them. I decided to investigation the matter for myself.

What I found was that, the assumption. It's a lethargic process of adding energy back to our batteries. It stemmed from allowing our batteries plates to become highly discharged before we attempt to charge them. What we think, is causing the problem. It's not actually causing the problem; but, we think it is because we don't actually understand the problem.

I find adding electrons back to a secondary cell battery isn't a lethargic process in and of itself. The problem is that, we're deep cycling the battery which is the problem. For instance; a deep cycle battery, it's made for deep cycling; ironically, deeper the discharged is. It'll take more time and energy to add electrons back to the battery because it's deep cycled.

The reason we deep cycle the battery because of the inconvenience of having to stop one cycling process in order to start the other, which is a problem itself as well. We could eliminate this lethargic process of adding electrons back to the battery if we could add them back to it while taking them out. This is one more reason why, I'm obsessed with this idea as well.

Sometimes; we overlook, a simple solution to a problem because we don't actually understand the problem; but, we think we do. For instance; since we could, add or subtract electrons from the same chemical composition of a battery plates. However; as more and more electrons, are subtracted from its plates. Then less and less electrons could be subtracted from its plates.

The reason is, it's due to the ever increasing chemical residue that is accumulating on the battery plates due to subtracting electrons from them. Wherein; one the key attribute of a secondary cell battery is that, we could add electrons back to its plates. In which; it diminishes, the chemical residue that is left on them from their discharging cycle as well.

Then the simple solution to the lethargic process of adding electrons back to the battery plates. It's to add electrons back to them before they become highly discharged which is the problem in the first place. Therefore; adding, electrons back to the battery while taking them out. Then we could kill two birds with one stone by adding energy back to our battery while taking them out.

If we assume, it'll be no benefits in adding energy back to our secondary cell batteries while taking it out because of the laws of physics. Then we're focusing on the wrong science based on my research. It seems like we haven't acquired enough information. To understand, what's required to add energy back to our secondary cell batteries while taking it out as well?

I found it's due to how we go about adding and subtracting energy from our secondary cell batteries, why we can't add energy back to them while taking energy out as well. It has little to do with the laws of physics in and of themselves; but, it has more to do with the laws of mechanics. But; most in the field of automotive battery technology, they don't believe me when I say this as well.

The reason we think that, we can't add electrical energy back to a secondary cell battery while using its chemical energy. We don't understand the difference between electrical and chemical energy. I find the only difference between them is how, the electrons will be delivered to a load source. In which; it involves, the type of electromotive force being used to deliver the electrons.

I find it could be light, heat, pressure, magnetic or a chemical electromotive force could be used to deliver the electrons to a load source. Unlike; a primary cell battery, a secondary cell battery's chemical composition. It doesn't generate its own electrons; but, its chemical reaction generates a potential difference between its negative and positive electrode to release electrons.

Therefore; a secondary cell battery, it's nothing more than a storage device for electrons. Then the idea, we can't add electrons back to it while taking them out. It shows a lack of understanding about the mechanics behind its cycling processes. Or we're focusing on the wrong science when it comes to its energy input and output process as well.

Here's why; although, the battery chemical reaction produces an electromotive force large enough to supply a usable amount of electrical current for general applications. However; we've to add electrons to the battery after, it has been assembled if it's going to do any type of electrical work; in other words, the battery is nothing more than a storage device for electrons.

Thus; the number of plates and the size of the plates in each cell in the battery, they'll determine the amount of electrons that could be stored in it; plus, it'll determine its total amp capacity as well. Regardless of the size of the plates and the number of plates in each cell, the residual effects of releasing electrons from them. It'll limit the amount of electrons that could be released from them as well.

Remember a lead acid battery electrical resistance is calculated by squaring current in amperes and multiplying it by the internal resistance of the battery. Thus; the amount of lead-sulfate has accumulated on its plates, it's compounded by their electrical resistance as well. Therefore; increasing, the amount of energy lost to heat during the discharging cycle of the battery as well.

The oxymoron behind the discharging cycle of the battery, it's the reason why we can't get back the same amount electrons that were put in. Since all the chemical compositions used in our secondary cell batteries today, they've the same problem. In which; it seems like, there's no such thing as a perfect chemical; composition to a longer-lasting battery as well.

Since adding electrical energy back to a secondary cell battery, it'll diminish its internal resistance; that is, created by its chemical composition. In which; it's responsible for, the oxymoron behind the discharging cycle of the battery. However; adding, electrical energy back to it will restore its chemical energy as well. Thus; adding, energy back to it while taking energy out should be our primary goal.

Here's why; because we can eliminate, the oxymoron that is associated with the discharging cycle of the battery if we've an intermittent charge on it. Wherein; it'll take less time and energy to add electrons back to its plates because they'll have less lead-sulfate build up on them. As a result; it'll eliminate, the long drawn out process of adding electrons back to them as well.

Understanding, the mechanical make-up and the mechanics behind the cycling processes of a lead acid automotive battery. I'm obsessed with the idea there's another option to increase the overall travel ranges of our vehicles on battery power. Without finding that, perfect chemical composition to a longer-lasting battery for them as well.

On a side note; regardless of the size of a battery plates and the number of plates in each cell, it'll not increase its voltage potential because it's derived from the type of chemical elements used to make up its chemical composition. In other words; one type of chemical composition, it might give you more or less voltage than another type of chemical composition would.

Let me give you another reason why, I'm obsessed with adding energy back to our batteries while taking it out as well. When a load source is connected to the terminals of a lead acid automotive battery, then electrons will flow from it to the load source; until, all of the battery electromotive force has been used up in order to power the load source.

When the battery electromotive force has been used up and then, it's said to be dead or fully discharged. However; it doesn't mean, all of the electrons in the battery have been used up. What it does mean is that, the battery chemical electromotive force has become so weak. It can't push out the remaining electrons because it could no longer overcome the resistance of the battery.

Thus; we can't get back, the same amount of electrons that were put in because of the residual effect of releasing electrons from the battery plates. Remember as more and more electrons are released from them. Then more and more of the chemical residue will accumulate on them; therefore, reducing their active areas to release more electrons from them.

In short; the battery voltage, it continue to weaken as more and more electrons are released from it. Since adding, electrons back to the battery will diminish the lead-sulfate buildup on its plates. Thus; decreasing, its internal electrical resistance as well as increasing its efficiency and overall capacity for its next chemical discharging cycle as well.

Understanding, how the chemical composition of the battery will behave during charging cycle of the battery. It shows there's a mechanical solution to increase its efficiency and overall capacity. If we find away to, have an intermittent charge on it while using it as well. This is why I'm obsessed with charging, the battery while discharging it to a load source as well.

Not only designing a lead acid automotive battery, so, electrons could enter and exit it at the same time is the idea here; but, how its chemical composition behaves during its charging cycle is part of the idea as well. Given that; adding electrons back to its chemical composition, it diminishes the lead-sulfate buildup on its plates and it restores their chemical energy as well.

Here's why; in order to restore, a lead acid automotive battery chemical energy. We use a charging source with a voltage potential higher than the normal battery voltage. For example;

if we've a 12-volt battery, then charging voltage must be greater than 12-volts in order to reverse the current flow back into the battery to charge it.

We normally use a magnetic electromotive force produced by a generated power source to charge the battery. Using solid-state or electronic circuitry to produce direct current, then we apply direct current across the terminals of the battery to add electrons back to it; thus, reversing the current flow back through it to replenish its chemical energy as well.

I found reversing the current flow back through the battery, it could mean two things. One is reversing the remaining electrons back through the battery; that is, left in it from its previous discharging cycle or two electrons, they're flowing in it from a charging source as well. Usually; it's the latter when we're talking about reversing, the current flow back into a battery.

Normally; when someone is talking about reversing, the current flow back through a battery. It means electrons are flowing into it from a charging source, opposed to electrons flowing out of it to a load source. Since a lead acid automotive battery chemical composition, it doesn't generate its own electrons. Then it'll need an internal and an external electromotive force to carry out its cycling processes.

Then the only difference, I found between the charge and the discharging cycle of the battery is that. The direction electrons are flowing between its negative and positive electrode during its cycling processes. Thus; the definition of its charge and discharging cycle, they only amounts to electrons flowing into it from a charging source or flowing out of it to a load source.

Not only does an oxymoron exist with the chemical structure of a lead acid automotive battery during its discharging cycle. Given that; as more and more electrons are removed from its plates, then less and less electrons could be removed from them. However; an oxymoron, it exists with the mechanical structure of the battery as well.

What I found during my research is that, the only reason the battery charge and discharging cycle. They can't exist at the same time because electrons, they've to go in it the same way they've to come out; in which, it's an oxymoron within itself. Given that; the cycling processes of the battery, they're a paradox within themselves because it could be charged or discharged.

However; we can't charge, the battery while discharging it because we've to charge it on the same terminal connection that it's discharged on. Since the battery, it'll need an internal electromotive force to discharge it and an external electromotive force to charge it. Then we'll find the process of adding electrons back to the battery will not stop the process of subtracting them from it.

It'll be a matter of which electromotive force will have the highest voltage potential between the charging voltage and the battery voltage. It'll determine which direction electrons will flow between the negative and the positive electrode within the battery. However; if the battery, it has more than two electrodes to reverse the current flow between.

Then the narrative changes concerning charging the battery while discharging it. Given that; it doesn't have to be charged on the same terminal connection that, it was discharged on. Since its chemical structure is two directional, then we don't have to stop subtracting electrons from its plates in order to add electrons back to them as well.

Once we change the narrative; that is, changing the mechanical structure of the battery from a one directional to a two directional system; so, electrons could enter and exit it at the same time. Then we don't have to stop its discharging cycle in order to start its charging cycle because electrons don't have to go in it the same way they've to come out as well.

During my research on the mechanical structure of a lead acid automotive starting battery, I talked to a mechanical engineer at Kettering University of Flint, Michigan. I was trying to get a better understanding about the mechanical structure of the battery and how, it played a role in its cycling processes; but, the engineer only wanted to talk about its amp capacity.

In which; it led me to believe that, the engineer didn't have enough knowledge about. How the battery mechanical structure will play a role in its cycling processes. He did give me his opinion about charging the battery while discharging it; in which, he thought it would be impossible; even if, electrons could enter and exit it at the same time as well.

I still listened to his comments about how to arrive at the total amp capacity of the battery as a tool to select the right battery for the right job. What I learned in the field of heating, cooling and refrigeration, it helped me to understand what the engineer was talking about when it comes to the total amp capacity of a secondary cell battery.

Here's the thing; it takes, one volt to move one ampere of current against one ohm of resistance. In which; 12 volts, it could move 12 amps against 12 ohms of resistance. But; 12 volts, it can't move 12 amps against 24 ohms of resistance because it'll not enough voltage to overcome the resistance; thus, it'll reflect the current away or electrons away.

The rate at which current will flow to a load source, it'll be based on two factors; the resistance of the load source and the voltage applied. Thus; the total amp capacity of a battery, it'll not matter if its voltage potential is less than the voltage needed to supply electrons to a load source. Given that; it'll not enough, voltage potential to supply electrons to the load source.

The total amp capacity of a battery, it'll determine by the amount of amps could be consumed from it at a given rate. For example; a 12-volt battery with an amp capacity of 480 amp hours, it means a load source requiring 12-volts and 1 amp to operate will have a run time of 480 amp hours. That is 480 amps divided by 1 amp to operate the load source.

On the other hand; if a load source, it requires 240 amps to operate. Then it'll have a run time of 2 amp hours using the same battery. It's important to understand, how to arrive at the total amp capacity of a battery for a particular application because it'll enable us to select the right battery for the right job as well.

While reflecting back on my conversations, I had with the engineer. I realized our conversation. It did included how the mechanical structure of the lead acid automotive starting battery played a role in its cycling processes. We didn't realize it at the time because neither one of us knew, how the battery mechanical structure played a role in its cycling processes as well.

Here's the thing; if a lead acid battery, it has three cells and they're 2-volts each with an amp potential of a 125-amps each. If the battery cells, they're connected in series with one another. Then its total amp capacity will be 750-amps; that is, 2-volts from each cell time the amp potential of one cell equaling 750 amps.

On the other hand; if the battery cells, they're connected in parallel with one another. Then it'll still have a total amp capacity of 750-amps because it'll be the amp potential of all of its cells, time's the voltage potential of one cell. That is 125 amps from each cell times the voltage potential of one cell, which equal to 750-amps as well.

Then how a battery cells, are interconnected with its terminal connections. It'll determine its total amp capacity as well. Likewise; if a battery cells, they're connected in series with its terminal connections. During its discharging cycle, its voltage potential will equal to the sum of all of its plates in each cell; but, its amp potential will only equal to its plates in one cell.

Now if a battery cells, they're connected in parallel with its terminal connections. During its discharging cycle, its voltage potential will only equal to the sum of its plates in one cell; but, its amp potential will equal to all of its plates in each cell. Therefore; the mechanical structure of a battery, it'll determine its total amp capacity as well.

What I did learn from the mechanical engineer is that, we just can't focus on the total amp capacity of a battery to determine if it could or couldn't power a load source. Given that; some load sources, they require more voltage than others. However; the mechanical structure of a battery, it tells us how we arrived at its total amp capacity as well.

Without changing, the contemporary design of a lead acid automotive starting battery; so, electrons could enter and exit it at the same time. It'll be disingenuous to assume it can't be charged while being discharged because of the laws of physics. It's like assuming the earth is flat, airplanes can't fly, sound can't be recorded and sea life doesn't exist in the deepest parts of the oceans.

In the past most people thought that, the earth was flat, airplanes couldn't fly, sound couldn't be recorded and sea life didn't exist in the deepest parts of the oceans as well. Given that; those in the scientific community, they told them so. However; some ventured outside the status quo and investigated the matters for themselves and then, our thinking changed.

Then we realize the earth was round, airplanes could fly, sound could be recorded and sea life does exist in the deepest parts of the oceans as well. After investigating, the mechanical and the chemical structure of lead acid automotive starting battery. I found its mechanical structure will determine <u>how</u> electrons will be added or subtracted from its plates.

However; I found that, the chemical structure of the battery. It'll determine if electrons <u>could</u> be added or subtracted from its plates. Also; I found, when we're talking about charging it while discharging it; somehow, we end up focusing on the wrong science. Maybe; we don't, know which science to focus on when it comes its cycling processes as well.

In the next chapter; let's explore, the electrical aspects of a lead acid automotive starting battery. Maybe; we'll find out which science, we should be focusing on when it comes to the cycling processes of the battery as well.

CHAPTER 5

ELECTRICAL ASPECTS OF AN AUTO BATTERY

WHILE EARNING, MY ASSOCIATE Degree in the field of heating, cooling and refrigeration. I learned there are three types of electrical circuit connections: a series, a parallel or a series-parallel circuit connection. These type of circuit connections, they're use to connect our load sources together; such as, electric motors, fans and safety switches when they're powered by the same power source.

To qualify as an electrical circuit connection, there must be a power source, a conducting source to carry the current and a load source to use the current. However; the average person, they don't know a lead acid automotive starting battery. It behaves like an electrical circuit during its cycling processes as well.

I only realized this myself when I started to carry out my experiments on the battery. Then I found it has all three attributes of an electrical circuit during its cycling processes; in which, it was surprising to me because I didn't think of it being such a device. After evaluating, its electrical aspects my thinking changed about its cycling processes as well.

Here's why; during the discharging cycle of the battery, its plates. They're the power source; but, its terminals, straps, tab connectors and its electrolyte solution. They're the conducting sources will carry the current to a load source. The amount of current flowing to the load source from the battery will be based on, how its plates and cells are connected with its terminals.

During the charging cycle of the battery, the charging source will be the power source for the battery plates; in which, they'll be the electrical devices using the current. The battery terminals, straps, tab connectors and its electrolyte solution will be the conducting sources. To carry the current to its plates based on, how they're interconnected with its terminals.

Surprisingly; I found that, a lead acid automotive starting battery has all three attributes of an electrical circuit during its cycling processes. For a moment, let's forget about the assumption.

It's due to the laws of physics why the battery can't be charged while being discharged as most in the field of automotive battery technology claims.

For a moment, let's focus on mechanics of the voltage and current and how they'll flow to and from our load sources. Such as; electric motors, ceiling fans, light bulbs and safety switches when they're powered by the same power source. Then it might seem conceivable to charge, a lead acid automotive starting battery while discharging it to a load source as well.

Given that; the battery plates, they behave like load sources during their charging cycle. Such as; electric motors, ceiling fans, light bulbs and safety switches consuming energy from the same power source as well. To understand why, it's possible to charge the battery while discharging it. We must understand there's a difference between its charge and discharging cycle as well.

During the discharging cycle of the battery, its plates act like a power source supplying energy to a load source. But; during its charging cycle, its plates act like load sources consuming energy from a power source. Not understanding the difference between them, it'll be difficult to comprehend why it'll be possible to charge the battery while discharging it as well.

Once we understand, there's a distinction between how electrons will flow out of a battery during its discharging cycle and how they'll flow into it during its charging cycle. Not only will we find it conceivable to charge it while discharging it; but, we'll understand how to design it; so, it could be charged while being discharged to a load source as well.

For instance; if we're using a deep cycle battery to power an electric motor and then, we attempt to charge the battery while it's still powering the electric motor. Then the battery will no longer be the power source for the electric motor; but, it'll become like the electric motor trying to consume energy from the charging source as well.

Thus; electrons will be distributed between the battery and the electric motor based on how, they're interconnected with one another in their relationship with the charging source. Then we'll realize the electrical connection made between them, it'll determine how electrons will flow between them. Since the same laws apply to our electrical devices, they'll apply to the battery plates as well.

When we're talking about charging, a secondary cell battery while discharging it to a load source. We must pay extra attention to the type of circuit connections made between them. However; we usually, end up focusing on the laws of physics. In which; I found that, we'll be focusing on the wrong science when we focus on those laws of physics.

If we focus on the mechanics behind, the charging cycle of a battery and not focus on its discharging cycle when we're talking about charge it while discharging it. Then we'll find ourselves focusing on, the right science. However; we've to separate voltage from current in order to understand why, it'll possible to charge the battery while discharging it as well.

Here's why; when it comes to a <u>series</u> circuit connection, the sum of the voltage drop throughout the circuit. It'll equal to the voltage flowing from the power source; but, current flowing from the power source will be the same throughout the circuit. When it comes to a <u>parallel</u> circuit connection, voltage is the same throughout each branch of the circuit.

However; current, it'll be divided between the branches of the circuit based on their individual resistance. Now; when it comes to a <u>series-parallel</u> circuit connection, then the laws governing both series and parallel circuitry will apply. Wherein; voltage and current, they'll flow to each branch in the circuit. Based on, how they're interconnected with one another in their relationship with a power source.

Since a secondary cell battery plates, they'll behave like load sources consuming energy during their charging cycle. Then the laws of series, parallel and series-parallel circuit connections will apply to them as well. However; voltage and current, they'll flow through those circuit connections differently. So; this is why, we've to separate voltage from current.

I found circuit connections; such as, series, parallel or series-parallel circuit connections. They'll determine how voltage and current will flow to and from a battery plates or to a load source. Such as; an electric motor, a fan or a safety switch consuming energy from a power source as well. Understanding this, then we'll not be focus on the laws of physics as well.

Such as; the electromagnetic laws of physics, the laws of electrochemistry, entropy, inertia or the laws of conservation of energy. When it comes to charging, a secondary cell battery while discharging it to a load source as well. We'll be to focusing on the connections made between the battery and the load source will in their relationship with a charging source.

Then we'll be focusing on the right science when it comes to charge a battery while discharging it to a load source. In other words; we'll not be focusing on, the electromagnetic laws of physics, the laws of electrochemistry, entropy, inertia or the laws of conservation of energy; but, we'll be focusing on the laws of series, parallel or series-parallel circuitry as well.

Whether; it'll be a series, a parallel or a series-parallel circuit connection made between a battery and a load source in their relationship with a charging source. Those circuit connections will determine how, voltage and current will flow between them. In which; I find that, it's the key to charging the battery while discharging it to the load source as well.

Here's why; voltage and current, they'll flow <u>to</u> a battery straps and tab connectors on its plates different than they'll flow <u>from</u> its tab connectors and straps on its plates connecting them together in parallel. I'll discuss this in more detail later on. My point here is that, we've to separate voltage from current; although, current is a combination of voltage electrons as well.

This is why we must have a clear understand about, the difference between voltage and current. If we're going to understanding, it's about the laws of mechanics when it comes to adding and subtracting electrons from our secondary cell batteries. If we don't understand this, then it'll seem so inconceivable or farfetched to charge them while discharging them to a load source as well.

Once we understand, it's about the laws of mechanics. Then it wouldn't seem so inconceivable or farfetched to charge our secondary cell batteries while discharging them to a load source as well. For instance; if a battery cells, they're connected in series with its terminal connections. Then the charging voltage and current, they'll flow to and from its plates.

The same way voltage and current will flow to and from our load sources; that is, connected in series with one another as well. Likewise; if a battery cells, they're connected in parallel with its terminal connections. Then the charging voltage and current will flow to and from its plates, the same way they'll flow to and from our load sources; that is, connected in parallel with one another as well.

Now; if a battery plates in each cell, they're connected in series-parallel with one another. Then the charging voltage and current, they'll flow to and from its plates. The same way voltage and current will flow to and from our load sources; that is, connected in series-parallel with one another as well.

Let me show you what I am alluding to, by analyzing how a lead acid automotive starting battery plates in each cell are interconnected with one another. In which; all the battery positive plates in each cell, they're interconnected with one another by a single strap. Likewise; all its negative plates in each cell, they're interconnected with one another by a single strap as well.

Thus; all of the battery positive plates in each cell, they're connected in parallel with one another. Likewise; all of its negative plates in each cell, they're connected in parallel with one another as well. However; its positive and negative plates in each cell, they're interconnected with one another by way of the electrolyte solution in each cell.

Therefore; the battery positive and negative plates in each cell, they're connected in series with one another by way of the electrolyte solution in each cell. Wherein; the battery plates in each cell, they're connected in series-parallel with one another by their straps, tab connectors and the electrolyte solution in each cell as well. See chapter 3.

Basically; how, the battery plates in each cell is constructed. We're using the laws of series, parallel and series-parallel circuitry to charge the battery plates in each cell during their charging cycle. Without having to, stop charging them in one cell in order to charge them in the next cell. All because of the type of connections made between its plates in each cell as well.

This is why, I say it's about the laws of mechanics and not about the laws of physics when it comes to charging a battery while discharging it to a load source or let's say adding and subtracting electrons from a battery plates as well. Based on, the conventional current flow concept that is voltage and current has to enter or exit the positive terminal of the battery in order to cycle it.

Let me describe how the charging voltage and current will behave flowing in the battery when they've to flow through the positive terminal connection of it. And then, flow to its strap and tab connectors connecting them together in parallel with its positive plates in that cell. Thus; the positive plates in that cell, they'll receive voltage and current based on the laws of parallel circuitry.

Since the positive plates, they'll be connected in parallel with the positive terminal connection by way of a strap and tab connectors in that cell; therefore; they'll receive the same amount of voltage flowing from the charging source. However; current flowing to those plates from the charging source, it'll be divided amongst them based on their individual resistance.

When the charging voltage and current, they're transferred from the positive plates to the adjacent negative plates in the cell with the positive terminal connection in it. The adjacent negative plates in that cell, they'll receive a voltage drop from the charging source because they're connected in series with the positive plates by way of the electrolyte solution in that cell.

The negative plates in the cell with the positive terminal connection in it, they'll receive the same amount of current as the positive plates in that cell. Given that; they're connected in series with one another, by way of the electrolyte in that cell. Thus; the charging voltage and current, they'll flow to and from those plates in that cell based on the laws of series-parallel circuitry.

Here's why; since the battery plates, they'll behave like load sources during their charging cycle and they'll be connected in series-parallel with one another by their straps and tab connectors and the electrolyte solution in that cell. Then the charging voltage and current will flow to and from them, the same way voltage and current will flow to and from our load sources as well.

We overlook a series-parallel circuit connection made between a lead acid automotive starting battery plates in each cell. Since we only think of the electrolyte solution in each cell as a medium that is ionized, by the charging current. What we don't realize is that, the electrolyte solution in each cell. It'll create a series circuit connection between the negative and the positive plates in each cell.

If we don't understand this, then we can't begin to comprehend why it's possible to charge the battery while discharging it to a load source. I found the electrolyte solution in each cell is the reason why electrons, they don't have to flow to and from the same tab connector on the same plate. In order to, flow to and from the same plate in each cell as well.

Wherein; we're able to charge, the battery plates in each cell without having to stop charging one plate in order to charge another; given that, they've only one tab connector each. So; it has little to do with, the laws of physics; but, it has more to do with the laws of mechanics when it comes to adding or subtracting electrons from the battery plates in each cell.

Likewise; when it comes to, charging a battery while discharging it to a load source. We can't focus on the laws of physics; but, we've to focus on the laws of mechanics as well. On the other hand; we've understand, there's a distinction between a battery charge and discharging cycle. Given that; voltage and current, they'll enter it differently than they'll exit it during its cycle processes.

In other words; during the charging cycle of a battery, voltage and current will flow to it different than they'll flow from it during its discharging cycle. Thus; we've to understand, the electrical aspects of a battery during its cycling processes. If not, then we can't begin to comprehend why it can't be charged while being discharged to a load source as well.

Understanding, the electrical aspects of a battery is just as important as understanding its mechanical and chemical aspects as well. If we don't understand these things, then we can't comprehend why it'll be possible to charge the battery while discharging it to a load source. It's a host of things we've to understand when it comes to concept of charging a battery while discharging it.

We just can't focus on one thing and think it's enough to determine if a battery could or couldn't be charged while being discharged to a load source. We've to focus on a lot of things concerning, the mechanical, the chemical and the electrical aspects of the battery in order to understand what's required to charge it while discharging it to the load source as well?

The first thing people want to talk about is the laws of physics when it comes to a conversation about charging, a lead acid automotive starting battery while discharging it. Given that; they think the laws of physics, are reasons it can't be charged while being discharged. The reason they think this because it's due to the lack of information about the battery as well.

Here's why; since the battery plates in each cell, they'll be connected in series-parallel with one another by their straps, tab connectors and the electrolyte solution in each cell. Then voltage and current will flow to its straps and tab connectors. It'll be different than voltage and current flowing from its tab connectors and straps connecting its plates together in parallel as well.

Using the conventional current flow concept again, let me repeat how the charging voltage and current will enter the positive terminal connection of the battery. So; when the charging current, it flows through the positive terminal. And then, reaches the strap and tab connectors connecting them together in parallel with the positive plates in that cell.

Then the charging current, it'll be divided amongst the positive plates based on their individual resistance as it flows across them. Remember each positive plate in the cell with the positive terminal connection in it. They'll receive the same amount of voltage flowing from the charging source because they'll be connected in parallel with the positive terminal connection.

Thus; how, voltage will flow to the positive plates connected in parallel with the positive terminal connection. It'll be different than current flowing to them from the charging source.

But; it'll be a twist, when voltage and current will flow <u>from</u> the positive plates through the electrolyte solution <u>to</u> the adjacent negative plates in the cell with the positive terminal in it.

Although; each positive plate in the cell with the positive terminal connection in it, they'll receive the same amount of voltage flowing from the charging source; but, the adjacent negative plates in that cell. They'll receive a voltage drop from the charging source because they're connected in series with the adjacent positive plates in that cell by way of the electrolyte solution.

When current, it leaves the positive plates in the cell with the positive terminal in it. And then, flow through the electrolyte solution to the adjacent negative plates. They'll receive the same amount of current as the adjacent positive plates in that cell because they're connected in series with one another by way of the electrolyte solution in that cell.

Although; current, it was divided amongst the positive plates in the cell with the positive terminal in it. Based on, their individual resistance when it was flowing across them from their strap and tab connectors connecting them together in parallel with the positive terminal. It'll be a twisted when voltage leaves the negative plates and flow to their tab connectors and strap.

It'll be a voltage drop at the strap as voltage flows from the negative plates to the tab connectors and strap connecting them together in parallel and a strap connecting them in series with the positive plates in the next cell as well. In other words; voltage, it doesn't combine at the strap connecting the negative plates together in parallel like current does.

However; when current, it leaves the negative plates in the cell with the positive terminal in it. And then, flow to their tab connectors and strap connecting them together in parallel and in series with the positive plates in the next cell. Then the sum of the current flowing across each plate, it'll be combined at the strap connecting them together in parallel.

Let me explain this in a different way; so, you'll have a better understanding how voltage and current will flow through each cell in a lead acid automotive starting battery during its charging cycle. Since it's one important aspect in understanding how to, charge the battery while discharging it to a load source as well.

When the charging voltage flows across, the positive plates connected in parallel with the positive terminal connection of the battery. Then the voltage coming from the charging source, it'll be the same across each positive plate in the cell with the positive terminal connection in it because they'll be connected in parallel with the terminal connection as well.

When voltage is transferred from the positive plates to the adjacent negative plates by way of the electrolyte solution in that cell, it'll be a voltage drop across the negative plates in that cell. Thus; the positive plates in the next cell, they'll receive a voltage drop from the negative plates in the previous cell because they're connected in series with one another by an electrical bus.

Therefore; voltage, it'll continue to drop evenly throughout each cell in the battery during its charging cycle. However; when the charging current, it's transferred from the positive plates to the adjacent negative plates in the cell with the positive terminal connection in it by way of the electrolyte solution in that cell as well.

Then the amount of current flowing across each negative plate in that cell, it'll be the same as the charging current flowing across each positive plate in that cell as well. Given that; the positive and the negative plates, they'll be connected in series with one another by way of the electrolyte solution in that cell as well.

However; the charging current, it'll regroup at the strap connecting the negative plates together in parallel in that cell and in series with the positive plates in the next cell as well. Thus; the next cell in the battery, it'll receive the same amount of current flowing from the charging source as well; so, the charging current will be the same throughout each cell in the battery.

Wherein; it has little to do with, the laws of physics; but, more to do with the laws of mechanics. What I found while exploring the electrical aspects of a lead acid automotive starting battery, it was very interesting because when the same plates have separate straps and tab connectors as well. Unusual things happen when voltage and current will flow to and from the plates.

I found voltage and current will flow <u>to</u> one strap and tab connectors connecting the plates together in parallel. It'll be in reverse of how voltage and current will flow <u>from</u> the other strap and tab connectors connecting the plates together in parallel as well. Let me try to make this very simple and clear because it's very important concerning charging a battery while discharging it.

For examples; if 20 amps of current were flowing across 10 plates connected in parallel with one another by a strap and tab connectors. Then 20-amps of current will be divided amongst the plates based on their individual resistance as current flows across them. When current leaves, each plate and reaches the other tab connectors and strap connecting them together in parallel.

Then the total current potential at the strap, it'll equal to 20 amps; although, 2 amps of current was flowing across 10 plates if they all had the same resistance; that is, 20 amps divided by 10. Wherein; current, it'll combine at the strap connecting the plates together in parallel; however, it'll be in reverse when current flows <u>to</u> the plates from their strap and tab connectors.

If 12-volts was flowing across, the same 10 plates; that is, connected in parallel with one another by a strap and tab connectors as well. Then voltage, it'll be the same across each plate as it flows across them from the strap and tab connectors connecting them together in parallel; but, when voltage leaves each plate and reaches the other tab connectors and strap.

Then the total voltage potential at the strap, it'll equal to 12-volts; that is, 12-volts flowing across one plate. Given that; voltage, it'll not combine at the strap when it's flowing from the plates to the strap connecting them together in parallel like current does; although, 12-volts was flowing across each plate when it flowed to them from the strap and tab connectors.

In essence; voltage, it'll be in reverse of how it'll flow to and from the strap and tab connectors connecting the battery plates together in parallel. Given that; voltage, it'll be the same at the strap as it was flowing across one plate; although, 12-volts was flowing across each plate from the strap and tab connectors connecting them together in parallel as well.

We've to understand voltage and current will flow differently to and from a strap and tab connectors connecting a battery plates together in parallel. After cutting off the top of the plastic casing of a lead acid automotive starting battery and leaving, its terminals, straps, tab connectors, plates and cells intact.

I was able to determine how electrons would flow through each cell in the battery by watching, the bubbles created by electrons flowing through the electrolyte solution in each cell. Also; I used, ammeters and voltmeters to determine how electrons will flow through each cell. Thus; I realized that, the connections made between them determine how electrons will flow between them as well.

When it comes to the amp capacity of an automotive battery, it means the same thing as voltage times current equals total power; however, we just say total amp capacity instead. Wherein; voltage, it's an electromotive force that moves electrons and current is the movement of electrons in one direction through or around a conducting source.

If we're going to understand why, it'll be possible to charge our deep-cycle automotive batteries while using them to power our vehicles. Then we've to separate voltage from current because we just can't focus on, the movement of current alone and think we're going to know what's going on when it comes to the idea of charging our batteries while using them as well.

However; there's a difference between, voltage and current because voltage will flow to and from a battery plates different than current will since it's the movement of voltage and electrons. Then current, it'll flow through or around a conducing source different than voltage will because it handles resistance different than current does due to electrons being involved.

Since like charges, they'll repel one another and follow the path of least resistance as well. Then current, it'll flow to and from a battery plates different than voltage will because it could flow through or around a conducing source without electrons; given that, it's a potential difference between two points because it could be the same at a battery and a load source; but, current doesn't.

Although; voltage, it could be the same at the battery and at the load source; however, it doesn't mean electrons are flowing to them at the same time. If the resistance at the battery is greater than the resistance at the load source, then electrons will be reflected back toward the load source. Thus; we'll not be able to charge, the battery while it has the load source on it.

Unless; a surplus of electrons, they'll be created at the load source in order to reverse the current flow back into the battery to charge it. Given that; current, it'll flow differently to the battery than it'll to the load source because its resistance to current is different than the battery resistance to current. See chapter 12.

Not only do we've to separate voltage from current; but, we've to understand voltage will flow through a series, a parallel or a series-parallel circuit connection different than current will. If we don't understand the difference, then we can't begin to comprehend why we can't charge our deep cycle automotive batteries while using them to power our vehicles as well.

Upon evaluating the electrical aspects of a lead acid automotive starting battery, I found we've to focus on a lot of things to understand what's required to charge our deep cycle automotive batteries while using them to power vehicles. Or the mechanics involved in adding or subtracting electrons from our automotive batteries' plates as well.

Here's why; having, a charge and a load source on the same terminal connection of a battery. Thus; it and the load source, they'll be in parallel with one another in their relationship with a charging source; therefore, voltage and current will flow between the battery and the load source will be subject to the laws of parallel circuitry as well.

Likewise; charging, a battery while using it to power a load source. Then voltage and current, they'll be subject to the laws of parallel circuitry as well. Given that; the battery, it has only one opening for electrons to enter or to exit it. Thus; we've to change it on the same terminal connection that, it's discharged on; therefore, the wrong mechanics are in play to charge it while using it as well.

This is why it's important to understand that, a parallel circuit connection made between two load sources. It doesn't work the same as a parallel circuit connection made between a battery and a load source. Remember if we've a charge and a load source on a battery at the same time, then the battery is no longer a power source for the load source.

However; the battery, it becomes like the load source trying to consume energy from the charging source as well. Since the mechanics of the charging voltage and current, they'll behave differently under different circumstances. Although; the battery plates, they'll behave like load sources consuming energy during their charging process as well.

But; there's a difference between, how a battery plates will consume energy compared to how a load source will consume energy. It has something do with the difference in their resistance to conduct current. We've to understand a secondary cell battery plates' resistance to current is different than a load source to current or electrons as well.

In other words; the resistance of the battery plates to current, they tend to reflect current away from them. However; a load source resistance to current, it's based on the number of turns that the wiring has to conduct current inside the load source. Although; voltage, it'll be the same across the battery and the load source; but, current will not.

Since current, it has to go in the battery the same way it has to come out. Then the battery will always be in parallel with the load source in their relationship with the charging source. Thus; current, it'll flow to the load source before it'll flow into the battery because their resistance to current is different and they'll be in parallel with one another as well.

If we don't understand, all the different circumstance involved. Then we might get the wrong impression, it's due to the mechanics of the current alone or the laws of physics. Why current, it'll flow to the load source before it'll flow into the battery. Given that; we'll be focusing on, the wrong science to figure out what's going on during the process as well.

If we take an in depth analysis of the mechanics; that is, taking place when there's a charge and a load source on a battery at the same time. Then we'll find the connections made between them, they'll be insufficient to charge the battery while it has the load source on it. Given that; voltage, it'll flow between them different than current will because of the connections.

Then we'll know that, we've to change the contemporary design of the battery in order to make the right connections between it and the load source in their relationship with the charging source. Given that; we'll know, the battery can't be charged while being discharged in its present state because it'll not work mechanically or otherwise.

Therefore; we've to figure out, what's required to carry out such a process without being impeded by the mechanics of the current or the laws of physics as well. Since a parallel circuit connection won't work, then there's only a series or a series-parallel circuit connection left to choose from; given that, it's only three types of circuit connections.

If we understand, the mechanics behind the charging cycle of a lead acid automotive starting battery. Then we'll know that, a series-parallel circuit connection will be the right circuit connection needed to charge it while discharging it to a load source. If electrons, they could enter and exit the battery at the same time as well.

Here's why; the battery plates in each cell, they're connected in series-parallel with one another by their straps, tab connectors and the electrolyte solution in each cell. So; we don't have to, stop charging one plate in order to charge another; although, they've only one tab connector each. Thus; the mechanics, behind the charging cycle of the battery.

It'll give us the information needed to understand what's required to charge it while discharging it to a load source. Without being impeded, by the mechanics of the current alone or the laws of physics as well. After taken, an in depth analysis behind the mechanics of the voltage and current when it comes to the charging cycle of the battery.

I found it boils down to mechanics or the procedure in how, electrons are added or subtracted from the battery plates. Given that; voltage and current, they'll adhere to the design of the battery as well as they'll adhere to the design of an electrical circuit connection; such as, a series, a parallel or a series-parallel circuit connection as well.

Remember each cell in a lead acid automotive starting battery is nothing more than a battery in and of itself. That is connected in series with one another and housed together in a plastic case to make up the whole battery as we know of today. When it's charged as a whole, we're simultaneously charging and discharging one battery while simultaneously charging and discharging another.

So; the question is, are we focusing on the right science when we suggest it's impossible to charge the battery while discharging it to a load source because of the mechanics of the current alone or the laws of physics? After exploring the mechanical, the chemical and the electrical aspects of the battery, I don't think so!

In the next chapter; let's briefly explore why, the lack of information about the mechanics behind the cycling processes of a lead acid automotive starting battery. It'll create a misconception about what we could or couldn't do with the battery; especially, if we don't understand its electrical aspect.

CHAPTER 6

LACK OF INFORMATION ABOUT AN AUTO BATTERY

AFTER EVALUATING THE MECHANICAL, the chemical and the electrical aspects of a lead acid automotive starting battery, it's hard to believe most people think that. It can't be charged while being discharged because of the mechanics of the current alone or the laws of physics as well. Wherein; this line of thinking, it's due to the lack of information about its cycling processes.

Although; some in the field of automotive battery technology, their assumptions might sound convincing at first. When they say that, it's impossible to charge the battery while discharging it because of the laws of physics. Once we start analyzing their assumptions, they seem like they stemmed from the lack of information about the battery itself.

Most think charging the battery while discharging it is unprecedented. Although; it goes through a simultaneous cycle process each time, it's charged. If people aren't familiar with the mechanics behind the charging cycle of the battery, then they won't know this. If they're familiar with it, then they don't see it as such a process because it's not defined as such.

When I point this simultaneous cycling process out, they conjure up all manner of laws of physics to justify their thinking. It's impossible to carry out such a process because of the laws of physics. In other words; skeptics, they're using rationalization to justify their thinking. Because they only focus on, what happens on the outside of the battery during its charging cycle?

In which; the skeptics, they can't formulate a valid opinion about charging the battery while discharging it if they only looking at it from the outside of the battery and not from the inside of it during its charging cycle. On the other hand; the lack of information given by those in the field of automotive battery technology, it has shaped our opinion about our batteries as well.

Given that; we haven't taken, an in depth analysis of the mechanics behind the cycling processes of our batteries for ourselves. To see why, they function the way that they do because most of the time. The scientific communities, they don't share all their information about their findings. Wherein; most findings, they're based on assumptions and not facts.

In some instances; the scientific communities, they use scientific jargon to explain their findings. In which; most in the general public, they don't understand the scientific jargon; thus, it keeps most of us in the dark about the world around us. It's no different when it comes to our battery technology as well.

For instance; most battery experts say that, we can't use an electronic alternator regulator control system to charge a deep-cycle automotive battery. Given that; it'll cook the battery based on the information received from the battery manufactures. After compiling and comparing data from many different sources about electronic alternator regulator control system.

It seems like there're only two reasons why, an alternator charging system can't charge a deep-cycle automotive battery. That is the battery is almost empty or almost full when an electronic alternator regulator control system throws. The full load of the alternator charging system across the battery this is what the battery manufactures, not telling us.

I found an alternator charging system will cook an automotive starting battery if it's almost empty or almost full as well. When; the electronic alternator regulator control system throws the full load of the alternator charging system across the battery as well. Are those battery manufactures telling us, the whole story when it comes to the alternator charging system?

After my investigation, it seems like it really doesn't matter if it's a deep-cycle battery or an automotive starting battery. What does matter if the battery is almost empty or almost full when an electronic alternator regulator control system, it throws the full load of the alternator charging system across the battery?

What I discovered during my research is that, the biggest difference between a deep cycle (lead acid) automotive battery and a lead acid automotive starting battery. The deep cycle battery plates are much thicker and they've more surfaces space than an automotive starting battery plates, other than that the batteries are pretty much the same.

It seems like if we set up a very shallow cycling process for a deep-cycle automotive battery under an alternator charging system. So; the battery isn't almost empty or almost full when an electronic alternator regulator control system throws, the full load of the alternator charging system across the battery as it does with an automotive starting battery.

Then it seems plausible that, we could charge the deep cycle automotive battery with an alternator charging system. On the other hand; due to the lack of information about why, an alternator charging system can't charge a deep cycle battery. It probably stemmed from the fact that, it was almost empty or almost full when the attempt was made to charge it as well.

I find the same reason we think a deep cycle battery can't be charged with an alternator charging system. It's the same reason we think the battery can't be charged while being discharged as well. It's due to the lack of information about the mechanics behind its cycling processes. This lack of information, it arrived from those in the field of automotive battery technology.

It seems like there's a disconnection between, how those in the field of automotive battery technology design a lead acid automotive starting battery. And how, they think its cycling processes will be carried out within the realm of physic as well. It seems like some or most, they're not telling us the whole story why it can't be charged while being discharged as well.

It really doesn't make any sense why we've to stop the discharging cycle of the battery in order to start its charging cycle. Because its plates in one cell, they could consume and store energy while supplying it to its plates in the next cell. Without having to, stop charging them in one cell in order to charge them in the next cell; although, they're separated by a non-conducting material.

On the other hand; some in the field of automotive battery technology, they claim it doesn't matter if the battery plates in one cell. They could consume and store energy while supplying it to its plates in the next cell. They still can't consume and store energy while supplying it to a load source because of the laws of conservation of energy.

Wherein; the reason some in the field of automotive battery technology put forth, it doesn't co-inside with the concept of charging the battery while discharging it to a load source. Given that; we'll be adding, energy back to the battery while taking it out to power the load source as well. Since the laws of physics, they don't contradict themselves as we know of.

Then it must be another explanation why some in the field of automotive battery technology would use. The laws of conservation of energy to discredit adding energy back to a battery while taking it out to power a load source. After exploring a lead acid automotive starting battery electrical aspect, I think it's due to the lack of information they've about the battery.

Given that; energy has to go in the battery the same way, it has to come out in order to charge its plates or for them to be discharged to a load source as well. In which; it might explain why, we think it can't be charged while being discharged to a load source because of the laws of physics. In reality, it's due to the laws of mechanics; but, disguised as the laws of physics.

For instance; the reasons, we could charge the battery plates in each cell. Without having to, stop charging them in one cell in order to charge them in the next cell. Because energy, it could flow in one direction from the battery plates in one cell to its plates in the next cell. Given how the battery plates and cells, they're interconnected with its terminal connections.

Since energy, it has to go in the battery the same way it has to come out in order to charge its plates or for them to be discharged to a load source as well. Then we can't charge the battery while discharging it to the load source because it would violate the laws of physics. Given that; energy, it has to go in the battery the same way it has to come out as well.

The lack of information about, energy having to go in the battery the same way energy has to come out. It'll create a false impression disguised as the laws of physics why, the battery can't be charged while being discharged to the load source. This false impression, it's create by the contemporary design of the battery; in which, it's the mechanics of it.

The contemporary design of a lead acid automotive starting battery, it'll make it impossible to charge it while discharging it to a load source. Given that; the laws of physics, they'll not allow it because like charges will repel one another and follow the path of least resistance as well. Some claim, even if energy doesn't have to go in the same way it has to come out.

We still couldn't charge the battery while discharging it to a load source due to the laws of conservation of energy. Given that; energy, it'll be lost to heat in both conversion processes of the battery; wherein, they'll cancel themselves out. Therefore; no gain, if we attempt to carry out both conversion processes of the battery at the same time.

I find this claim comes from the lack of information about what happens on the inside of a lead acid automotive starting battery during its charging cycle. Remember the battery plates in each cell, they could consume and store energy while supplying it to its plates in the next cell. So; all of the plates in each cell, they could be charged during their charging cycle as well.

Although; energy loss to heat, it'll occur during the electrolysis process of the battery plates; but, it'll be made up by the charging source as well. Then the claim we can't charge the battery while discharging it because energy lost to heat, it'll occur in both of its conversion processes. It seems flawed because it's not telling, the whole story about the process itself.

If we're adding, energy back to the battery while taking it out. Then any energy lost to heat during the process, it should be made up by the charging source as well. Then claim, it'll be no benefits in adding energy back to the battery while taking it out because of the laws of conservation of energy is flawed. They've no bearing on whether the battery is charging or not.

Given that; any energy lost to heat during the process of adding energy back to the battery while taking it out, it should be made up by the charging source as well. Then claim it'll be no benefits in adding energy back to the battery while taking it out because of the laws of conservation of energy, it shows a lack of information about those laws of physics as well

If we're adding energy back to the battery while taking it out, then the laws of conservation of energy will not apply because we're adding energy back to the battery as well. Thus; it doesn't make any sense to discredit, the process because energy lost to heat occurs in both of its conversion processes as well.

On the other hand; the claim, it'll be no benefits in adding energy back to the battery while taking it out. The claim itself, it doesn't actually depict what's happening on inside the battery during a simultaneous cycling process of it. Although; energy, it'll be lost to heat during the simultaneous cycling process; but, it'll be made up by the charging source as well.

It seems like those who are using, the laws of conservation of energy to discredit a simultaneous cycling process of a secondary cell battery. They're trying to pull the wool over our eyes about what we could or couldn't do with our batteries. In other words; it doesn't matter if we change their contemporary, we still can't charge them while discharging them.

If we don't venture outside the status quo, then we might never know if those in the field of automotive battery technology are telling us the truth. If there's another option to increase, the overall travel ranges of a vehicle on battery power. Without finding that, perfect chemical composition to a longer-lasting battery for it.

Wherein; it's strange that, we could charge or discharge the same battery; but, we can't charge it while discharging it because of the laws of physics. If we understand how, the mechanical structure of a lead acid automotive starting battery comes together and how it'll play a role in its cycling processes.

Then we'll know those in the field of automotive battery technology, they're trying to pull the wool over our eyes. Given that; we'll know that, it'll have little to do with the laws of physics why a lead acid automotive starting battery can't be charged while being discharged; but, it'll have more to do with the laws of mechanics why it can't.

Here's why; the battery plates in each cell, they're connected in series-parallel with one another by their straps, tab connectors and the electrolyte solution in each cell. So; electrons, they could flow from a strap to a tab connector on one plate. And then, flow across that plate through the electrolyte solution to an adjacent plate.

Without electrons flowing to and from, the same tab connector on the same plate. Thus; the battery plates in one cell, they're simultaneously charged and discharged amongst themselves in each cell. Since the plates in one cell, they're connected in series with the plates in the next cell. Then they're simultaneously charged and discharged to its plates in the next cell as well.

Based on the information about the charging cycle of the battery, it seems like it's not due to the laws of physics why it can't be charged while being discharged. In which; it seems like, it has more to do with mechanics because of how electrons have to enter and exit each cell within the battery during its charging cycle as well.

Wherein; electrons, they could flow in one direction from one plate to the next within the battery and to its plates in the next cell. Without having to, flow to and from the same tab connector on the same plate in order to flow to and from the same plate. Using mechanics, we don't have to stop charging the battery plates in one cell in order to charge them in another.

It seems like those in the field of automotive battery technology, they're focusing on the wrong science. When they suggest that, it's due to the laws of physics why we've to stop the discharging cycle of the battery in order to start its charging cycle. Actually; it's due to the laws of mechanics, disguised as the laws of physics because we don't have enough information about the battery.

Having the lack of information about the mechanical structure of a battery and how, it'll play a role in its cycling processes. We might get the wrong impression why we've to stop its discharging cycle in order to start its charging cycle. Wherein; it might be a disconnection between how we think it's designed and how, we think its cycling processes will be carried out as well.

Obtaining information about the mechanics behind the charging cycle of a lead acid automotive battery, it seems like those in the field of automotive battery technology. They're not telling us the whole story why the battery can't be charged while being discharged. They're just telling us that, it's due to the laws of physics with lingering questions why.

It seems like they're speculating because their claim contradicts the mechanics behind the charging cycle of a lead acid automotive starting battery. Since we can't see beyond the plastic casing of the battery, then we can't compare what's happening on the outside of it to what's happening on the inside of it during its cycling processes as well.

Then we're inept to ascertain; whether, it's due to the laws of physics why the battery can't be charged while being discharged due to the lack of information. Thus; we're left at mercy of those in the field of automotive battery technology to give us, the whole story; wherein, they have been lacking in this area concerning the mechanics behind the battery.

What I discovered is that, the interior design a lead acid automotive starting battery. It allows electrons to be simultaneously added and subtracted from its plates during their charging cycle. Because of how, they're interconnected with its terminal connections. But; its exterior design, it'll not allow electrons to be simultaneously added and subtracted from its plates.

In other words; we can't simultaneously add and subtract electrons from the battery plates in order to power a load source because of how, they're interconnected with its terminal connections as well. The lack of information given by those in the field of automotive battery technology about the battery, it's disturbing and it seems disingenuous because it doesn't add up.

Here's why; remember each cell in a lead acid automotive battery, it's nothing more than a battery in and of itself. That is connected in series with one another and housed together in a plastic case to make up the whole battery as we know of today. When it's charged as a whole, we're simultaneously charging and discharging one battery to another.

So; it begs the question, are those in the field of automotive battery technology telling us the truth? When they say that, we can't charge the battery while discharging it to a load source because of the laws of physics. What's the big difference between the battery plates in one cell consuming and storing energy while supplying it to its plates in the next cell?

Compared to, the battery plates consuming and storing energy in one cell while supplying it to a load source as well. I find the answer is mechanics because energy has to go in the battery the same way that, energy has to come out in order to charge its plates or for them to be discharged to the load source; but, those in the field of automotive battery technology haven't mention this.

Then the next question is why the battery plates in one cell, they could be simultaneously charged and discharged to its plates in the next cell. Without being impeded, by the laws of physics and the answer is mechanics as well. Remember energy could flow in one direction from the battery plates in one cell to its plates in the next cell as well.

Wherein; the process of energy flowing in one direction, it'll allow the battery plates in one cell to be simultaneously charged and discharged to its plates in the next cell. Without being impeded, by the laws of physics. The lack of information given by those in the field of automotive battery technology, it makes it hard to figure out why it can't be charged while being discharged as well.

However; it'll not hard to figure out, why the battery can't be charged while being discharged. If we take an in depth analysis of its mechanical structure and how, it'll play a role in its cycling processes as well. On the other hand; some in the field of automotive battery technology, they'll add more confusion to why it can't be charged while being discharged as well

They'll use the laws of entropy, inertia or the laws of conservation of energy to discredit a simultaneous cycling process of the battery. Wherein; those assumptions, they contradict the very essence of what those laws of physics mean. If we're adding energy back to the battery while taking it out, then those laws of physics will not apply to the concept of charging it while discharging it.

Wherein; the battery, it'll not fall under the category of a closed or a perpetual motion system; in which, those laws of physics depict because energy isn't introduced back into the system to keep its energy from running down. Therefore; the idea, it's either due to the laws of entropy, inertia or the laws of conservation of energy is absurd in and of itself as well.

Here's why; those assumptions, they contradict the foundation of the laws of entropy, inertia or the laws of conservation of energy as well. Because those assumptions, they don't convey what those laws of physics depict. In other words; if energy, it's not added back to a closed or a perpetual motion system. Then its energy will eventually run down under those laws of physics.

I could tell you within a couple of paragraphs why the laws of entropy, inertia or the laws of conservation of energy will not apply to a simultaneous cycling process of a secondary cell battery. However; you probably wouldn't believe me, given the conventional wisdom we possess about the cycling processes of our secondary cell batteries as well.

Wherein; it requires me to, explain line up on line and precept up on precept why the laws of entropy, inertia or the laws of conservation of energy. They'll not apply to a simultaneous cycling process of our secondary cell batteries. If we're adding, energy back to our batteries while taking it out as well.

It's a host of things, other than the laws of physics will prevent our secondary cell batteries from being charged while being discharged. Likewise; a host of things, they've to come into play in order to charge our batteries while discharging them as well. Therefore; if I don't keep repeating myself, line up on line and precept up on precept while writing my book.

Then I probably couldn't overcome the conventional wisdom, we possess about the cycling processes of our secondary cell batteries; so, you probably wouldn't believe why we could charge them while discharging them as well. I believe the lack of information given by those in the field of automotive battery technology is vital information needed.

Without this vital information, we're left at mercy of those in the field of automotive battery technology. About we could or couldn't do with our secondary cell batteries within the realm of physics. Likewise; without this vital information, we don't know if finding a perfect chemical composition to a longer-lasting battery is the key in revolutionizing the electric vehicle as well.

On the other hand; I discovered that, a lead acid automotive battery plates. They behave like load sources consuming energy during their charging process; such as, electric motors and ceiling fans. In which; this type of information, it's vital to understand why the battery could or couldn't be charged while being discharged to a load source as well.

Wherein; this type of information, it's lacking for those in field of automotive battery technology as well. Also; I found that, voltage and current will flow to and from a secondary cell battery plates to a load source. Based on, how they're interconnected with one another in their relationship with a power source as well.

In other words; if a battery plates, they're connected in series, in parallel or in series-parallel with a load source. Then voltage and current, they'll flow to and from the battery to the load source according to how they're interconnected with one another. This type of information, it's not given by those in the field of automotive battery technology as well.

While earning, my Associate Degree in Science; in the field of heating, cooling and refrigeration. It helped me to understand how voltage and current will flow through the body of a battery based on its design. Taking a lead acid automotive starting battery apart piece by piece, it gave me insight into why it functions the way that it does.

Wherein; the design concept of the battery, it's the deciding factor in determining how voltage and current will enter or exit it. I found a parallel circuit connection made between it and a load source in their relationship with a charging source. It'll be one of the hosts of things will preventing the battery from being charged while being discharged to the load source as well.

Notwithstanding; the battery, it has only one opening for electrons to enter or to exit it. The contemporary design of the battery, it'll allow an insufficient circuit connection to exist between it and a load source in their relationship with the charging source. In which; this vital information, it hasn't been revealed by those in the field of automotive battery technology as well

So; no matter how we slice it, line up on line or precept up on precept. It all boils down to the lack of information that, we've about our secondary cell batteries; in which, it shapes our opinions about their cycling processes. Wherein; it's a disconnection between how our batteries are designed and how, their cycling processes will be carried out within the realm of physic as well.

Remember on the inside of a lead acid automotive starting battery. Its plates in each cell, they're connected in series-parallel with one another by their straps, tab connectors and the electrolyte solution in each cell. And each cell, it's connected in series with its terminal connections as well; thus, electrons could flow from a strap to a tab connector on one plate.

And then, electrons could flow across that plate through the electrolyte solution to an adjacent plate. Without flowing to and from, the same tab connector on the same plate. Thus; the battery plates in each cell, they're simultaneously charged and discharged amongst themselves in each cell because of how they're interconnected with one another in each cell.

Since each cell, it's connected in series with one another by way of an electrical bus. Then the plates in one cell, they're simultaneously charged and discharged to the plates in the next cell. Without having to, stop charging them in one cell in order to charge in the next cell. Thus; all of the battery plates in each cell, they could be charged at once during their charging cycle.

Looking at the battery from the outside of its plastic casing, it's obvious why we can't carry out a simultaneous cycling process of it because of the laws of physics. If we comprehend, what's happening on the inside of the battery during its charging cycle? Then it'll be obvious why, it'll be possible to carry out a simultaneous cycling process of it because of the laws of physics as well.

Wherein; the lack of information, about the mechanics behind the cycling processes of the battery. It'll create a misconception about what we could or couldn't do with it within the realm of physic because of its contemporary design as well. Remember if a battery cells, they're connected in series with its terminal connections.

Then electrons, they could pass through the battery plates in one cell in order to flow to its plates in the next cell. Thus; its plates in one cell, they could consume and store electrons while supplying them to its plates in the next cell as well. However; if the battery cells, they're connected in parallel with its terminal connections.

Then electrons, they can't pass through the battery plates in one cell in order to flow to its plates in the next cell. But; electrons, they could flow to its plates in each cell based on their individual resistance. Thus; its plates in one cell, they can't consume and store electrons while supplying electrons to its plates in the next cell as well.

Having vital information about, what'll happen on the inside of a battery during it cycling processes. If its cells, they're connected in series or in parallel with its terminal connections. It'll shed light on what we could or couldn't do with it within the realm of physic. Given that; its design, it's the deciding factor in determining how electrons will enter or exit it as well.

Focusing on the mechanics behind the charging cycle of a lead acid automotive starting battery, we'll find that. The electromagnetic laws of physics will not pose a problem when it comes to charging its plates in one cell while discharging them to the next cell as well. Likewise; we'll find that, neither the laws of entropy, inertia nor the laws of conservation of energy will pose a problem as well.

Although; energy lost to heat, it'll occur during the charging cycle of the battery. However; we're still able to charge, its plates in each cell while discharging them to the next cell. Without having to, stop charging them in one cell in order to charge them in the next cell; even though, they're separated by a non-conducting material called plastic as well.

What's so fascinating, about the concept of simultaneously charging and discharging a battery to a load source. It'll be no different than electrons passing through the battery plates in one cell in order to flow to its plates in the next cell. It seems simple to understand and yet, we end up focus on the wrong science because of the lack of information as well.

When I've conversations about a simultaneous cycling process of a battery, it seems like its mechanical structure seems irrelevant to those whom I'm talking to. Given that; we've to charge it on the same terminal connection that, it was discharged on. Wherein; it seems like, it'll be one reason why it couldn't be charged while being discharged as well.

However; those who I've a conversation with, they'll blame it all on the laws of physics why a battery can't be charged while being discharged; wherein, one reason should be obvious! If we've to charge it on the same terminal connection that, it was discharged on. Given that; electrons, they've to go in it the same way they've to come out in order to cycle it as well.

If there's a charge and a load source, on the same terminal connection of a battery. Then they'll be in parallel with one another and then, the amount of current flowing into the battery. It'll be dictated by the amount of resistance at the load source. This type of information, it's not given by those in the field of automotive battery technology.

Remember if a battery cells, they're connected in series with its terminal connections. Then it'll allow each cell in the battery to receive the same amount of current flowing from the charging source. But; it'll be a voltage drop, across each cell in the battery during its charging cycle. This type of information, it'll be very helpful if we're talking about charging it while discharge it.

On the other hand; having, no information about how voltage and current will flow between a battery and a load source in their relationship with a charging source. It'll be hard to grasp why, the battery can't be charged while being discharged to the load source. If you don't understand that, voltage will flow in and out of a battery different than current will.

If the battery and the load source, they're connected in parallel with one another in their relationship with the charging source. Then the parallel connection between them, it'll allow the charging current to follow the path of least resistance between them. However; the charging voltage, it'll be the same across the battery and the load source at the same time.

Not having this type of information, it'll be hard to understand why the battery couldn't be charged while being discharged to the load source as well. We just can't focus on one thing because there's a host of things, other than the laws of physics will prevent a battery from being charged while being discharged to a load source as well.

If we can't comprehend why those hosts of other things, they'll prevent a battery from being charged while being discharged to a load source. Then we're inept to determine why the battery can't be charged while being discharged to the load source as well. The reason I say inept because there're other reasons, other than the laws of physics in play as well.

Here's why; although, a lead acid automotive starting battery plates in each cell. They're separated by a non-conducting material called plastic. We're still able to charge its plates in each cell without having to stop charging them in one cell in order to charge them in the next cell. Without being impeded by the laws of entropy, inertia or the laws of conservation of energy as well.

Wherein; it has little to do with the laws of entropy, inertia or the laws of conservation of energy. But; it has more to do with the laws of mechanics why, we could charge the battery plates in each cell without having to stop charging them in one cell in order to charge them in the next; although, they're separated by a non-conducting material called plastic as well.

If we don't have this type of information about, the mechanics behind the charging cycle of the battery. Then we can't take in consideration, the design of the battery is the reason why we don't' have to stop charging its plates in one cell in order to charge them in the next cell as well. Wherein; we'll end up focusing, on the laws of physics instead of the laws of mechanics.

If we equate, what we know about a lead acid automotive starting battery. Then it'll be hard to believe, it'll be no benefits in adding energy back to it while taking energy out as well. Remember each cell in the battery is nothing more than a battery in and of itself; that is, connected in series with one another and housed in a plastic case to make up the whole battery as we know of today.

When the battery is charged as a whole, we're simultaneously adding and subtracting energy from one battery while simultaneously adding and subtracting energy from another. Then the idea it'll be no benefits in adding energy back to the battery while taking energy out. It could only stemmed from the lack of information about its mechanical make up as well.

In the next chapter; let's talk about misconceptions or why the lack of information about a secondary cell battery. It'll create misconceptions about what we could or couldn't do with it because of the lack of information about it as well.

CHAPTER 7

MISCONCEPTIONS ABOUT AN AUTO BATTERY

OUR PERCEPTIONS ABOUT, WHAT we could or couldn't do with our secondary cell batteries. It stems from misconceptions because of the lack of information that, we've about our secondary cell batteries as well. We think there's no other way to carry out their cycling processes; but, to stop one process in order to start the other; in which, it's nothing more than a misconception as well.

We mainly blame it, on the laws of physics why we've to stop one cycling process of our batteries in order to start the other; although, we've figured out how to charge them while they've a load source on them. We devised ways for a charging source to create a surplus of electrons at a load source in order to reverse the current flow back into our batteries to charge them.

However; we're only charging, our batteries and not charging them while discharging them. Given our misconceptions, we've about their cycling processes since we believe that. The laws of physics, they're the reasons we've to stop their discharging cycles in order to start their charging cycles. In which; I think that, we really haven't thought this thing through.

Those in the field of automotive battery technology and the field of electrochemistry, they're suggesting our ability to add and subtract electrons from the same chemical composition of a battery. It's a phenomenon or a prodigy in some respect since their skepticism implies that, it's impossible to add electrons back to a battery while taking them out as well.

What the skeptics failed to realize is that, their assumption is a misconception because a twenty-five year old French physicist named Gaston Plante'. In which; he made, the first secondary cell battery or the first lead acid automotive battery. Thus; he brought together, the mechanical, the chemical and the electrical aspects of the battery in such a manner.

So; we could charge or discharge, the same battery; in which, he advanced our battery technology in mid 1800s. Although; it was a phenomenon or a prodigy in a sense because it went beyond the primary cell battery made by Allesandro Volta in the late 1700s. No doubt there were skeptics in his time when they thought, he couldn't put energy into a device.

Likewise; in Gaston Plante' time skeptics thought that, he couldn't add energy back to a device once it was taken out. In which; the skeptics thought, it couldn't be done. Wherein; one man, he brought together our mechanical and chemical know how to create energy in a device to allow us to have portable energy.

While another man, he brought together our mechanical, chemical and electrical know how to add energy back to a device once it was taken out to advance our battery technology in the mid 1800s. We were wrong then and now to think we can't bring together our mechanical, chemical and electrical knowhow to charge our batteries while discharging them as well.

You would think those in the field of automotive battery technology today, they would have the skills or the technical knowhow to go beyond what Allesandro Volta and Gaston Plante' accomplished more than a century ago. In this point in history, we should be able to charge our secondary cell batteries while discharging them as well.

However; our misconceptions about our secondary cell batteries, they've kept us searching for that perfect chemical composition to a longer-lasting chemical reaction for them for more than a century. In order to, increase their efficiencies and overall capacities thus, increasing the overall travel ranges of our vehicles on battery power as well.

We're given many scientific explanations by those in the scientific community why things work or don't work; but, it doesn't mean their explanations are the only explanations. What I found was that, their explanations are based on their understanding on how things work or don't work. Contrary to popular belief, scientists don't know everything.

Imagine where the world would be today if no one ventured outside the status quo to investigate matters for themselves. We probably still be assuming that, the earth is flat, we'll have no airplanes, no sound recording and we'll continue to assume sea life doesn't exist in the deepest parts of the oceans as well.

If we don't venture outside the status quo, then misconceptions could shape our perceptions as well. Since we could charge or discharge the same battery, then to assume it can't be charged while being discharged due to the laws of physics. It would be a mistake for us to make such an assumption without investigating matter first as well.

Here's why; the assumption in and of itself, it'll be based on the contemporary design of our secondary cell batteries. Given that; energy, it has to go in them the same way energy has to come out in order to cycle them. Then the assumption, they can't be charged while being discharged is nothing more than a misconception created by their contemporary designs as well.

For example; we could replenish, a secondary cell battery chemical energy with electrical energy; however; if we use the wrong type of electrical current. Then we might get the wrong impression, the battery chemical energy can't be replenish with electrical energy. If we're using alternating current instead of direct current to replenish, its chemical energy as well.

The nature of the alternating current, it'll not allow the battery chemical energy to be replenished. Here's why; each time the current, it alternates within a circuit. It'll create a void or a gap within the circuit. Thus; electrons, they'll flow back and forth within the void or the gap to fill the void or the gap as the charging system rotates because of the nature of the current.

If we attempt to, charge the battery with the alternating current. Then electrons will not flow in one direction long enough to breakdown the resistance of the battery plates in order to charge them. If we don't understand this, then it might create a misconception that we can't replenish the battery chemical energy with electrical energy as well.

Wherein; it doesn't mean that, the battery chemical energy can't be replenished with electrical energy. What it does mean is that, we're using the wrong type of current to replenish the chemical energy of the battery. If we use direct current to replenish, its chemical energy. Then we'll be using, the right current to replenish its chemical energy.

Given that; the charging electrons, they'll not have a void or a gap to flow in and out of within the circuit between the battery and the charging source. Since the charging electrons, they'll be able to flow in one direction toward the battery plates to break down their resistance to charge them. And then, we'll be able to charge the battery plates.

Here's my point; it's not due to the laws of physics, why a secondary cell battery can't be charged while being discharged as we believe. However; it's a misconception created by, the contemporary design of the battery. We haven't acquired enough information about our secondary cell batteries. To understand, what's required to charge them while discharging them?

I discovered if current could flow in one direction from a battery plates in one cell to its plates in the next cell. Then its plates in one cell, they could consume and store energy while supplying it to its plates in the next cell as well. We think the mechanics of the current alone, they'll determine how they'll enter or exit the battery; in which, it's a misconception.

However; the mechanics of the current, they're submissive to the design of a battery as well as they're submissive to a circuit connection made between load sources as well. A parallel circuit connection made between a battery and a load source in their relationship with a charging source. It'll create a misconception the battery can't be charged while it has the load source on it.

Unless; the charging source, it'll create a surplus of electrons at the load source in order to reverse the current flow back into the battery to charge it; however, it's due to current having to go in the battery the same way it has to come out as well. I found voltage will flow through a parallel circuit connection different than current will. See chapter 12.

Wherein; the mechanics of the voltage and current, they'll behave differently flowing between a battery and a load source because of the parallel circuit connection made between them. Although; voltage, it'll be the same across the battery and the load source. But; current, it'll not because of the parallel circuit connection made between them. See chapter 12 as well.

If we think, it's due to the mechanics of the current alone. Then it's a misconception because we failed to realize that, a parallel circuit connection made between a battery and a load source. It doesn't work the same as a parallel circuit connection made between two load sources or two batteries, which is one of our misconceptions as well.

It seems like we don't understand or overlooking, current will flow through a circuit connection different than voltage will. If we don't understand or acknowledge this, then we're leaving ourselves open for misconceptions. About what we could or couldn't with our secondary cell batteries because of the laws of physics as well.

We can't begin to comprehend why it'll be possible or not possible to charge our secondary cell batteries while discharging them as well. On the other hand; we just can't look at the mechanics of the current alone and then, determine they'll prevent a secondary cell battery from being charged while being discharged; but, we've to look at the structure the current is flowing through as well.

It'll be a host of things coming together, other than the mechanics of the current alone will prevent a battery from being charged while being discharged. Let me go into more details about what will happen when there's a charge and a load source on a battery at the same time. First of all; the charge and the load source, they'll be on the same terminal connection of the battery.

Since the battery, it has only one opening for current to enter or to exit it. Then it'll be in parallel with the load source in their relationship with the charging source. In which; this parallel circuit connection, it'll allow certain aspects of the current to come into play; such as, like charges will repel one another and follow the path of least resistance as well.

Since current, it's the movement of electrons. Then it can't simultaneously enter and exit the same terminal connection of the battery travelling on the same path. Given that; electrons, they'll repel one another and follow the path of least resistance as well. However; if we think it's due to the mechanics of the current alone, then it's nothing more than a misconception as well.

However; it's part of a host of things, like the contemporary design of the battery. Given that; current, it has to go in the battery the same way it has to come out. Wherein; this is one of the things, along with a list of others will prevent the battery from being charged while being discharged; in which, we've to get familiar with those hosts of things as well.

Here's the deal, we're either charging a secondary cell battery or discharging it to a load source; but, we can't do both. Likewise; if we think, it's due to either the laws of electrochemistry, entropy, inertia or the laws of conservation of energy. Why it can't be charged while being discharged to the load source, then it's still a misconception as well.

The plates in each cell in a lead acid automotive starting battery, they could consume and store energy while it's supplied to the plates in the next cell. Although; the plates in each cell, they're separates by a non-conducting material called plastic. Then the theories, it's either due to the laws of electrochemistry, entropy, inertia or the laws of conservation of energy are flawed as well.

We could take energy from a battery plates in one cell, so, it could be added to its plates in the next cell. Without the laws of electrochemistry, entropy, inertia or the laws of conservation of energy being an issue. Given that; the battery plates, they're connected in series-parallel with one another in each cell and each cell is connected in series with its terminal connections as well.

Then it seems like, we're focusing on the wrong science when we consider those laws of physics. They'll be an issue when it comes to charging a secondary cell battery plates while using them to power a load source as well. When those laws of physics, they're not an issue when taking energy from the battery plates in one cell, so, it could be added to its plates in the next cell as well.

It's a misconception to think, either the laws of electrochemistry, entropy, inertia or the laws of conservation of energy. They'll be a reason why a secondary cell battery can't be charged while being discharged to a load source. However; in reality, it's due to the laws of motion and the effect of forces on bodies or the design of the battery.

On the other hand; if we don't understand, the mechanics of the current will be submissive to the design of a secondary cell battery as well as the connections made between it and a load source in their relationship with a charging source as well. Then we're subject to misconceptions about what we could or couldn't do with the battery as well.

Here's why; the connections made between, a lead acid automotive starting battery plates in each cell. They're the reasons we don't have to stop charging its plates in one cell in order to charge them in the next cell; in which, it's due to the laws of motion and the effect of forces on bodies or due to the design of the battery.

The mechanical structure of a battery, it's the deciding factor in determining how its cycling processes will be carried out. Wherein; it's a part of a host of things will prevent, it from being charged while being discharged as well. If we focus on, the mechanical structure of the battery. It'll help us understand the simplicity in charging it while discharging it as well.

Since the laws of physics, they're vast and complex. And sometimes, they're misinterpreted or misapplied because of misconceptions. We've to simplify the approach to understand, why it'll be possible to charge a secondary cell battery while discharging it. If we think we've to focus on the vast and complex laws of physics, then it's a misconception as well.

All we've to do is focus on, how electrons will flow through the body of a battery based on its design. In other words; how, electrons will enter the battery from a charging source and how, electrons will exit the battery to a load source. Also; we've to separate, voltage from current because voltage will enter or exit the battery different than current will.

Although; current, it's a combination of voltage and electrons. However; focusing on the mechanics of the voltage and current alone, it'll make our task much easier as well. If not, then we're subject to misconceptions. Given that; the laws of electrochemistry, entropy, inertia or the laws of conservation of energy, they're much complex laws of physics to interpret as well.

Here's why; if there's a parallel circuit connection made between a battery and a load source in their relationship with a charging source. It's one of a host of things will prevent, the battery from being charged while being discharged to the load source other than the battery having only one opening for current to enter or to exit it as well

For instance; the parallel circuit connection, made between the battery and the load source in their relationship with the charging source. It'll allow current to flow into the battery or flow to the load source based on, the type of resistance they've to conduct current. If the battery resistance is greater than the load source resistance, then current will flow to the load source.

Here's why; current will be following the path of least resistance between, the battery and the load source. Likewise; if the resistance at the load source is greater than the resistance at the battery, then the charging current will flow into the battery. The parallel connection made between them, it makes it difficult to charge the battery while it has the load source on it.

Understanding that, it'll be difficult to charge the battery while it has the load source on it because of the connections made between them, and the difference in their resistance to current. Then it's no need to focus on, the vast and complex laws of physics to figure this out. All we've to do is focus on the connections made between them in their relationship with the charging source.

Realizing, voltage will response differently to resistance than current will because electrons are involved. See chapter 12. Although; voltage, it'll be the same across the battery and the load source when they're connected in parallel with one another. However; current, it'll not be the same across them. In which; it's something that, we've realize as well.

For instance; voltage, it'll flow through a parallel circuit connection different than current will. Then we'll find that, the mechanics of the voltage and current alone. They're not the deciding factor in determining how they'll flow between the battery and the load source in their relationship with the charging source.

Given that; the battery resistance to current, it'll be different than the load source resistance to current. Then the different circumstance, it'll determine how current will flow between them in their relationship with the charging source because of the parallel connection made between them.

Then voltage and current will flow between the battery and the load source in their relationship with the charging source. It'll be assisted by the parallel circuit connection made between them. Thus; it has more to do with, the type of connection made between them than it does with the mechanics of the voltage and current alone.

Knowing it's the connections made between, the battery and the load source in their relationship with the charging source. Then it'll be no big misconception about why, we can't charge the battery while it has the load source on it. Unless; a surplus of electrons, are created at the load source in order to reverse the current flow back into the battery to charge it.

Likewise; it'll be no big misconception about why, we can't charge the battery while discharging it to the load source as well. Given that; it'll be all about, the connections made between the battery and the load source in their relationship with the charging source as well. It's not about the mechanics of the voltage and current alone or the laws of physics as well.

Another big misconception is that, we think a secondary cell battery is nothing more than a storage device for electrons. If we realize, it's a storage device for electrons. However; its internal structure, it creates the electrical circuit connections needed. To transfer electrons between its plates and cells because of the connections, they've with its terminal connections.

Not only will voltage and current will flow to and from the battery plates in each cell based on, how they're interconnected with its terminal connections. However; voltage and current, they'll flow to and from the battery plates to a load source based on how they're interconnected with one another as well.

There're a host of things, we've to focus on other than the mechanics of the voltage and current alone or the laws of physics. If we're going to figure out, is it possible or not to charge a secondary cell battery while discharging it to a load source as well. Given that; it's not all about, the mechanics of the voltage and current alone or the laws of physics as well.

Here's why; when there's a charge and a load source on, the same terminal connection of a battery. We can't see the electrical connections made between its plates and cells in their relationship with its terminal connections. All we could see are the connections made from the outside of its plastic casing; thus, we can't determine how voltage and current will flow through it.

In other words; we can't determine how, voltage and current will enter the battery flowing from the charging source or exit the battery flowing to the load source. When there's a charge and a load source is on the battery at the same time, all because of its plastic casing. Until; we take, an in depth analysis of the mechanics taking place inside of its plastic casing.

Then we'll never know, what's taking place within the battery during its cycling processes. Given the misconception, we've about the way its cycling processes will be carried out within the realm of physic as well. In other words; we can't see beyond, its plastic casing to see if. The laws of physics or its design will determine how, its cycling processes will be carried out.

If we don't understand, how the mechanical structure of the battery will play a role in its cycling processes. Then we're subject to misconceptions about, how voltage and current will flow through the body of the battery. Thus; subject to misconceptions about, what we could or couldn't do with the battery because of the mechanics of the voltage and current as well.

For instance; if a battery cells, they're connected in series with its terminal connections. Then each cell in the battery, they'll receive the same amount of current flowing from the charging source; however, each cell will receive a voltage drop from the charging source. The reason this will happen because of the connections made between the battery cells and its terminals.

If we don't understand, the reason why each cell in the battery will receive the same amount current or each cell will receive a voltage drop from the charging source. Then we're subject to misconceptions about, how voltage and current will flow through the body of the battery. In which; our misconceptions, they'll be created by a lack of information we possess.

Given that; we'll not know, how voltage and current will flow through a series, a parallel or a series-parallel circuit connection. If we re-visit chapter 3 again, it'll help explain why we've misconceptions about. What we could or couldn't do with our secondary cell batteries because of the lack of information about, how their mechanical structures will play a role in their cycling processes.

If we don't understand that, the design of a battery is the deciding factor in determining how voltage and current will enter or exit it. Then we're inept to blame it on, the mechanics of the current alone or the laws of physics why the battery can't be charged while it has a load source. Not along why, the battery can't be charged while being discharged to the load source as well.

If we don't understand, the mechanics of the voltage and current will be submissive to the design of a battery as well as they're submissive to a circuit connection between it and a load source in their relationship with a charging source as well. Then we're subject to misconceptions about, what we could or couldn't do with the battery within the realm of physic as well.

On the other hand; if we understand, how the mechanics of the voltage and current will behave flowing through the body of a battery during its cycling processes. Then it'll be simple to understand why it can't be charged while being discharged to a load source as well. Given that; we figured out, how to charge a secondary cell battery while it has a load source on it.

Although; the battery, it has only one opening for electrons to enter or to exit it and it'll be in parallel with the load source in their relationship with the charging source as well. How we achieved this goal because we manipulated, the mechanics of the current with the charging source. Allowing, it to produce more electrons than needed to power the load source.

Therefore; any additional electrons flowing toward the load source, they'll be reflected back into the battery to charge it; given that, it's in parallel with the load source as well. However; the battery, it has only one opening for electrons to enter or to exit it. Wherein; it creates this misconception, we've to stop one cycling process in order to start the other as well.

When it comes to charging the battery while discharging it to the load source, then logic tells us creating a surplus of electrons at the load source. In order to, reverse the current flow back into the battery to charge it will not work. Wherein; logic will tell us that, we've to find another method to manipulate the mechanics of the current in order to carry out this process.

During my research, I found it's possible to manipulate the mechanics of the current with the design of a battery as well. Given that; voltage, it'll force electrons in and out of the battery based on its design. Then we could get around the mechanics of the current that, poses a problem when there's a charge and a load source on the battery at the same time.

We don't have to stop the discharging cycle of a battery in order to start its charging cycle. If electrons, they don't have to go in the battery the same way they've to come out. Our big misconception is that, we're focusing on the contemporary design of the battery. In which; it tricks us in believing, we've to stop its discharging cycle in order to start its charging cycles.

Here's why; if a battery plates in each cell, they're connected in series-parallel with one another by their straps, tab connectors and the electrolyte solution in each cell. And each cell, it's connected in series with its terminal connections as well. Then its plates in one cell, they could be simultaneously charged and discharged to its plates in the next cell as well.

Although; the battery plates in each cell, they'll be separated by a non-conductive material called plastic. And yet, we wouldn't have to stop charging its plates in one cell in order to charge them in the next cell. All because its plates in each cell, they're connected in series-parallel with one another. And each cell, it's connected in series with its terminal connections as well.

Then it's simple to understand why, we could charge the battery plates in each cell. Without having to, stop charging them in one cell in order to charge them in the next cell; although, they're separated by a non-conducting material called plastic as well. We don't have to focus on the mechanics of the current alone or the laws of physics to figure this out as well.

Then the reason is clear why, we can't charge a secondary cell battery while discharging it to a load source; given that, the wrong mechanics are in play. If we're focusing on, anything other than the design of the battery and the connections made between it and a load source. In their relationship with a charging source, then we're focusing on the wrong science.

Likewise; if we don't understand that, voltage and current will enter or exit a battery differently based on its design. Then we're to subject misconceptions about, what we could or couldn't do with the battery because of the mechanics of the voltage and current alone, not because of the laws of electrochemistry, entropy, inertia or the laws of conservation of energy as well.

This is why I went outside, the status quo to conduct my own investigations. Since it seems like some in the field of automotive battery technology, they're speculating about what we could or couldn't do with our secondary cell batteries within the realm of physic. In which; I found the laws of physics have little to do with how our batteries cycling processes, are carried out as well.

Here's why; I found that, each cell in a lead acid automotive starting battery. It's nothing more than a battery in and of itself; that is, connected in series with one another and housed together in a plastic case to make up the whole battery as we know of today. When it's charged as a whole, we're charging and discharging one battery while charging and discharging another.

Then it must be a misconception thinking that, we can't charge the battery as a whole while discharging it to a load source as well. Since scientists, they don't always get it right the first time because of unknown variables may exist. Sometimes scientists, they actually have to carry out the experiments for themselves in order to get it right.

For instance; for centuries, we've recognized Pluto as a planet in our solar system; but, in the twentieth century. Those in the International Astronomical Union, they came to the conclusion that. The primitive stargazers were wrong to assume Pluto was a planet. It's nothing more than a chunk of ice based on, light reflections in their modern day telescopes and other cosmic events.

In which; modern day astronomers' conclusions, they caused an up roar in the scientific community of astronomy. However; in the twentieth-first century, NASA sent a space probe to Pluto. After reaching its orbit, not only did they find Pluto was planet as the primitive stargazers believed; but, it has moons orbiting it as well.

Sometimes; we've to carry out the experiments for ourselves in order to get it right because if we do not, then misconception becomes our perception. What's ironic about our misconceptions, why we can't charge a lead acid automotive starting battery while discharging it. We don't perceive each cell in the battery as nothing more than a battery in and of itself.

When we charge the battery as a whole, we're charging and discharging one battery while charging and discharging another. Given that; energy, it could enter and exit each battery flowing in one direction; thus, energy flowing in one direction. It's the key in charging a battery while discharging it to a load source as well.

It seems like those in the field of automotive battery technology, they're focusing on the wrong science. When they say that, it's impossible to add energy back to a lead acid automotive starting battery while taking it out and it'll be no benefits in it due to the laws of physics as well. It's hard to believe their assumptions based on the mechanical make-up of the battery as well.

Given that, the battery plates in each cell, they're connected in series-parallel with one another by their straps, tab connectors and the electrolyte solution in each cell. Therefore; electrons, they could flow from a strap to a tab connector one plate. And then, flow across that plate through the electrolyte solution to an adjacent plate without flowing to and from the same tab connector.

Thus; one plate, it's simultaneously charged and discharged to an adjacent plate by way of the electrolyte solution in each cell because of how they're interconnected with one another. Since each cell, it's connected in series with the terminal connections of the battery. Then it makes it possible to simultaneously charge and discharge its plates in one cell to its plates in the next cell.

Wherein; it seems like, the design of the battery is the deciding factor. In determining how, its cycling processes will be carried out; that is, one process at a time or both processes simultaneously. Thus; it seems like, it's a misconception to assume it'll be impossible add energy back to the battery while taking it out and it'll be no benefits in it as well.

On the other hand; it seems like, it's has been a serious miscalculation about what we could or couldn't do with our secondary cell batteries because of the laws of physics. In the next chapter; let's see why misconceptions, they cause us to focusing on the wrong science when it comes to the energy input and output process of our secondary cell batteries as well.

CHAPTER 8

FOCUSING ON THE WRONG SCIENCE

AFTER OBSERVING, THE MECHANICS behind the charging process of a lead acid automotive starting battery. I found its plates in one cell, they're simultaneously charged and discharged to its plates in the next cell by way of an electrical bus. So; all of its plates in each cell, they could be charged during their charging process without having to stop charging them in one cell in order to charge them in the next.

Although; the battery plates in each cell, they're separated by a non-conducting material called plastic. I find neither the laws of electrochemistry, entropy, inertia nor the laws of conservation of energy were an issue. When the battery plates in one cell, they were simultaneously charged and discharged to its plates in the next cell as well.

Although; energy, it was lost to heat during the process; but, the battery plates were still charged. Then claiming, it'll be impossible charge a lead acid automotive starting battery while discharging it and it'll be no benefits in it as well. It's nothing more than a misconception stemmed from focusing on, the wrong science as well?

What science, are those in the field of automotive battery technology talking about? When they say that, it'll be impossible add energy back to a battery while taking it out and it'll be no benefits in it as well. They can't be talking about, the laws of electrochemistry because we already adding and subtracting energy from the same battery plates during its charging cycle.

Likewise; those in the field of automotive battery technology, they can't be talking about the laws of entropy, inertia or the laws of conservation of energy. If we're adding, energy back to a battery while taking it out as well. Then what science they're talking about when they say, it'll be impossible add energy back to a battery while taking it out and it'll be no benefits in it as well.

Wherein; I found the laws of physics, they don't determine how energy is added or subtracted from a battery plates. Based on, the mechanics behind the charging cycle of a lead acid automotive starting battery. Then the question is, why it'll be impossible to add energy back to a battery plates while taking it from them and it'll be no benefits in it as well.

If we consider, what happens on the inside of a lead acid automotive starting battery during its charging cycle when energy is transfer from its plates in one cell to its plates in the next cell. Then we'll realize that, the laws of electrochemistry, entropy, inertia or the laws of conservation of energy. They'll not be an issue when it comes to adding energy back to it while taking energy out as well.

Since the interior design of the battery, it allows its plates in each cell to consume and store energy while supplying it to its plates in the next cell during its charging cycle. Given that; its plates, they're connected in series-parallel with one another by their straps, tab connectors and the electrolyte solution in each cell. And each cell, it's connected in series with its terminal connections as well.

If we consider the mechanics behind, the charging cycle of the battery. Then it'll be clear why its plates in each cell, they could consume and store energy while supplying it to its plates in the next cell. Without having to, stop charging them in one cell in order to charge them in the next cell; although, they're separated by a non-conducting material called plastic as well.

Then the answer to why, we think it'll be impossible to add energy back to the battery while taking it out and it'll be no benefits in it because of the laws of physics. It's nothing more than a misconception because we're focusing on the wrong science. Our misconceptions, they were established by the mechanical structure of the first lead acid automotive battery more than a century ago.

Gaston Plante' made the first secondary cell battery in 1859; thus, he established how energy. It'll be added or subtracted from the battery plates by designing it with one opening for electrons to enter or to exit it; although, it has two terminals. But, they only equal to, one opening for electrons to enter or to exit it; therefore, it can't be charged while being discharged as well.

However; I found, it's only to the extent electrons have to go in the battery the same way they've to come out. Given that; like charges, they'll propel one another and follow the path of least resistance as well. Then how could anyone come to the conclusion that, it'll be impossible to add energy back to the battery while taking it out and it'll be no benefits in it as well.

If they know anything about, the mechanics behind the charging cycle of the battery. Then they wouldn't be blaming it on the laws of electrochemistry, entropy, inertia or the laws of conservation of energy. Why it'll be impossible to add energy back to the battery while taking it out and it'll be no benefits in it as well.

If we consider, how energy has to enter or exit a lead acid automotive starting battery during its cycling processes. Then we could come to a valid conclusion why it'll be impossible to add energy back to it while taking energy out and it'll be no benefits in it without being the laws of electrochemistry, entropy, inertia or the laws of conservation of energy into the conversation as well.

Here's why; since energy, it has to go in the battery the same way it has to come out. Then the battery, it can't be charged while being discharged and it'll be no benefits in it because we can't carry out the process. If we compare the exterior design of the battery to its interior design, then we'll see why we've to stop its discharging cycle in order to start its charging cycle as well.

Since we've to charge the battery on the same terminal connection that, it was discharged on; wherein, it poses a problem. It has little to do with the laws of electrochemistry, entropy, inertia or the laws of conservation of energy why, we've to stop its discharging cycle in order to start its charging cycle; proven by, the mechanics behind its charging cycle.

Here's why; the interior design of the battery, it'll allow its plates in one cell to be simultaneously charged and discharged to its plates in the next cell. Without having to, stop charging them in one cell in order to charge them in the next cell. One could conclude that, the exterior design of the battery prevents it from being charged while being discharged as well.

Given that; electrons, they've to go in the battery the same way they've to come out. Wherein; its interior design, it allows electrons to flow in one direction from its plates in one cell to its plates in the next cell. Without having to, stop charging them in one cell in order to charge them in the next cell; so, it's all about mechanics and not about the laws of physics.

Then the big question is that, are we focusing on the wrong science? When we assume that, we've to stop the discharging cycle of the battery in order to start its charging cycle because of the laws of physics. If we focus on the concept of a run capacitor or a power transformer, then we wouldn't be so sure it's due to the laws of physics as well.

Here's why; I found that, we don't have to stop the discharging cycle of a lead acid automotive starting battery in order to start its charging cycle. Only to the extent that, electrons have to go in it the same way they've to come out in order to cycle it. If we're focusing on the wrong science, we might get the impression it's due to the laws of physics as well.

Focusing on, how electrons will enter and exit each cell in a lead acid automotive starting battery during its charging cycle. Then we might understand, it's not due to the mechanics of the current alone or the laws of physics. Why we've to stop its discharging cycle in order to start its charging cycle; but, it'll be due to the mechanics involved in carrying out the process.

Here's why; the plates in each cell in a lead acid automotive starting battery, they're connected in series-parallel with one another by their straps, tab connectors and the electrolyte solution in each cell; in which, it allows them to be simultaneously charged and discharged amongst themselves in each cell and to the plates in the next cell as well.

Given that; each cell in the battery, they're connected in series with one another by way of an electrical bus. Thus; we've series-parallel circuit connections, connected in series with other series-parallel circuit connections. Therefore; the battery plates in one cell, they could be simultaneously charge and discharge to its plates in the next cell as well.

Wherein; all of the battery plates in each cell, they could be charged during their charging cycle. Without having to, stop charging them in one cell in order to charge them in the next cell. Given that; its plates in each cell, they're connected in series-parallel with one another by their straps, tab connectors and the electrolyte solution in each cell as well.

Therefore; the charging electrons, they could flow from a strap to a tab connector on a plate. And then, flow across that plate through the electrolyte solution to an adjacent plate. Without flowing to and from, the same tab connector on the same plate. Thus; electrons, they could flow in one direction from the battery plates in one cell to its plates in the next cell as well.

It has little to do with the mechanics of the current alone or the laws of physics, why we don't have to stop charging the battery plates in one cell in order to charging them in the next cell. It's due to mechanics or the procedure in how electrons are added or subtracted from its plates in each cell. The mechanics behind its charging cycle gives credence to this conclusion.

I've heard many different opinions why the battery can't be charged while being discharged. In which; it's based on, a person level of understanding about the mechanics behind its cycling processes; wherein, they're based on conventional wisdom and not facts. It's no evidence showing we've to stop its discharging cycle in order to start its charging cycle.

Other than energy having to go in the battery the same way it has to come out. Likewise; there's no evidence showing, it'll be no benefits in a simultaneous cycling process of the battery because of the laws of physics. Other than energy having to go in the battery the same way energy has to come out as well.

On the other hand; having incomplete information about, the concept of charging the battery while discharging it; wherein, it could cause us to focus on the wrong science. If we don't understand, what'll happen on the inside of the battery during its cycling processes as well? There's a disconnection between what happens on the inside of it and what happens on the outside of it.

Wherein; some in the field of automotive battery technology, they can't see beyond the exterior design of the battery to determine how its cycling processes will be carried out. Thus; some in the field of automotive battery technology, they're inept to determine if it's possible or not to carry out a simultaneous cycling process of the battery and will it be a benefit in it or not as well.

What I'm trying to say is that, we could replenish the lead acid automotive starting battery chemical energy with electrical energy as well. Then it's no reason to think, it'll be no benefits in adding energy back to the battery while taking energy out as well. Unless; some in the field of automotive battery technology, they don't understand the mechanics behind its charging cycle.

During my research, I found some people believe if we attempt to carry out a simultaneous cycling process of the battery. Its endothermic and exothermic chemical reaction will take place at the same time; therefore; no gain. If we attempt to charge it while discharging it; given that, they'll cancel themselves out because one absorbs heat and the other releases heat.

I found the lack of understanding about the cycling processes of the battery. It'll create a false perception that its endothermic and exothermic chemical reactions will take place at the same time if we attempt to charge it while discharging it. Some in the field of automotive battery technology, they subscribed to this type of ideology about the cycling processes of the battery.

However; I found that, there's a problem with this type of ideology because it doesn't add up with the mechanics behind the charging cycle of the battery. Here's why; its plates in each cell, they could store energy while it's supplied to its plates in the next cell during their charging cycle; although, energy lost to heat will occur during their electrolysis process as well.

Nevertheless; energy lost to heat, it'll have no bearing on whether the battery plates in each cell is charging or not. Then the idea of energy being lost to heat in both conversion processes of the battery, it'll render a simultaneous cycling process of it void. It seems ridicules if we're charging the battery while discharging it as well.

If you understand, what happens on the inside of the battery during its charging cycle? Then you would assume during a simultaneous cycling process of it, any energy lost to heat would be made up by the charging source as well. Thus; any energy lost to heat during the simultaneous cycling process, it wouldn't have any bearing on whether the battery plates are charging or not.

It seems like some in the field of automotive battery technology, they're focusing on the wrong science to ascertain; whether, it'll be a benefit or not in adding energy back to the battery while taking it out. Wherein; it seems like some in the field of automotive battery technology, they need to take an in depth analysis behind its charging cycle to answer this question as well.

Adding energy back to the battery while taking it out to power a load source, it seems like it'll be no different than simultaneously adding and subtracting energy from the battery plates in one cell to its plates in the next cell. Although; energy loss to heat, it'll occur during the process; but, it shouldn't have no bearing on whether the battery is charging or not as well.

I found it'll be up to, the capability of the charging source to add energy back to the battery during its charging cycle; likewise, it'll be up to the capability of the charging source to add energy back to the battery while it's discharging to a load source as well. The endothermic and the exothermic chemical reaction taking place within the battery during its cycling processes.

Wherein; I found that, the endothermic and the exothermic chemical reaction. They're nothing more than indicators of the intensity of heat being created by energy lost to heat.

This might help explain, why the battery gets hot during its charging cycle and not during its discharging cycle because one chemical reaction, it absorbs heat and the other releases heat.

Although; energy lost to heat, it'll occurs when a lead acid automotive starting battery plates in one cell, are simultaneously charged and discharged to its plates in the next cell. In which; energy lost to heat, it'll not be a factor during their charging cycle as well. We've to consider this when it comes to a conversation about charging the battery while discharging it as well.

If we don't consider energy lost heat, not being a factor during the charging cycle of the battery. Then we're focusing on, the wrong science if we think. It'll be no benefits in a simultaneous cycling process of the battery. Because energy lost to heat, it'll occur in both of its conversion processes as well.

On the other hand; if we think, there's nothing. We could do about the inherent energy lost to heat in both conversion processes of our secondary cell batteries. Then we're focusing on, the wrong science as well. If we think eliminating, the energy lost to heat in both conversion processes of our batteries. It'll increase the overall travel range of a vehicle on battery power as well.

If we were focusing on the right science, then we'll know not to burden ourselves with the inherent energy lost to heat in both conversion processes of our batteries. We only need to eliminate the inherent energy lost to heat in their discharging cycles. If we're trying to, increase the overall travel ranges of our vehicles on battery power as well.

Here's why; only, the amount of energy lost to heat in the discharging cycle of our batteries become relevant. When it comes to, the overall travel ranges of our vehicles on battery power. If we're adding energy back to our batteries while taking it out, then the amount of energy lost to heat during their discharging cycles become irrelevant as well.

Remember the internal electrical resistance of a lead acid automotive starting battery. It's calculated by squaring current in amperes and then multiplying it by the internal resistance of the battery. In which; its electrical resistance, it's compounded by its internal resistance during its discharging cycle.

We could minimize the effects of energy lost to heat during the discharging cycle of the battery by adding energy back to it while taking energy out as well. Given that; adding energy back to the battery, it decreases its internal resistance; therefore; decreasing, its electrical resistance during its discharging cycle as well.

The factors that affect the efficiency and overall capacity of a lead acid automotive battery, they're all interrelated. If we charge one factor, then we change them all. So; with an intermittent charge on the battery while using it, then we could keep its internal resistance as low as possible with the intermittent charge on it while using it as well.

Not only decreasing the internal resistance of the battery. But; decreasing, the affects its internal resistance has on its efficiency and overall capacity as well. Given that; adding, electrical energy back to the battery, it diminishes its internal resistance; thus, minimizing its electrical resistance when it comes to its next chemical discharging cycle as well.

Thus; less internal resistance, it equals less electrical resistance; therefore, increasing the overall travel range of a vehicle on battery power; given that, it'll have less energy loss to heat as well. Mechanically; if we could add energy back to the battery while taking energy out, then we could do something about the inherent energy lost to heat in both of its conversion processes as well.

For example; as more and more energy is released from the battery, then more and more of the lead-sulfate will accumulate on its plates. Their electrical resistance will increase exponentially proportional to the amount of lead-sulfate that has accumulated on them. Thus; with an intermittent charge on them, then it'll translate into less lead-sulfate build up on them as well. See chapter 4.

In which, it translates into less, energy lost to heat during the discharging cycle of the battery and more energy to power a load source as well. In short; if we've an intermittent charge on the battery while using it to power a load source, then the amount of energy lost to heat during the chemical discharging cycle of the battery becomes irrelevant as well.

Therefore; the amount of energy needed to power the load source for a certain period of time becomes irrelevant as well. Wherein; having, an intermittent charge on the battery will extend its chemical energy because it'll increase its efficiency and overall capacity for its next chemical discharging cycle; thus, extending the run time of a load source on battery power as well.

Then the idea we can't do anything about the inherent energy lost to heat in a lead acid automotive battery, it's because we're not focusing on the right science. Since its plates in each cell, they're connected in series-parallel with one another by their straps, tab connectors and the electrolyte solution in each cell and each cell is connected in series with its terminals as well.

Then the battery plates in each cell, they could consume and store energy while it's supplied to its plates in the next cell. If any energy lost to heat occurs during the charging cycle of the battery, it'll be made up by the charging source as well. Thus; it rules out, the idea it'll be impossible to add energy back to the battery while taking it out and it'll be no benefits in it as well.

Focusing on the mechanics behind the charging cycle of a lead acid automotive starting battery, we can't use the laws of entropy, inertia or the laws of conservation of energy. To discredit, a simultaneous cycling process of the battery if we're adding energy back to it while taking energy out. If so, then we're focusing on the wrong science as well.

Given that; the battery plates in one cell, they could be simultaneously charged and discharged to its plates in the next cell. Without the laws of entropy, inertia or the laws of conservation of energy being an issue. Although; the battery plates in each cell, they're separated by a non-conducting material called plastic as well.

Then adding energy back to the battery while taking it out to power a load source; wherein, the laws of entropy, inertia or the laws of conservation of energy will not be an issue; however, some people assume they will. If we're focusing on those the laws of physics, then we're focusing on the wrong science when it comes to a simultaneous cycling process of the battery as well.

If we comprehend why, the battery plates in each cell. They could consume and store energy while it's supplied to its plates in the next cell. Then we'll understand why, we don't to stop charging its plates in one cell in order to charge them in the next cell as well. Then we'll know that, the laws of entropy, inertia or the laws of conservation of energy will be the wrong science to focus on.

Since a lead acid automotive starting battery plates in one cell, they could consume and store energy while supplying it to the plates in the next cell. It should be an indictment against our misconceptions it can't be charged while being discharged to a load source because of the laws of entropy, inertia or the laws of conservation of energy as well.

Likewise; it should be a confirmation that, the battery could be charged while being discharged to a load source without being impeded by the laws of entropy, inertia or the laws of conservation of energy as well. Understanding the mystery behind, why we could charge or discharge the same battery; but, we can't charge it while discharging it; in which, it's a paradox within itself.

It doesn't make any sense to focus on the laws of entropy, inertia or the laws of conservation of energy when it comes to a conversation about charging a lead acid automotive battery while discharging it as well. Given that; its plates in one cell, they could consume and store energy while it's supplied to its plates in the next cell without the laws of physics being an issue.

Therefore; it had to be another explanation, other than the laws of physics preventing the battery from being charged while being discharged. I had to cut off the top of, the plastic casing of the battery. That housed its plates and cells and left them intact, so, I could see. What would happen on the inside of it during its cycling processes based on, its internal design. See figure T-4.

After cutting off the top of the casing of the battery, I was able to observe how electrons would flow through it. By observing, the bubbles created by the electrons when they flowed through the electrolyte solution in each cell during its cycling processes. Wherein; it helped me to understand, the paradox behind why we could charge or discharge it; but, we can't charge it while discharging it.

Instead of focusing on the laws of electrochemistry, entropy, inertia or the laws of conservation of energy, I focused on the contemporary design of the battery. Wherein; it revealed that, we could charge or discharge it; but, we can't charge it while discharging it because energy has to go in it the same way energy has to come out as well.

We could all agree that, we could charge or discharge the same battery. But; what we don't agree on is that, why we can't charge it while discharging it. In figure T-4; it shows, an inside view of a lead-acid automotive starting battery to help shed some light on this argument.

Figure T-4

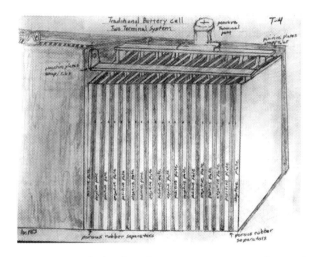

(Inside view of a lead acid automotive starting battery)

If we focus on the mechanical structure of the battery and how it'll play a role in its cycling processes, then we'll find it's a matter of mechanics and not a matter of physics why it can't be charged while being discharged. For one moment; let's imagine that, a lead acid automotive starting battery is a house with only one outer door for us to enter or to exit it.

The straps connecting the battery's cells together in series, they represent the inter doors of the house; so, we could enter or exit each room in it. However; each cell in the battery, they represent the rooms in the house; so, we could reside in them. But; once we enter the house through the outer door, then we've to go through the same door in order to exit the house as well.

Thus; in order to exit the house, we've to travel back through the same doors and rooms to exit through the outer door. In which; I find, it's not wise to build a house this way in case of an emergency because you would have only one way out. Likewise; I find that, it's not wise to build a second cell battery with only one opening for energy to enter or to exit it as well.

Especially; when it comes to the hybrid vehicles, we've to depend more heavily on their gasoline engines than their electric motors to propel them. Since we can't use their electric motors to, propel them while charging their batteries because energy has to go in them the same way it has to come out. See chapter 25 and 26 as well.

With an all-electric vehicle, we've to travel as far as possible on a single battery charge because of the inconvenience having to stop using the vehicle in order to charge its batteries. Those in the field of automotive battery technology, they know it's a problem when it comes to their energy input and output process of our batteries because they're the architect of their designs.

Here's why; after analyzing the mechanics, behind the cycling processes of a lead acid automotive starting battery. I found during its charging cycle, voltage and current could flow through all of its cells flowing in one direction. On the other hand; during its discharging cycle, voltage could flow from all of its cells flowing in one direction to a load source.

However; current, it could only flow in one direction from one cell in the battery to a load source during its discharging cycle. Thus; I found that, voltage flows from all of the battery plates in each cell to increase its voltage and current potential for its plates in one cell during its discharging cycle to power in order to a load source.

On the other hand; current, it could only flow from the plates in a cell with a terminal connection in it during the discharging cycle of the battery; whether, it's the negative or the positive terminal connection. But; it all depends on, what concept you subscribe to; whether, it's the conventional current flow or the electron current flow concept. See chapter 12.

Since a lead acid automotive starting battery positive terminal connection, it's connected in parallel with its positive plates at one end of its cells. Likewise; its negative terminal connection, it's connected in parallel with its negative plates at the other end of its cell. Thus; current, it either has to flow through a negative or a positive terminal to enter or to exit the battery.

Focusing on the right science, then it's a matter of mechanics and not a matter of physics. Why the battery, it can't be charged while being discharged to a load source. If we think, it's due to the laws of electrochemistry, entropy, inertia or the laws of conservation of energy; thus, we're focusing on the wrong science as well.

There's a disconnection between how we think the laws of physics will play a role in charging the battery while discharging it and how, they'll actually play a role in it. For more than a century, we design our secondary cell batteries in such a manner. It defeats the purpose of them being rechargeable batteries because they've to be charged on the same terminal they're discharged on.

So; it's not entirely due to the mechanics of the current alone or the laws of physics why, our batteries can't be charged while being discharged. However; it's due to, the wrong mechanics in play to charge them while discharging them because of their mechanical structures. I know it's hard for some in the field of automotive battery technology to believe.

Given that; they think, it has something to do with the mechanics of the current alone or the laws of physics. Wherein; it's only to the extent that, current has to go in our batteries the same way it has to come out. It seems like those in the field of automotive battery technology, they're focusing on the wrong science when it comes to the energy input and output process of our batteries.

If like charges have to travel the same path and flow in the opposite direction of one another in order to, simultaneously enter and exit our batteries. Then it'll be impossible to charge them while discharging them since like charges will repel one another and follow the path of least resistance as well.

However; it doesn't mean that, like charges can't simultaneously enter and exit our batteries. What it does mean is that, like charges can't simultaneously enter and exit the same terminal connection of our batteries. There's a big difference between focusing on the wrong science than focusing the right science when it comes to the energy input and output process of our batteries.

Those who think, they know about the cycling processes of our batteries. They'll immediately invoke the laws of physics into a conversation about charging them while discharging them. Without realizing, they're focusing on the wrong science; wherein, it becomes nothing more than a distraction or bewilderment for most people.

When I bring up the idea of charging our batteries while discharging them, most skeptics will overlook the part where we're adding energy back to our batteries while taking it out. On the other hand; they'll overlook, the part where our batteries will need to have separate terminals for charging while discharging in order to carry out such a process as well.

Here's why; if we didn't have to charge our batteries on the same terminal connections that, they were discharged on. Then the electromagnetic laws of physics will no longer support the idea. That our batteries can't be charged while being discharged because like charges, they'll repel one another and follow the path of least resistance as well.

Likewise; those assumptions, it'll be no benefits in a simultaneous cycling process of our batteries because of the laws of entropy, inertia or the laws of conservation of energy; even if, energy could enter and exit our batteries at the same time. Those assumptions, they'll be thrown out the window if we're adding energy back to our batteries while taking it out as well.

I know some laws of physics will prevent our batteries from being charged while being discharged; however, they're not the primary reasons as well. After taken an in depth analysis of, the mechanics behind the cycling processes of a lead acid automotive starting battery. I realize that, we're focusing on the wrong science when it comes to its energy input and output process.

Here's why; remember each cell in the battery, it's nothing more than a battery in and of itself. That is connected in series with one another and housed together in a plastic case to make up the whole battery as we know of today. When it's charged as a whole, we're charging and discharging one battery while simultaneously charging and discharging another as well.

It doesn't make any sense to assume a lead acid automotive starting battery as we know of today. It can't be charged while being discharged to a load source because of the laws of physics. Unless; we're focusing on, the wrong science when it comes to the energy input and output process of the battery as well. See chapter 10.

We don't need to work in the field of automotive battery technology or the field of electrochemistry to figure out. Why a lead acid automotive starting battery, it can't be charged while being discharged to a load source once we observe the inner workings of the battery. Then we'll realize, it's due to the wrong mechanics in play to charge it while discharging it.

In the next chapter; let's explore the reasons why the laws of physics, they're not the reason in and of themselves. Why our secondary cell batteries, they can't be charged while being discharged to a load source as well.

CHAPTER 9

THE LAWS OF PHYSICS

IT ALREADY HAS BEEN established that electrons, they could be simultaneously added and subtracted from a lead acid automotive starting battery plates during its charging cycle. So; it's no need to argue that, the laws of electrochemistry or the electromagnetic laws of physics are the reasons the battery can't be charged while being discharged to a load source as well.

Likewise; it's no need to bring, the laws of entropy, inertia or the laws of conservation energy into a conversation. About why the battery, it can't be charged while being discharged to a load source. If we're adding energy back to the battery while taking it out, then those laws of physics will not apply because the battery wouldn't be considered a closed system under those laws.

Here's why; one part of the laws of entropy states that, the measure of a system thermal energy per unit temperature is unavailable for doing useful work. Because work is obtained from ordered molecular motion, the amount of entropy is also a measure of the molecular disorder, or the randomness of a system; stated by a German physicist, Rudolf Claudius in 1850.

The laws of Inertia; Newton's first law of motion states an object at rest, it stays at rest. And an object in motion, it strays in motion with the same speed and the same direction; unless, acted upon by an unbalance force. The first law of conservation of energy, it states the total amount of energy remains constant in an isolated system.

In which; it implies, energy can neither be created nor destroyed; but, it can be changed from one form to another. The second law of conservation of energy is about the quality of energy, it states as energy is transferred or transformed more and more of it is wasted. The third law of conservation of energy is a branch of physical science.

It deals with the relation between heat and other forms of energy such as; mechanical, electrical, or chemical and by the extension of the relationships between all forms of energy. Looking at the laws of entropy, inertia or the laws of conservation of energy, it's no reason to assume it'll be no benefits in a simultaneous cycling process of a battery because of those laws physics.

I haven't figured out why some in the field of automotive battery technology. They would use the laws of entropy, inertia or the laws of conservation of energy to discredit a simultaneous cycling process of a battery; especially, if we're adding energy back to it while taking energy out; wherein, it doesn't make any sense to use them as a counter argument.

The laws of physics, they're vast and complex; sometimes, they're miss-interpreted or miss-applied; so, not to miss-interpret or miss-apply them. I'm going to use the mechanical structure of a secondary cell battery and how, it'll play a role in its cycling processes. To show, we're focusing on the wrong science when it comes to its energy input and output process as well.

Focusing on the mechanical, chemical and electrical aspect of a lead acid automotive starting battery, it'll show its contemporary design will impede it from being charged while being discharged. Wherein; it's conceivable that, we could charge or discharge it; but, it's inconceivable it could be charged while being discharged because of the laws of physics.

What're the laws of physics any ways, I found they're nothing more than the understanding of how matter and energy will work under different circumstances. Some laws of physics, they're named after certain people like; Boyle's Law, Charles' Law, Newton's law or Dalton's Law because they discovered certain ways, how matter and energy will work.

However; it doesn't means, it's the only way matter and energy will work. What it does mean is that, different circumstances determine how matter and energy will work under those circumstances. If we're going to use the laws of physics to discredit a simultaneous cycling process of a battery, then we'll be focusing on the wrong science to make that determination.

Those who are skeptical about charging a lead acid automotive starting battery while discharging it, they say I'm wrong to assume the battery. It goes through a simultaneous cycling process each time it's charged. The skeptics say the reason each plate in the battery is charged during its charging cycle because the charge is across the whole battery at once.

What I discovered while carrying out my experiments on, a lead acid automotive starting battery is that. The skeptics, they're wrong to assume the charge is across the whole battery at once because it doesn't add up with the mechanics behind its charging cycle. Let me explain this in a step by step process to show the charge will not be across the whole battery at once.

For instance; since the battery plates in each cell, they're connected in series-parallel with one another. And each cell, it's connected in series with its terminal connections as well. Then the charge can't be across the whole battery at once; given that, the charging electrons have to pass through its plates in one cell in order to flow to its plates in the next cell.

Therefore; the battery plates in one cell, they've to store electrons while they're flowing to the plates in the next cell as well. Thus; the battery plates in one cell, they're simultaneously charged and discharged to its plates in the next cell. So; all of its plates in each cell, they could be charged during their charging cycle as well.

There's no other way to explain this since the battery plates in each cell, they're connected in series-parallel with one another by their straps, tab connectors and electrolyte solution in each cell. And each cell, it's connected in series with its terminal connections as well. Therefore; the charging electrons, they've to flow from a strap to a tab connector on one plate.

Then electrons, they'll flow across that plate through the electrolyte solution to an adjacent plate. Without flowing to and from, the same tab connector on the same plate. Thus; fixed electrons on one plate, they could reflect free electrons coming from the charging source toward an adjacent plate by way of the electrolyte solution in that cell as well.

Thus; the plates in one cell, they're simultaneously charged and discharged amongst themselves. Since the plates in one cell, they're connected in series with the plates in the next cell by an electrical bus. Then fixed electrons on the plates in the previous cell, they'll reflect free electrons coming from charging source toward the plates in the next cell as well.

As a result; the battery plates in one cell, they're simultaneously charged and discharged to its plates in the next cell. Although; the plates in each cell, they're separated by a non-conductive material called plastic. Wherein; electrons, they've to pass through the battery plates in one cell in order to flow to its plates in the next cell as well.

Therefore; the charge, it's not across the whole battery at once. However; after electrons, they've pass through the plates in one cell in order to flow to the plates in the next cell. Then the charge is across, the whole battery at once. If we've incomplete information about the process, then we might get the wrong impression about the process itself as well.

For instance; we could charge or discharge, the same battery; but, we can't charge it while discharging it. Given that; energy has to go in it, the same way energy has to come out in order to charge its plates or for them to be discharged as well. If we don't understand, how the mechanical structure of the battery will play a role in its cycling processes.

Then we might get the wrong impression about, what we could or couldn't do with the battery because of the laws of physics. All because energy has to go in the battery, the same way it has to come out. However; some in the field of automotive battery technology, they claim even if energy doesn't have to go in the battery, the same way it has to come out.

We still couldn't charge the battery while discharging it because it'll be no benefits in it due to energy lost to heat in both of its conversion processes. Thus; they'll cancel themselves out; therefore, no gain because of the laws of conservation of energy; in which, it sounds ridiculous if we understand what those laws of physics mean or depict.

Since the third law of thermodynamics states that, energy loss to heat in a closed or a perpetual motion system. It can't be used to do any more work than it already has done; that is, create heat. So; if the system, it's going to do any more work? Then energy, it has to be added back to the system with an outside power source in order to do more work.

Therefore; if we're charging the battery while discharging it, then we haven't violated the laws of conservation of energy. Given that; we're adding energy back to the battery with an outside power source, so, it could do more work. Thus; it seems like some in the field of automotive battery technology, they're deliberately misinterpreting those laws of physics as well.

Here's why; although, a lead acid automotive starting battery plates. They go through an electrolysis process during their charging cycle; in which, energy loss to heat does occur. However; the plates in each cell, they'll consume and store energy while it's supplied to the plates in the next cell. And yet the plates in each cell, they're still charged as well.

It seems unlikely the laws of entropy, inertia or the laws of conservation of energy. They'll be an issue when it comes to a simultaneous cycling process of the battery. Given that; any energy loss to heat during the process, it'll be made by the charging source as well; thus, it seems ridiculous to them to discredit a simultaneous cycling process of the battery as well.

If we solicit what the laws of entropy, inertia or the laws of conservation of energy means? And then, try to apply them to the concept of a simultaneous cycling process of a battery. It'll not add up if we're adding energy back to the battery while taking it out. Wherein; those laws of physics, they refer to a closed or perpetual system where energy isn't added back.

In short; if we're adding, electrons back to the battery with an outside power source to do more work. That is to store electrons on the battery plates after they've been consumed electrons due to the electrolysis process. Wherein; the laws of entropy, inertia or the laws of conservation of energy will not be an issue because we're adding electrons back to the battery plates.

Therefore; we don't have to worry about, energy loss to heat affecting the charging cycle of the battery plates because the charging source. It'll make up any energy lost to heat during their charging cycle. Thus; if we're charging the battery plates while discharging them to a load source, then any energy lost to heat will be made up by charging source as well.

Then you've to conclude that, it includes the electrolysis process of the battery plates and the energy being consumed by the load source as well. So; it seems like, only two things we've to worry about when it comes charging the battery while discharging it to the load source. The ability of the charging source and the connections made between them.

If we're looking beyond, the ability of the charging source and the connections made between it and a battery in their relationship with a load source. Then we're focusing on the wrong science to ascertain; whether, it's possible or not to add electrons back to the battery while discharging it to the load source.

Remember a secondary cell battery chemical composition. It doesn't generate its own electrons; but, it only generates an electromotive force to release electrons from it. Thus; we've to conclude that, a charging source could add and subtract electrons from the battery at the same time as well. Then the definition of the charge and the discharging cycle of the battery only mean two things.

One is electrons, are flowing into the battery or flowing out of it. However; it could mean electrons, they're flowing in and out of each cell in it as well. Here's why; if a battery plates in each cell, they're connected in series-parallel with one another by their straps, tab connectors and the electrolyte solution in each cell and each cell is connected in series with its terminals as well.

Then electrons have to pass through the battery plates in one cell in order to flow to its plates in the next cell because of the series and series-parallel circuit connections made between them. Thus; the battery plates in one cell, they're charging while they're discharging to its plates in the next cell by way of an electrical bus as well.

In essence; during the charging cycle of a lead acid automotive starting battery plates, a charge and a discharging process is taking place at the same time. There's no other way to explain this; that is, one definition of the charge and the discharging cycle of the battery means electrons are flowing in and out of each cell in it as well. Based on, the mechanics behind its charging cycle.

Let me explain in this fashion why one definition of the charge and the discharging cycle of the battery mean electrons are flowing in and out of each cell as well. Given that; the charging electrons, they could flow from a strap to a tab connector on one plate. And then, flow across that plate through the electrolyte solution to an adjacent plate as well.

Without electrons flowing to and from, the same tab connector on the same plate. Then fixed electrons on one plate, they could reflect free electrons coming from the charging source toward an adjacent plate by way of the electrolyte solution in each cell. Since the plates in one cell, they're connected in series with the plates in the next cell by an electrical bus as well.

Then fixed electrons on the plates in one cell, they'll reflect free electrons coming from charging source toward the plates in the next cell as well. Therefore; a charge and a discharging process of the battery plates, are taking place at the same time during its charging cycle as well. Then the idea a battery charge and discharging cycle, they can't exist at the same time because of the laws of physics.

Given that; they're different processes, it doesn't hold up based on the mechanics behind the charging cycle of a lead acid automotive starting battery. Wherein, it has little to do with the laws of physics, when it comes to simultaneously charging and discharging the battery plates in one cell to its plates in the next cell. It has more to do with the laws of mechanics than anything else.

So; I find that, we can't use the laws of entropy, inertia or the laws of conservation of energy to discredit a simultaneous cycling process of a lead acid automotive starting battery. Given that; it's a matter of mechanics, how electrons will flow in or out of it. Not a matter of physics, how electrons will flow in or out of it based on the mechanics behind its charging cycle as well.

Some people want to believe that, the charge and discharging cycle of a lead acid automotive starting battery. It can't exist at the same time because of the laws of physics. However; if electrons, they could enter the battery on one terminal and exit it on another terminal connection. Then its charge and discharging cycle, they could exist at the same time.

Since it'll eliminate the process of having to charge the battery on the same terminal connection that, it was discharged on. Then it'll be a matter of mechanics and not a matter of physics why its charge and discharging cycle can't exist at the same time. Wherein; the process of charging it while discharging it will be nothing more than an extension of its charging cycle.

On the other hand; it'll be hard to believe, the laws of electrochemistry will prevent a battery plates from consuming and storing energy while it's supplied to a load source as well. Given that; it'll not be consistent with, the mechanics behind the charging cycle of a lead acid automotive starting battery plates during their charging cycle as well.

Since energy could be simultaneously added and subtracted from the battery plates in one cell to its plates in the next cell without being impeded by its chemical composition. Although; its plates in each cell, they're separated by a non-conducting material called plastic. Then what the different between its plates consuming and storing energy while it's supplied to a load source?

I found the only difference is that, energy has to go in the battery the same way it has to come out in order to charge its plates or for them to be discharged to a load source as well. After taken, an in depth analysis of the mechanics behind the cycling processes of a lead acid automotive starting battery. Its four reasons it can't be charged while being discharged.

The number one reason is that, it has only one opening for electrons to enter or to exit it. Reason two is that, it and a load source will be in parallel with one another in their relationship with a charging source. Reason three is that, the battery resistance is greater than the resistance at the load source.

Reason four is that, a load source is consuming electrons faster than they could break down the resistance of the battery plates to charge them. We could conjure up all manner of laws of physics to justify why, a secondary cell battery can't be charged while being discharged. However; it always boils down to, how we go about adding and subtracting electrons from it.

Somehow, we always assume it's due to the laws of physics why a battery can't be charged while being discharged because we're focusing on the wrong science. It could be one thing or it could be a combination of things, other than the laws of physics; but, we use them as escape goats to justify our ineptness to identify those things preventing it from being charged while being discharged.

We just can't eliminate one thing and expect to charge a battery while discharging it; but, we've to eliminate all those underline causes as well. Although; the design concept of a secondary cell battery, it's the main reason it can't be charged while being discharged. I find there're other extenuating circumstances involved, other than laws of physics as well.

For example; let's consider, the design concept of an automotive starting battery with top and side post terminal connections. See figure B-1 in chapter 10. Since the battery, it has two negative terminals in one cell and two positive terminals in another cell as well. Wherein; it seems like electrons, they could enter and exit it at the same time.

Upon carrying out, a number of experiments on the battery. It wasn't the case electrons could enter and exit it at the same time because they could not. I found it was still necessary to stop its discharging cycling in order to start its charging cycle. We just can't focus on one thing and think it's possible or not possible to charge the battery while discharging it.

Although; the battery, it has four terminal connections. I found a host of things came into play to prevent it from being charged while being discharged, other than the laws of physics. Not only do we've to focus on its exterior structure; but, we've to focus on its interior structure to see how it plays a role in its cycling processes as well.

Here's why; the top posts terminal connections of the battery, they're connected to their respective plates by a strap and tab connectors. Its side post terminal connections, they're only connected to their respective top posts terminal connections by a strap. Thus; its respective top and side posts terminal connections, they only equal to one terminal connection. See figure B-1 in chapter 10.

If we don't take an in depth analysis of the mechanical structure of the battery and see how, it'll play a role in its energy transfer process. Then we might blame it on the laws of physics why it can't be charged while being discharged as well. But; if we take an in depth analysis of its mechanical structure, then we'll find it's not due to the laws of physics as well.

However; we'll find that, it's due to electrons having to go in the battery the same way they've to come out; although, it has two negative terminals in one and two positive terminals in another cell as well. Wherein; it seems like, electrons could enter and exit the battery at the same time; but, they can't base on the connections made between its respective terminal connections.

Although; you could have a charge on, one terminal and a load source on the other respective terminal connection. It still would be like having a charge and a load source on the same terminal connection of the battery. Thus; we still couldn't charge it while discharging it to the load source because electrons still have to go in the same way they've to come out as well.

I've been advocating it'll be possible to charge a battery while discharging it if it had separate terminals for charging and discharging; however, an automotive starting battery with top and side post terminal connections. It gives no evidence it could be charged while being discharged;

even if, it had a charge on one terminal and a load source on the other respective terminal as well.

Here's why; the two respective terminal connections of the battery, they only equal to one terminal connection because of how they're interconnected with one another. Therefore; the battery, it still has only opening for electrons to enter or to exit it. Wherein; we just can't eliminate one factor and think that, we could or couldn't charge the battery while discharging it.

We've to find and eliminate all those contributing factors will prevent the battery from being charged while being discharged as well. In other words; if a battery, it has a charge on terminal and a load source on another terminal connection. It doesn't necessarily mean it could be charged while being discharged to the load source as well.

I found there are other extenuating circumstances will prevent the battery from being charged while being discharged to a load source. It could be related to how the battery multi-terminals will be interconnected with its plates and where those terminals will be located amongst its cells; in which, it'll determine if it could be charged while being discharged to the load source as well.

What I'm saying, other than the laws of physics. The design of a battery could prevent it from being charged while being discharged as well. If we're focusing on, the mechanical structure of an automotive battery with top and side posts terminal connections. Its top posts terminals, they're interconnected with their respective plates by way of a strap and tab connectors.

However; the battery side posts terminals, they're only interconnected with their respective top post terminals by way of a strap. I found the design concept of the battery is the reason it can't be charged while being discharged to a load source. Although; the battery, it has two negative terminals in one cell and two positive terminals in another as well.

If we take an in depth analysis of the mechanical structure of the battery, then we'll know why it can't be charged while being discharged. If not, then we'll blame it on the laws of physics why it can't be charged while being discharged; although, it has two negative terminals in one cell and two positive terminals in another cell as well.

If we don't take an in depth analysis of the mechanical structure of the battery, then we'll not know electrons. They still have to go in the same way they've to come out in order to charge the battery plates or for them to be discharged to a load source as well. Although; it's a charge on, one terminal and a load source on the other respective terminal of the battery as well.

However; we can't expect to charge, the battery while discharging it to the load source if electrons have to go in the same way they've to come out as well. If we know this, then we're half way there. We don't need to focus on the vast and complex laws of physics to figure this out. All we've to do is focus on the laws of mechanics to figure this one out.

Here's why; if a battery plates in each cell, they're separated by a non-conducing material called plastic and yet, its plates in one cell. They're simultaneously charged and discharged to its plates in the next cell by way of an electrical bus. Without being impeded, by the laws of physics. Then we need to focus on the laws of mechanics and not the laws of physics.

It doesn't make any sense to blame it on the laws of physics why a battery can't be charged while being discharged to a load source. If a battery plates in each cell, they could be charged without having to stop charging them in one cell in order to charge them in the next cell; although, they're separated by a non-conducting material called plastic as well.

Here's why; if a battery plates in each cell, they're connected series-parallel with one another by their straps, tab connectors and the electrolyte solution in each cell. Then electrons, they could flow from a strap to a tab connector on one plate. And then, flow across that plate through the electrolyte solution to an adjacent plate without flowing from the same tab connector.

If the battery cells, they're connected in series with its terminal connections as well. Then electrons, they could flow in and out of each cell flowing in one direction without having to stop charging the plates in one cell in order to charge them in the next cell; although, they're separated by a non-conducting material called plastic as well.

So; we don't have to focus on the vast and complex laws of physics to figure out why a battery can't be charged while being discharged to a load source. Because the laws of physics, they've little to do with it. All we've to do is focus on the design of the battery and the connections made between it and the load source in their relationship with a charging source.

This is why I've came to the conclusion that, we'll be focusing on the wrong science if we're focusing on the laws of physics when it comes to a conversation about charging a battery while discharging it to a load source. The laws of physics, they don't determine how electrons will be added or subtracted from the battery plates; but, its design will.

If we decipher the mechanics behind, a lead acid automotive starting battery's charging cycle correctly. Then it'll show certain laws of physics, they could be manipulated by its design as well. Given that; the laws of mechanics, they'll determine how electrons will be added or subtract from the battery plates during their charging cycle.

Here's why; remember if the battery plates in each cell, they're connected in series-parallel with one another. And each cell, it's connected in series with its terminal connections as well. It'll allow its plates in each cell to consume and store electrons while supplying them to its plates in the next cell. In order to, charge all of its plates in each cell during their charging cycle.

Although; energy, it's lost to heat during the process. Neither the laws of entropy, inertia nor the laws of conservation of energy will be an issue. When energy is taken from the battery plates in one cell, so, it could be added to the plates in the next cell as well. Even though; the plates in each cell, they're separated by a non-conducting material called plastic as well.

What's the difference between simultaneously charging and discharging, the battery plates in one cell to its plates in the next cell? Without the laws of entropy, inertia or the laws of conservation of energy being an issue. Compared to, simultaneously charging and discharging them to a load source. I find the answer to be mechanics and not physics.

Remember if a battery cells, they're connected in <u>series</u> with its terminal connections. Then we can't expect to power a load source while charging the battery as well. If the charging voltage and current, they've to flow through every cell in the battery before reaching the load source. Given that; it'll be a voltage drop before voltage, it reaches the load source.

Although; the charging current, it'll be the same throughout each cell in the battery. However; the load source, it might not receive the proper voltage needed to supply electrons to it. Likewise; if a battery cells, they're connected in <u>parallel</u> with its terminal connections. Then we can't expect to charge the battery while powering a load source as well.

If the charging voltage and current, they've to flow through every cell in the battery before reaching the load source as well. Although; there will be no voltage drop before voltage, it reaches the load source. However; the charging current, it'll not flow across the battery plates to charge them if their resistance is greater than the resistance at the load source.

Thus; the charging current, it might flow directly to the load source and not flow across the battery plates in order to charge them. Therefore; they might not get, the proper amount of current needed to charge them as well. So; we've to ask ourselves, is it due to the laws of mechanics or is it due to the laws of physics? The answer is the laws of mechanics.

If we understand that, voltage and current will enter or exit a battery based on its design. Then we'll find that, it's a matter of mechanics and not a matter of physics. Wherein; it'll be no need to focus on, either the laws of entropy, inertia or the laws of conservation of energy. When it comes to a conversation about charging a battery while, discharging it to a load source as well.

If we consider, the automotive starting battery with top and side posts terminal connections. It doesn't matter if it has a charge on one terminal and a load source on the other respective terminal. It still can't be charged while being discharged to the load source as well. Wherein; it doesn't mean that, it's due to the laws of physics as well.

I found it was more to it than having a charge on one terminal and a load source on the other respective terminal. However; I found it was where those terminals, they're located amongst the battery cells and how they're interconnected with its plates in that cell as well. In which; it'll determine if the battery could or couldn't be charged while, being discharged to the load source as well.

On the other hand; thinking about, the design concept of an automotive battery with top and side posts terminal connections. You might assume it has two openings for current to enter or to exit it at the same time. Given that; it has two positive terminals in one cell and two negative terminal connections in another cell as well.

When it comes to current entering the battery, it really doesn't have two openings for current to enter it at the same time. Given that; its top and side posts terminal connections, they're interconnected by a strap and not by their respective plates. Thus; the charging current, it could only flow to the battery plates from one of its respective terminals to its respective plates.

Then current, it could only flow across the strap that joins the two respective terminals together when current is entering battery from both of its respective terminals. Likewise; current, it could exit the battery from both of its respective terminals as well. Thus; current, it could only flow back and forth across the strap that joins the two respective terminals together.

When there's a charge on, one terminal and a load source on the other respective terminal. Then the charging current, it can't flow from one of the battery respective terminal to its respective plates to charge them. Given that; the charging current, it could only flow back and forth across the strap that joins the two respective terminals together as well.

Therefore; the charging current, it'll only flow toward the load source when the charge is on one respective terminal and the load source is on the other respective terminal of the battery. I found it had little to do with the laws of physics why it couldn't be charged while being discharged to the load source; but, it had more to do with the mechanical structure of the battery.

If we're not focusing on, the mechanical makeup of the battery and how it'll plays a role in its energy transfer process. Then we might get the wrong impression, it's due to the laws of physics why it can't be charged while being discharged to the load source. In the next chapter; let me explain why, it's not due to the laws of physics; but, it's due to the wrong mechanics are in play.

CHAPTER 10

THE WRONG MECHANICS IN PLAY

I FIND IT'S EASY for some to discredit the possibility of a simultaneous cycling process of a lead acid automotive starting battery; even if, energy could enter and exit it at the same time. Since we can't see the fact that, we shouldn't have to stop its discharging cycle in order to start its charging cycle since we don't have to stop charging its plates in one cell in order to charge them in the next.

Although; the battery plates in each cell, they're separated by a non-conducting material called plastic. Some people, they'll conjure up all manner of laws of physics to justify their assumptions why a battery can't be charged while being discharged. While overlooking, the fact that the battery isn't designed for charging while discharging as well.

I didn't realize our batteries weren't designed for charging while discharging; until, I started exploring the mechanical structure of a run capacitor and a power transformer. After exploring, their mechanical structures and how they played a role in their energy transfer process. Then I realized the wrong mechanics were in play to charge our batteries while discharging them as well.

I found if we eliminate the obvious, then we'll find it has little to do with the laws of physics in and of themselves why our batteries can't be charged while being discharged; but, it'll have more to do with the wrong mechanics in play. In figure B-7, I'm going to use a simple illustration to explain. Why the wrong mechanics, are in play to charge our batteries while discharging them.

I'm going to use a lead acid automotive starting battery to show, why it can't be charged while being discharged because the wrong mechanics are in play. In other words; the battery, it's not designed to the right specification for charging while discharging.

Figure B-7

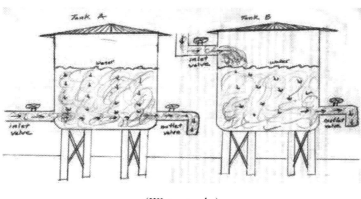

(Water tanks)

I'm going to use the same examples I used in my book, "Miracle Auto Battery" to give a clear understanding why, the battery can't be charged while being discharged because one picture is worth a thousand words. In figure B-7, it shows two water tanks; A and B. Tank-A; it'll illustrate, what'll happen when there's a charge and a load source on the same terminal of a battery.

Tank-B, it'll illustrate what'll happen when a battery truly has separate terminals for charging and discharging; but, we'll talk about water tank-B in chapter 13. I hope tank-A, it'll help simplify. Why we're focusing on, the wrong science when it comes to a conversation about charging our batteries while discharging them to a load source.

Wherein; it's not due to the laws of physics; but, it's due to the contemporary design of our batteries why they can't be charged while being discharged to a load source. In figure B-7, tank-A has an inlet and an outlet valve at the same depth and near the bottom of it. Thus; it'll represent, what'll happen when there's a charge and a load source on the same terminal of a battery.

For instance; if we're going to simultaneously add and subtract water from tank, then its inlet and outlet valves will be open at the same time. Therefore; it'll create, a path for the incoming water to flow directly out of the tank because its inlet and outlet valves, they're at the same depth and they're open at the same time as well.

As a result; the wrong mechanics, are in play to simultaneously added and subtracted water from the tank. Although; water, it'll be flowing in and out of both valves at the same time; but, the process will only be subtracting water for the tank and not adding water to it. Wherein; it's not due to the laws of hydrodynamics, it's due to other circumstances involved.

Such as; the weight of the water, gravity and where the two valves are located within the tank in their relationship with one another. If water is already in the tank, then it'll be trying to get out through the open outlet valve because of the weight of the water already in the tank and gravity; especially, where the outlet valve is located in its relationship with the inlet valve.

Since both valves, they're at the same depth and they'll be open at the same time. Then it'll allow the incoming water to flow directly out of the tank, along with the water that is already in the tank as well. Given that; the incoming water, it'll be flowing through one valve and out the other. It'll create suction allowing water already in the tank to flow out of it as well.

Instead of adding water back to the tank, we'll be subtracting water from it; so, it'll end up empty because of its design concept. Since its outlet valve, it'll be open at the same time as its inlet valve. Then the path of least resistance for the incoming water, it's to flow out of the outlet valve because of where the outlet valve is located within the tank in its relationship with the inlet valve.

It has little to do with the laws of hydrodynamics; but, it has more to do with those extenuating circumstances preventing water from being simultaneously added and subtracted from the tank. Wherein; it has more to do with its design than anything else, not the laws of hydrodynamics why water can't be simultaneously added and subtracted from it.

The laws of hydrodynamics, they're nothing more than the behavior of water under different circumstances. In other words; because of other extenuating circumstances involved, the laws of hydrodynamics. They'll not allow water to be simultaneously added and subtracted from the tank because of where its inlet and outlet valve is located within it.

Since the incoming water, it could flow directly out of the tank before it could begin to fill it because both of its valves will be open at the same time. Then the water has a direct path to the open outlet valve; wherein, it's the main reason why. Water can't be added back to the tank while it's taken out as well.

We face a similar scenario when there's a charge and a load source on the same terminal connection of a battery. It'll be like having an inlet and an outlet valve open at the same time as well. Wherein; the charging current, it'll have a direct path to the load source because of the parallel circuit connection made between it and the battery.

Therefore; the charging current, it doesn't have to flow into the battery to charge it if its resistance is greater than the resistance at the load source. In reality, it has little to do with the mechanics of the current alone or the laws of physics as well. It has more to do with the design of the battery and the connections made between it and the load source.

The connections made between the battery and the load source, it'll allow electrons already stored in the battery to flow out of it. Anything short of creating, a surplus of electrons at the load source with the charging source to reverse the current flow back into the battery to charge it; wherein, it'll end up empty if we attempt to charge it while it has the load source on it.

Based on the electron current flow concept, current could only enter or exit the negative terminal connection of a battery to cycle it. Since it has only one negative terminal connection, then it can't be charged while being discharged; thus, it's not entirely due to the mechanics of the current alone or the laws of physics; but, it's mainly due to its design as well.

If we wanted to charge the battery while discharging it, then logic should tell us it'll need more than one negative terminal connection. If we haven't figured this out yet, then we're inept to explain why it can't be charged while being discharged; likewise it's more to it than it having more than one negative terminal to charge it while discharging it as well.

If we learned anything about, the mechanics behind the cycling processes of a secondary cell battery. Then we'll know, a host of things have to come in to play in order to charge it while discharging it. Those things, includes having more than one negative terminal and where they'll be located within the battery in their relationship with one another as well.

Likewise; if we learned anything about the mechanics of the voltage and current, then we'll know they'll be submissive to the design of the battery as well as they'll be submissive to the design of an electrical circuit connection as well. Then we'll know that, the design of the battery is the deciding factor in determining how voltage and current will enter or exit it as well.

We don't have to focus on the vast and complex laws of physics to figure out why a secondary cell battery can't be charged while being discharged to a load source. All we've to do is focus on the design of the battery and the connections made between it and the load source in their relationship with a charging source; in which, it's more easily to understand as well.

This is why it's important to understand, the mechanical structure of a battery and how it'll play a role in its cycling processes as well. See chapter 3. On the other hand; we've to understand, how voltage and current will flow through different types of circuit connections; such as, series, parallel or series-parallel circuit connections as well.

If we're going to understand, what's required to charge a battery while discharging it to a load source? Then we can't end up focusing on, the wrong science or the wrong mechanical when it comes to a conversation about charging the battery while discharging it to the load source. Wherein; we might think, it's impossible to carry out such a process as well. See chapter 5.

I found in order to charge a secondary cell battery while it has a load source on it. The charge and the load source will be on the same terminal connection of the battery because of its contemporary design. Thus; it and the load source, they'll be in parallel with one another in their relationship with the charging source as I discovered.

If you know anything about a parallel circuit connection, then you'll know the wrong mechanics are play to charge the battery while it has the load source on it. Since it has, only one opening for current to enter or to exit it and it'll be in parallel the load source. These combinations of things, they'll prevent the battery from being charged while it has the load source on it.

Given that; the amount of resistance at the load source, it'll determine the amount of current will flow into the battery because of the parallel circuit connection made between them. The parallel circuit connection made between them in their relationship with the charging source. It'll allow the charging current to follow the path of least resistance between them.

In other words; current, it'll flow to the battery or flow to the load source; but, it depends on which one has the least resistance. In which; the parallel circuit connection made between them, it'll create this dilemma. However; allowing, the charging source to create a surplus of electrons at the load source to reverse the current back into the battery will solve that dilemma as well.

Since any additional electrons flowing from the charging source toward the load source, they'll be reflected back into the battery to charging it. Wherein; the charging source, it'll create more electrons than needed to power the load source. In which; it'll create, electrical resistance at the load source to reverse the current flow back into the battery to charge it.

So far, it's the only way we could charge a battery while it has a load source on it; although, they'll be in parallel with one another; thus, using the mechanics of the current to work in our favor. When we're talking about charging a battery while discharging it to a load source, we're facing similar scenarios when they're connected in parallel with one another as well.

However; we've to do more than allowing a charging source to create a surplus of electrons at a load source in order to reverse the current flow back into a battery to charge it because it has only one opening for current to enter or to exit it. In which; it's the main reason why, it can't be charged while being discharged to a load source in the first place.

A battery having only one opening for current to enter or to exit it and it'll be in parallel with a load source in their relationship with a charging source. It'll present a unique problem when it comes to charging the battery while discharging it to the load source. Given we've to charge the battery on the same terminal connection that, it's discharged on as well.

When we're having, a conversation about charging a battery while discharging it to a load source. We've to understand the concept behind the process itself. Wherein; it has little to do with the mechanics of the current alone or the laws of physics; but, it has more to do with the design concept of the battery itself. See chapter 3.

If we understand, the concept of charging a battery while discharging it to a load source. Then we end up focusing on the wrong science because we don't understand that, the wrong mechanics are in play to charge the battery while discharging it to the load source. You would think those in the field of automotive battery technology, they would this as well.

Given that; they design our batteries with, only one opening for current to enter or to exit them; plus, they'll be in parallel with a load source in their relationship with a charging source as well. In which; these combinations of things, they'll pose a problem when it comes to charging our batteries while discharging them to a load source as well.

You would think those in the field of automotive battery technology, they would have known creating a surplus of electrons at a load source. It wasn't the only solution to charging a battery while it has a load source on it as well. Given how they design a battery, so, we don't have to stop charging its plates in one cell in order to charge them in the next cell as well.

On the other hand; it makes you wonder, what type of battery those in the field of automotive battery technology used to come to the conclusion. It'll make no difference; even if, electrons could enter and exit it at the same time. We still couldn't charge it while discharging it to a load source because of the laws of physics as well.

There's only one automotive battery on the market; so far, that resembles a battery with two openings for electrons to enter or to exit it at the same time. It's an automotive starting battery with top and side posts terminal connections; also, a modern day cell phone battery with its multi-terminal connections as well.

Although; it seems like, they've more than one opening for energy to enter or to exit them at the same time; but, they do not. If we don't take, an in depth analysis of their mechanical structures and see how they'll play a role in their cycling processes. Then we might get the wrong impression why they can't be charged while being discharged because of the laws of physics as well.

Let's start with an automotive starting battery with top and side posts terminal connections. To see why the wrong mechanics are play to charge it while discharging it; although, it has two positive terminals in one cell and two negative terminals in another cell as well. See figure B-1.

Figure B-1

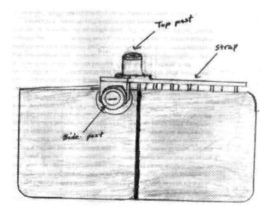

(Automotive battery w/ top and side post connections)

The top and side posts terminal connections of an automotive starting battery, it allows it to be retro-fitted for older or newer model vehicles. Given that; most of the newer model vehicles, their cable connections. They're designed for side posts terminal connections to reserve space for complex equipment or electrical devices added to the newer model vehicles.

For these reasons, many battery manufactures started designing our automotive starting batteries with top and side posts terminal connections or universal terminal connections. Their top post terminal connections, they allow cables to be fitted on top of them. Their side post terminal connections, they allow cables to be screwed on the side of them.

Although; automotive starting battery with top and side posts terminal connections, it has two negative terminals in one cell and two positive terminals in another cell as well. However; the battery, it can't be charged while being discharged. Wherein; it gave me insight into why, the wrong mechanics were play to charge it while discharging it as well.

I found one respective terminal connections of the battery was connected to its respective plates by a strap and tab connectors. However; the other respective terminal, it was only connected to its respective terminal by a strap. Wherein; the two respective terminals, they only equal to one terminal connection; although, there're two of them as well.

When there's a change on, one terminal and a load source on the other respective terminal connection of the battery. Then it'll be like having, the charge and the load source on the same terminal connection of the battery. We'll not know this if we don't take an in depth analysis of its mechanical structure to see how it'll play a role in its energy transfer process.

If we don't an in depth analysis of its mechanical structure, then we might get the wrong impression it can't be charged while it's being discharged because of the laws of physics as well. This false impression, it'll become an erroneous belief because it'll be based on a false premise. That the battery has multi-terminals for electrons to enter or to exit it as well; but, it doesn't.

Here's the deal; although, the battery appears to have multi-terminals for electrons to enter and to exit it at the same time. But; it doesn't because of how its multi-terminals, are interconnected with its plates. In which; the wrong mechanics, are in play to charge it while discharging it. Given how its respective terminals are interconnected with one another as well.

The top posts terminal connections of the battery are connected to their respective plates by a strap and tab connectors. However; the side posts terminal connections of the battery, they're only interconnected with their respective top posts terminal connections by a strap and not by their respective plates; so, the wrong mechanics are in play to charge it while discharging it.

Wherein; the mechanics of the current, they'll dictate whether electrons are flowing into the battery or flowing toward the load source. It'll be based on which one has the most resistance to current. We could blame it on the mechanics of the current alone or the laws of physics; but, in reality. The wrong mechanics in play, they're dictating the direction of the current.

For instance; the modern day cell phone batteries, they're good examples of a battery having multi-terminal connections and yet, they still can't be charged while being discharged because of their mechanical structures as well. In figure 2A, it shows the design concept of some modern day cell phones batteries with multi-terminal connections.

If you notice, the batteries might have three or four terminal connections. Some in the field of battery technology, they say the design concept of modern day cell phone batteries with multi-terminal connections. Those terminals, they're used for thermistors to provide overcharge protection for the batteries.

Figure 2A

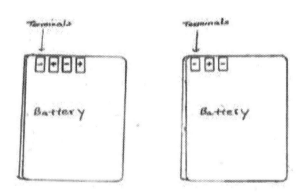

(Examples of some modern day cell phones' batteries)

Nevertheless; I find that, modern day cell phone batteries. They've multi-terminal connections for current to enter or to exit them at the same time. In which; it changed, my perspective about charging a battery on the same terminal connection it was discharged on. I found energy doesn't have to go in the battery the same way it has to come as well.

Although; our cell phones batteries, they've more than one opening for current to enter or to exit them; but, they can't be charged while being discharged because of the wrong mechanics are in play. It's not due to the mechanics of the current alone or the laws of physics. But; the batteries, they're not designed to the right specification for charging while discharging.

Wherein; some of the multi-terminal connections for the cell phones' batteries in particular, they're used for thermistors to provide overcharge protection for the batteries and not to charge them while discharging them. Thermistors are types of resistors that are sensitive to temperature changes; thus, their resistance changes when the temperature of a battery changes.

Thus; hotter a battery gets, a thermistor increases its resistance to limit the amount of current flowing into the battery; therefore, providing over charge protection for it during its charging process. Although; a thermistor, it detects thermo energy; but, energy has to have more than one way to exit the battery in order for the thermistor to work.

Therefore; energy, it has to flow to and from the battery at the same time. If a thermistor, it's going to provide overcharge protection for the battery. So; you could assume that, its multi-terminals will allow energy to enter and to exit it at the same time as well. As a result; it gave me, a new perspective on how to cycle a lead acid automotive starting battery as well.

Understanding the purpose of a cell phone's battery multi-terminals, it helped me to understand. We've have to change the exterior design of a lead acid automotive starting battery in order to charge it while discharging it. Likewise; an automotive starting battery with top and side posts terminal connections, it gave me more insight into it as well.

A cell phone's battery with its multi-terminal connections and an automotive starting battery with its top and side post terminal connections; although, they can't be charged while being discharged. But; they gave me a glimpse into why, our batteries can't be charged while being discharged because the wrong mechanics are in play to carry out such a process.

On the other hand; an automotive starting battery with top and side posts terminal connections, it proves one thing tradition could be broken. When it comes to the design concept of our automotive batteries, they could have more than two terminal connections and still work according to their specifications as well.

If we think, it's inconceivable or farfetched to charge our automotive batteries while discharging them because of the laws of physics. Then we're focusing on, the wrong mechanics because in figure B-1. It showed the design concept of an automotive starting battery with top and side posts terminal connections to help illustrate why it's not due to the laws of physics.

The wrong mechanics are in play to charge the battery while discharging it; although, it has multi-terminal connections as well. I found it's not limited to having a charge and a load source on the same terminal connection of it; but, it could have a charge on one terminal and a load source on another. And yet, it still can't be charged while being discharged as well.

I found a battery could have multi-terminal connections. And yet, electrons still have to go in it the same way they've to come out as well. If we take an in depth, analysis of the mechanics structure of an automotive starting battery with top and side posts terminal connections. It'll verify electrons still go in it the same way they've to come out as well.

Although; the battery, it has two positive terminals in one cell and two negative terminals in another cell as well. I found it takes more than having a charge on one terminal and a load source on another respective terminal connection to charge a battery while discharging it to a load source. It also depends on how those terminals are interconnected as well.

Remember water tank-A scenario in figure B-7, it illustrated what'll happen when the wrong mechanics are in play to simultaneously add and subtract water from the tank. We could apply this to an automotive starting battery with top and side posts terminal connections; given that, the wrong mechanics in play to simultaneously add and subtract electrons from it as well.

Although; the battery, it could have a charge on one terminal and a load source on the other respective terminal. And yet, it still can't be charged while being discharged to the load source. Given that; one terminal, it's connected to its respective plates by a strap and tab connectors; but, the other respective terminal is only connected to its respective terminal by a strap.

Thus; the respective terminals of the battery, they'll only equal to one terminal; given that, they're joined by a strap and not by their respective plates. Therefore; the two respective terminal connections, they'll be connected in parallel with one another by a strap; as a result, if the resistance at the battery is greater than the resistance at the load source.

Then the charging current, it'll not come in contact with the battery plates in order to charge them; given that, it could flow through one respective terminal and flow across the strap that joins them together. And then, flow out one respective terminal to the load source without coming in contact with the battery plates to charge them as well.

It'll have little to do with the mechanics of the current alone or the laws of physics. Why the charging current, it'll not come in contact with the battery plates to charge them; but, it'll have more to do with the wrong mechanics in play. Given how, the two respective terminals are interconnected with one another by a strap and not by their respective plates.

Thus; current, it could only flow to and from one respective terminal to its respective plates. However; one respective terminal connection, current can't flow to and from it to its respective plates without flowing through the other respective terminal connection; given that, it's only connected to its respective terminal by a strap that joins them together in parallel.

Therefore; when there's a charge on one terminal and a load source on the other respective terminal connection, then the charging current. It'll flow through one terminal and flow across the strap joining the two respective terminals together and out the other to the load source without coming in contact with the battery plates to charge them.

Since the two respective terminal connections, they'll be in parallel with one another by way of a strap. Then it's like having, a charge and a load source on the same terminal connection of the battery as well. Although; the battery, it has two positive terminals in one cell and two negative terminals in another; but, they only equal to one positive and one negative terminal.

Then the reason why the battery, it can't be charged while being discharged to the load source because of the mechanics of the current or the laws of physic. It'll be based on the same premise as having a charge and a load source on the same terminal connection of a battery; wherein, the wrong mechanics are in play to charge it while discharging it the load source as well.

We've to understand that, a secondary cell battery behave differently during its discharge cycle than it does during its charging cycle because of the different mechanics involved. We've to understand these different mechanics if we going to understand why we've to stop its discharging cycle in order to start its charging cycle as well.

During the battery discharging cycle, it acts like a power source. However; during its charging cycle, it acts like a load source consuming electrons from a power source. For example; if we've two electric motors and one is connected to, the top negative and positive post terminal connections of an automotive battery with top and side posts terminal connections.

The other electric motor, it'll be connected to the side negative and positive post terminal connections of the battery; thus, it'll be no voltage drop for either motor. Because they'll receive the same amount of voltage flowing from the battery; given that, they'll be connected in parallel with one another in their relationship with the battery.

However; current, it'll flow to the motors based on their individual resistance and voltage applied by the battery. But; if we disconnect, one electric motor from its terminals and connected a charging source to them. We'll still have two load sources connected in parallel with one another because once a charge is on the battery, then it's no longer a power source.

Given that; the battery, it becomes like the motor trying to consume energy from the charging source as well. Although; the battery and the motor, they'll receive the same amount of voltage flowing from the charging source. It'll be a twist when it comes to them receiving current from the charging source because it'll flow to them based on their individual resistance.

The twisted is that, current will flow between the battery and the motor based on the resistance between them in their relationship with the charging source. In other words; if the resistance at the battery is greater than the resistance at the motor, then the charging current will only flow to the motor. And not, flow into the battery at all because the wrong mechanics are in play.

Here's why; remember a load source resistance to current, it'll be based on the number of turns that the wiring has inside of it to conduct current. But; a battery resistance to current, it'll act like a non-conducting material that will reflect current away from it. Thus; the charging current, it'll flow toward the load source before flowing into the battery because it's the path of least resistance.

Since the battery and the load source, they'll be in parallel with one another. Then it's like having an inlet and an outlet valve of a tank open at the same time and they're at the same depth of the tank as well. Let's go back to the example where we disconnected one motor from one set of terminals and then connected a charging source to them in order to get a better understanding.

The laws of parallel circuitry, they still will be in play because the battery and the motor will be in parallel with one another in their relationship with the charging source. However; we'll get a different result, compared to when there were two electric motors on the battery at the same time; given that, both motors received current at the same time.

However; I find that, a parallel circuit connection made between a battery and a load source. It doesn't work the same as a parallel circuit connection made between two load sources; such as, the two electric motors because the wrong mechanics are in play when it comes to a parallel connection made between a battery and a load source. See chapter 12.

There're other extenuating circumstance we've to consider when it comes to charging a battery while it has a load source on it because of the mechanics of the current. Not only do we've to understand voltage will enter or exit a battery different than current will; but, voltage and current will flow through a battery differently based on which cycling process is carried out as well.

Here's why; if a battery cells, they're connected in series with its terminal connections. Then its total voltage potential, it'll equal to all of its plates in each cell; but, its current potential will only equal to its plates in one cell during its discharging cycle. However; during its charging cycle, voltage flowing from the charging source will be evenly divided amongst its cells.

But; current flowing from the charging source, it'll be the same across each cell in the battery during its charging cycle. If a battery cells, they're connected in parallel with its terminal connections. Then its total current potential, it'll equal to all of its plates in each cell. But; its total voltage potential, it'll only equal to its plates in one cell during its discharging cycle.

During its charging cycle, voltage will be the same across each cell coming from the charging source. But; current, it'll be divided between the battery cells based on the individual resistance of the plates in each cell. In view of these facts, it seems like the design of a battery is the deciding factor in determining how voltage and current will enter or exit it as well.

It has little to do with the mechanics of the voltage and current alone. Why a battery, it can't be charged while being discharged; but, it has more to do with the wrong mechanics being in play. In the next chapter; let's see why focusing on, the wrong science and mechanics will lead to erroneous beliefs about what we could or couldn't do with a battery as well.

CHAPTER 11

ERRONEOUS BELIEFS ABOUT AN AUTO BATTERY

IT WAS NON-RECHARGEABLE BATTERIES before there were rechargeable batteries. I bet when the first rechargeable was made people were saying the same thing then as people, are saying today about charging it while discharging it. That it can't be done because of the laws of physics as well. Those who are skeptical of these ideas, their skepticism is created by erroneous beliefs as well.

If the skeptics knew that, each cell in a lead acid automotive starting battery is nothing more than a battery in and of itself. That is connected in series with one another and housed together in a plastic case to make up the whole battery as we know of today. Then they might be less skeptical about charging it while discharging it to a load source as well.

Given that; each time, the battery is charged as a whole. We're simultaneously charging and discharging one battery while simultaneously charging and discharging another. I believe the reason why some skeptics, they think it's impossible to simultaneously charge and discharge the battery as a whole because of the lack of information about its cycling processes.

The mechanics behind the charging cycle of the battery, it shows it's possible to charge it while discharging it and it'll be a benefit in it as well; wherein, the process will be nothing more than an extension of its charging cycle as well. Here's why; a secondary cell battery, it doesn't generate its own electrons like a primary cell battery does.

Thus; a secondary cell battery, it's nothing more than a storage device for electrons; therefore, it'll need an external electromotive force to add electrons to it and an internal electromotive force to subtract electrons from it. Wherein; the process of adding electrons back to the battery, it'll not stop process of subtracting electrons from it as well.

I've compiled data from books written or supported by those in the field of automotive battery technology. To come to a logical conclusion, it's possible to add electrons back to a lead acid automotive starting battery while taking them out and it'll be a benefit in it as well. If we change, its contemporary design; so, electrons could enter and exit it at the same time as well.

Then the idea that we've to stop the discharging cycle of the battery in order to start its charging cycle, it's nothing more than an erroneous belief. Since this idea, it's not consistent with the mechanics behind its charging cycle as well. Given that; its plates in each cell, they could consume and store electrons while supplying them to its plates in the next cell as well.

Given how the battery plates and cells are interconnected with its terminal connections. We don't have to stop charging its plates in one cell in order to charge them in the next cell; although, they're separated by a non-conducting material called plastic. It seems like its exterior design prevents it from being charged while being discharged to a load source.

The idea of the battery can't be charged while being discharged a load source. It's nothing more than an erroneous belief handed down from one generation to the next because no one ventured outside the status quo to change the contemporary design of the battery; so, it doesn't have to be charged on the same terminal connection it was discharged on as well.

I'm going to share with you what I've discovered during my analysis of a lead acid automotive starting battery charging cycle. In hope it'll clear up some of the erroneous beliefs about what we could or couldn't do with it because of the laws of physics. Wherein; those erroneous beliefs, they're not consistent with the mechanics behind its charging cycle as well.

Those erroneous beliefs behind, why we can't charge the battery while discharging it because of the laws of physics. They're nothing more than myths handed down from one generation to the next. For instance; some who are skeptical of the process, they assume it'll be due to the laws of entropy or the laws of inertia because the battery energy will eventually run down.

Wherein; it seems like, it's nothing more than erroneous beliefs because we'll be adding energy back to the battery while taking it out; most likely, the battery energy will not run down if we're adding energy back to it while taking energy out as well. Thus; the skeptics assumptions, are flawed because they're overlooking energy being added back to the battery as well.

The mechanics behind charging cycle of the battery, they don't support the skeptics' assumptions. Nor the laws of entropy or the laws of inertia; given that, they'll not apply if we're adding energy back to the battery while taking it out as well. When it comes to adding energy back to the battery while taking energy out, I find it's about mechanics and not about physics.

In which; it's about, how we go about adding or subtracting energy from the battery. If we think it's about the laws of physics, then it's nothing more than an erroneous belief because the laws of physics. They don't determine how energy will be added or subtracted from the battery plates because only the laws of mechanics do.

There are a host of things will come into play to prevent a battery from being charged while being discharged, other than the laws of physics as well. For instance; the connections made between, the battery and a load source in their relationship with a charging source. It could be one reason, other than the laws of physics as well.

Not to mention that the battery has only one opening for energy to enter or to exit it. In which; it's the main reason why, it can't be charged while being discharged as well. It's a host of things will prevent the battery from being charged while being discharged, other than the laws of physics; so, we just can't focus on one thing when it comes to charging it while discharging it as well.

On the other hand; some skeptics say that, the battery can't be charged while being discharged because its charge and discharging cycle. They're two different processes; therefore, they can't exist at the same time because of the laws of physics, I found this is an erroneous belief as well. The only reason they're two different processes because of its contemporary design.

Wherein; energy, it has to go in the battery the same way energy has to come out in order to cycle it. For instance; in order to charge the battery, the charging voltage as to be higher than the normal battery voltage to reverse the current flow back into it to charge it. If energy, it has to go in the battery the same way it has to come out as well.

Then the charge and the discharging cycle of the battery, they can't exist at the same time if energy has to go in the same way it has to come out. So; is it due to the laws of physics or is it due to the contemporary design of the battery. In which; I found that, it's due to the contemporary design of the battery why they can't exist at the same time as well.

However; if energy, it doesn't have to go in the battery the same way energy has to come out. Then it'll be no conflict between its charge and discharging cycle because it'll come down to the strength of the electromotive forces being used to add or to subtract energy from it. Whichever electromotive force has the highest voltage potential will prevail over the other.

The reason it's true a battery charge and discharging cycle, they can't exist at the same time because energy has to go in it the same way energy has to come out. Therefore; it's due to the fact that, it has to be charged on the same terminal connection it was discharged on. Some in the field of electrochemistry, they'll chuckle when I say this because they think its utter non-sense.

However; they think, I know nothing about the chemistry of the lead acid automotive starting battery. What they failed to realize, all I need to know could electrons be added or subtracted from its chemical composition if so, then it's all about mechanics. Thus; all I've to do is that, change its mechanical structure; so, it'll not prevent it from being charged while being discharged.

What some in the field of electrochemistry, they failed to realize is that. The concept of charging a lead acid automotive starting battery while discharging it to a load source. It's not about the laws of electrochemistry; given that, we could already add or subtract electrons from the same chemical composition of its plates during their charging cycle; so, it's about mechanics.

Remember in order to replenish the chemical energy of a lead acid automotive starting battery with electrical energy, it's a threefold process. First; the charging voltage, it must be higher than the normal battery voltage to reverse the current flow back into it to charge it. In order to, restore the active areas of its plates; so, they could release more electrons.

Secondly; the charging electrons, they've to diminish the lead-sulfate build up on the battery plates that accumulated on them during their discharging cycle. And third; the remaining electrons, that wasn't used up during the electrolysis process of the battery plates. They'll be stored on its plates to restore their chemical energy as well. See chapter 4.

The assumption it's about the chemistry of the battery why, it can't be charged while being discharged is nothing more than an erroneous belief. It shows the lack of understanding about the mechanics behind its charging cycle as well. Since the battery plates in each cell, they're connected in series-parallel by their straps, tab connectors and the electrolyte solution in each cell.

Then electrons could flow from a strap to a tab connector on one plate and then, flow across that plate through the electrolyte solution to an adjacent plate by way of the electrolyte solution. Without flowing to and from, the same tab connector on the same plate; thus, free electrons will be jolting with one another for space on each plate in each cell as well.

Since each cell, it's connected in series with one another by way of an electrical bus. Then electrons could flow in one direction from the plates in one cell to the plates in the next cell without having to stop charging them in one cell in order to charge them in the next cell; although, they're separated by a non-conducting material called plastic as well.

Although; the electromagnetic laws of physics, they do have a role in the charging cycle of the battery. However; it's only to the extent that, like charges will repel one another and follow the path of least resistance. But; the charging cycle of the battery is mostly carried out by mechanics, in which, it has little to do with the laws of electrochemistry as well.

The laws of electrochemistry, they don't determine how electrons will be added or subtracted from the battery. They only determine if electrons could be added or subtracted from its plates. However; the laws of mechanics or the design of the battery, it's determine how electrons will be added or subtracted from its plates. See chapter 3.

I find it's no need to focus on the laws of electrochemistry when it comes to charging the battery while discharging it. See chapter 4. On the other hand; some people assume, it'll be a lethargic process of adding electrons back to a lead acid automotive starting battery because of the laws of physics; thus, we couldn't add them back fast enough to charge it while discharging it.

Given that; electrons, they'll flow directly toward a load source and not into the battery when there's a charge and the load source is on it at the same time. I found this assumption is an erroneous belief because it stemmed from the initial charging cycle of the battery, given it plates will be empty during their initial charging cycle as well.

Remember the chemical composition of a lead acid automotive starting battery doesn't generate its own electrons. Therefore; electrons, they've to be added to the battery when it's first assembled. So; it'll take awhile to store, electrons on its plates due to their surface space and their electrical resistance as well. See chapter 4.

However; the amp potential of the battery, it's based on how many plates in each cell and the size of the plates in each cell as well. In which; it'll determine, how many electrons will be stored on its plates as well. At a certain point during their initial charging cycle, it'll become harder and harder to add electrons to them due to their surface space and electrical resistance as well.

The idea it'll be a lethargic process of adding electrons back to the battery because the laws of physics is an erroneous belief stemming from its initial charging cycle. Here's why; after its initial charging cycle, I found only if its plates are highly discharged before we attempt to charge them, it'll be a lethargic process of adding electrons back to them as well.

Since the chemical reaction that takes place within a lead acid battery, it'll release two electrons for every sulfate-ion radical that has bonded to its lead plates during their chemical discharging cycle. The lead and the sulfate-ions will bond together to create lead-sulfate. As more and more electrons are released from, then more of the lead-sulfate will accumulate.

In which; lead-sulfate, it'll try to resist electrons from being released or stored on the battery plates. Thus; the accumulation of the lead-sulfate on its plates, it'll slow down the rate at which electrons could be added or subtracted from them. Therefore; deeper the chemical discharge is, it'll take more time and energy to add electrons back to the battery plates as well.

Given the time, it'll take to diminish the lead-sulfate build up on the battery plates. This is why some people think, it'll be a lethargic process of adding electrons back to its plates because of the laws of physics; in which, it's due to other extenuating circumstances involved because we try to go as long as possible before recharging our batteries due to inconvenience well.

Although; a deep cycle battery, it's built for deep cycling. However; it'll become, a lethargic process of adding electrons back to it because of the duration of its discharging cycle. Wherein; there's a correlation between, its discharge and charging cycle as well. Given the amount of time and energy, it takes to add electrons back to the battery during its charging cycle.

It's exponentially proportionally to the duration of the discharging cycle of the battery as well. I found in order for a deep cycle battery to become more efficient, we've to stop deep cycling it; in which, it sounds ironic because it's built for deep cycling as well. If we have an intermittent charge on it while discharging it, we could shorten the duration of its charging cycle as well.

With an intermittent charge on the battery, we could keep the chemical residue buildup on its plates as low as possible; thus, it'll take less time and energy to add electrons back to its plates. Then we wouldn't have to worry about the lethargic process of adding electrons back to them while taking electrons from them to power a load source as well.

With a very shallow cycling process of the battery, we could shorten the duration of its charging cycle using its chemical electromotive force to discharge it and then, use a charging source to discharge it as well. Thus; we could add, electrons back to the battery quicker compared to allowing it to become highly discharged before we attempt to charge it.

The lethargic process of adding electrons back to a battery occurs because of the mechanical process of having to stop its discharging cycle in order to start its charging cycle. Since electrons, they've to go in it the same way they've to come out. There's a combination of things will slow down the process of adding electrons back to it, other than the laws of physics as well.

I found if we could add electrons back to the battery while taking electrons out, then we could avoid the lethargic process of adding them back to it. It's the whole purpose of adding them back to it while taking them out. There're a host of things, other than the laws of physics will reinforce our perceptions why it'll be a lethargic process of adding electrons back as well.

If we could charge the battery while discharging it, then we wouldn't have to worry about the lethargic process of adding electrons back to it. Wherein; it'll debunk the theory its charge and discharging cycle can't exist at the same time, along with the theory electrons couldn't be added back to it fast enough to charge it while discharging it to a load source; thus, killing two birds with one stone as well.

Likewise; the theory if we attempt to, carry out both conversion processes of the battery at the same time. It'll be no benefits in it because of the laws of conservation of energy due to energy loss to heat in both conversion processes of the battery; therefore; no gain. This is an erroneous belief because we'll be charging it while discharging it as well.

Thus; any energy loss to heat, it'll be made up by the charging source as well. Then the theory it'll be no benefits in a simultaneous cycling process of the battery is debunk as well. Here's why; the theory, it doesn't account for the energy being added back to the battery while it's taken out as well.

Remember if a lead acid automotive starting battery plates in each cell, they're connected in series-parallel with one another by their straps, tab connectors and the electrolyte solution in each cell. Then its plates in each cell, they're simultaneously charged and discharged amongst themselves and to its plates in the next cell as well.

The theory it'll be no benefits in a simultaneous cycle process of the battery because of the laws of the laws of conservation of energy. It doesn't add up with the mechanics behind the charging cycle of the battery; although, energy will be loss to heat during its charging cycle; but, it'll be made up by the charging source as well.

Although; the battery plates in each cell, they're separated by a non-conducting material called plastic. However; by using an electrical bus, we're still able to take energy from the battery plates in one cell, so, it could be added to its plates in the next cell. Even though; energy loss to heat, it'll occur during the process; but, we're still able to charge all the plates in each cell.

It'll be illogical to assume, it'll be no benefits in adding energy back to the battery while taking it out to power a load source because of the laws of physics. Based on, the mechanics behind the charging cycle of a lead acid automotive starting battery because it'll sound like an erroneous belief. Then the question is, are we focusing on the right science and mechanics?

Given that; it seems like, the only reason it'll be no benefits in adding energy back to battery while taking it out to power a load source because we're unable to carry out such a process. Since energy, it has to go in the battery the same way it has to come out as well. Then it'll be impossible to add energy back to it while taking energy out to power a load source as well.

Since electrons, they've to go in the battery the same way they've to come out in order to cycle it. Then it'll prevent it from be charged while being discharged; given that, like charges will repel one another and follow the path of least resistance as well. Then to say it's due to the laws of physics is disingenuous because actually, it's due to the design of the battery.

Given that; like charges, they'll repel one another and follow the path of least resistance as well. It'll have little to do with why the battery can't be charged while being discharged. If electrons, they didn't have to go in it the same way they've to come out in order to charge its plates or for them to be discharged to a load source as well.

Wherein; the assumption, it's due to the laws of physics why the battery can't be charged while being discharged. It'll be nothing more than erroneous belief conjured up by someone, who haven't taken an in depth analysis of the mechanics behind the charging cycle of the battery to make the determination is it due to the laws physics or the design of the battery.

We view a battery; as though, it's a closed system; given that, we can't add energy back to it while taking energy out because we're taking it out. In which; it reinforces our erroneous beliefs that, we've to stop one process in order to start the other because of the laws of physics; but, it's only true to the extent energy has to go in the same way it has to come out.

This process of energy having to go in the same way it has to come out, it creates an erroneous belief there's no other way to carry out the cycling processes of a lead acid automotive starting battery; but, to stop its discharging cycle in order to start its charging cycle. In which, it contradicts, the mechanics behind the charging cycle of the battery as well.

Remember the battery plates in each cell are connected in series-parallel by their straps, tab connectors and the electrolyte solution in each cell. Thus; electrons, they could flow from a strap to a tab connector on one plate. And then, flow across that plate through the electrolyte solution to an adjacent plate without flowing to and from the same tab connector on the same plate.

Since each cell in the battery is connected in series with its terminal connections by an electrical bus, then we don't have to stop charging its plates in one cell in order to charge them in the next cell; although, they're separated by a non-conducting material called plastic. Thus; electrons, they could flow in one direction from its plates in one each cell to them in the next cell as well.

Based on, the mechanics behind the charging cycle of a lead acid automotive starting battery. It'll debunk the theory its charge and discharging cycle can't exist at the same time because of the laws of physics. Given that; the battery plates in one cell, they're charged and discharged to its plates in the next cell as well.

Likewise; it'll debunk, the theory it'll be no benefits in adding energy back to the battery while taking it out as well. Given that; its plates in one cell, they could consume and store energy while supplying energy to its plates in the next cell; although, they're separated by a non-conducting material called plastic as well.

If we look at, the mechanics behind the charging cycle of a lead acid automotive starting battery. All those assumptions about why it can't be charged while being discharged because of the laws of physics. They're nothing more than erroneous beliefs stemmed from the lack of information about the mechanics behind its cycling processes as well.

If we don't take an in depth analysis of the mechanics behind, the cycling processes of the battery. Then assuming its charge and discharging cycle can't exist at the same time, it's like assuming sea life doesn't exist in the deepest parts of the oceans without investigating the matter first; thus, we all know how that assumption turned out as well.

Given that; the assumption, it was an erroneous belief because it was based on sea life as we knew of at the time. If we don't venture outside, the status quo to change the contemporary designs of our secondary cell batteries. Then our assumptions about why, they can't be charged while being discharged becomes nothing more than an erroneous belief as well.

We've to ask ourselves do those assumptions about why we can't charge our secondary cell batteries discharging them because of the laws of physics hold true. If we don't have to charge our batteries on the same terminal connections that, they were discharged on; given that, we could charge or discharge them as well.

Given that; a lead acid automotive starting battery plates in each cell, they could consume and store energy while supplying it to the plates in the next cell. Without having to, stop charging them in one cell in order to charge them in the next cell. We've to put it to the test to see if, we've to stop its discharging cycle in order to start its charging cycle as well.

If we look beyond the exterior design of the battery, we might find its charge and discharging cycle could exist at the same time. Without being impeded, by the laws of physics as well. Wherein; it seems like, it's a matter of mechanics and not a matter of physics when it comes to adding or subtracting electrons from the battery plates as well.

Although; energy, it'll be lost to heat during the electrolysis process of the battery plates during their charging cycle. It has no bearing on whether they're charged or not; so, it must be an erroneous belief to assume. It'll be no benefits in adding energy back to the battery while taking it out to power a load source because it seems like, it'll be no difference between them.

Let's me explain why energy loss to heat in both conversion processes of a lead acid automotive starting battery. It'll have no bearing on whether a simultaneous cycling process of it will be beneficial or not. Here's why; energy loss to heat, it comes in two forms during the energy transfer process of the battery.

First; energy, it's lost to heat from the electrical resistance of the battery plates during their discharging cycle; in which, it's irreversible. Secondly; energy loss to heat, it occurs during the charging cycle of the battery; in which, it's reversible. Thus; when the battery is discharged, its chemical reaction is endothermic or it absorbs heat.

However; when the battery is charging, its chemical reaction is exothermic or it releases heat. But; the endothermic chemical reaction of the battery, it balances out the heat released by its electrical resistance during its discharging cycle; so, it doesn't seem to get hot during its discharging cycle; but, energy is still lost to heat during its discharging cycle as well.

The energy lost to heat during its discharging cycle will limit the amount of energy could be used to power a load source. In which; it affects, the overall capacity of the battery as well. However; when energy is loss to heat during its charging cycle, it rises its temperature and all that heat is energy lost; but, it'll be made up by the charging source as well.

Thus; we're still able to add, energy back to the battery plates during their charging cycle. Subsequently; the amount of energy lost to heat, it'll be exponentially proportional to the amount of energy will be cycled through the battery during its cycling processes. Wherein; it'll be based on, the amount of lead-sulfate accumulation on the battery plates as well.

It'll take more time and energy to add electrons back to the battery plates due to the amount of lead-sulfate accumulation on them during their discharging cycle; so, more energy will be lost to heat during the charging cycle as well. In other words; the amount of time, it takes to discharge its plates. It'll be exponentially proportionally to the time it takes to charge them as well.

However; when it comes to the assumption, it'll be no benefits in a simultaneous cycling process of the battery because energy will be lost to heat in both of its conversion processes. It doesn't accurately portray what'll happen during a simultaneous cycling process of it. Given that; it doesn't account for, the energy being added back to it while energy is taken them as well.

Wherein; energy lost to heat, amongst the battery plates during their charging cycle. It becomes reversible because it'll be made up by the charging source as well. This is what, the assumption. It'll be no benefits in a simultaneous cycling process of the battery leaves out because it's not accounting for the energy being added back to the battery while it's taken out as well.

One could assume any energy lost to heat during a simultaneous cycling process of the battery, it'll become reversible because it'll be made up by the charging source as well. On the other hand; those who are skeptical of charging the battery while discharging it, they think it'll be using its chemical electromotive force to discharge it while it's being charged as well.

I found this is an erroneous belief because during a simultaneous cycling process of the battery. It'll not be using its chemical electromotive force to discharge it while it's being charged as well. We'll only be using the electromotive force created by a charging source to charge the battery while discharging it as well.

Then the assumption energy lost to heat in both conversion processes of the battery, they'll cancel themselves out if we attempt to carry them out at the same time; therefore, no gain. It's nothing more than an erroneous belief because any energy loss to heat during the process, it'll be made up by the charging source as well.

So; it's disingenuous to use, the laws of entropy, inertia or the laws of conservation of energy to discredit a simultaneous cycling process of battery because energy is lost to heat in both of its conversion processes as well. What's amazing about using those laws of physics to discredit a simultaneous cycling process of battery, they become a contradiction as well.

Here's why; the battery, it'll not be considered a closed system under those laws of physics if we're adding energy back to it while taking energy out as well. Wherein; it'll show a lack of knowledge or understanding, about those laws of physics if we're using them to discredit a simultaneous cycling process of the battery because it'll be a depiction of them as well.

Maybe; we're combining, what we know about an endothermic or an exothermic chemical action of the cycling processes of the battery. Thus; assuming, its endothermic and exothermic chemical action will take place at the same time. If we attempt to charge it while discharging it, they'll cancel themselves out because one it absorbs heat and the other releases heat.

There're three things wrong with this kind of assumption; number one is that, energy will be added back to the battery while it's taken out. Number two; the process, it'll only consist of using a charging source to charge the battery while discharging it. Number three is that; any energy loss to heat during the process, it'll be made up by the charging source as well.

The question is, do we actually understand the mechanics behind the cycling processes of a lead acid automotive starting battery or we just speculating? If we're just speculating, then it explains why we've all those erroneous beliefs about what we could or couldn't do with it because of the laws of physics as well.

In the next chapter; let's focus on, the mechanics of the current and then we might find out why all those erroneous beliefs came into existence about what could or couldn't do with our batteries due to the laws of physics because we've been focusing on the wrong science as well.

CHAPTER 12

THE MECHANICS OF THE CURRENT

THE REASONS WE'VE SO many erroneous beliefs about what we could or couldn't do with our batteries because of the laws of physics. It's due to the lack of understanding about the mechanics of the current. We've to separate voltage from current if we're going to understand why we can't charge our batteries while discharging them because of the laws of physics as well.

Throughout my book, I've been separating voltage from current to help shed some light on the difference between them. Remember voltage is an electromotive force that moves electrons and current is, the combination of voltage and electrons moving through and around a conducting source. Given that; voltage, it'll behave different than current will.

We've to know the difference between voltage and current if we're going to understand why a battery can't be charged while being discharged. If not, then we won't understand voltage will enter or exit a battery different than current will; but, both will enter or exit the battery based on its design as well.

Likewise; voltage and current, they'll flow through a battery differently based on which cycling process will be carried out as well. If we just focusing on, the mechanics of the voltage and current alone. Then we'll find the cycling processes of a battery is carried out by the laws of motion and the effect of forces on bodies and not by the laws of physics as some may believe.

For more than a century, our perceptions about why we've to stop the discharging cycle of a battery in order to start its charging cycle because of the laws of physics have become the norm. It shouldn't be any surprise those in the field of automotive battery technology will have doubt about charging a battery while discharging it as well.

On the other hand; we can't expect to charge a battery while discharging it if energy has to go in it the same way energy has to come out as well. Wherein; it's safe to assume, it can't be charged while discharged because of the laws of physics as well. In contrast, it's safe to assume. It'll be possible to charge it while discharging it because of the laws of physics as well.

Since like charges, they'll repel one another and follow the path of least resistance when they travel through the body of a battery as well. Those in the field of automotive battery technology, they'll determine how a battery cycling processes will be carried out. Let's forget about the laws of physics for a moment because they're vast and complex as well.

Let's focus on the mechanics of the voltage and current alone something we could easily fathom. Remember voltage is the potential difference between two points, which moves electrons. Wherein; the movement of electrons, it's called current and it does the electrical work. In figure 3, the arrows shows how voltage and current will flow within a battery based on its design.

Figure 3

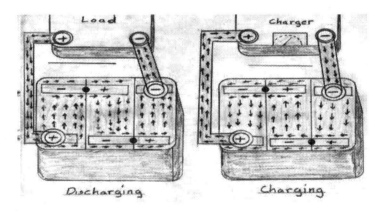

(Cycling processes of a battery)

If we use the electron current flow concept and then, during the discharging cycle of a lead acid automotive starting battery. Voltage and current will flow from its negative terminal to a load source and then, voltage will return back to the battery through its positive terminal. But; current, it'll not return back to the battery because it'll be used by the load source.

However; during the charging cycle of the battery, the charging voltage and current will flow through its negative terminal to replenish its chemical energy. Thus; the charging current, it'll break down the chemical residue that has accumulated on the battery plates during their discharging cycle; in which, it'll restore the battery electromotive force.

When the chemical residue, it's released back into the battery electrolyte solution while storing electrons on its plates as well. Thus; the charging voltage, it'll return back to its source through the positive terminal of the battery based on the electron current flow concept. On the other hand; it depends on who we're talking to, a chemist or an electrician.

Given that; an electrician will say that, current will flow to and from the positive terminal of a battery. But; a chemist will say that, current will flow to and from the negative terminal of a battery. Thus; we end up with two concepts, the electron current flow and the conventional current flow concept. See figure R-1.

Figure R-1

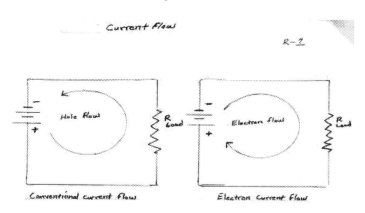

(The conventional and the electron current flow concepts)

When we're talking about A/c versus D/c for A/c, the black wire means hot and the white wire means neutral. And for D/c, the black wire means neutral and the red wire means hot. Based on a chemist's view during the discharging cycle of a battery, electrons will flow from the negative terminal of the battery to a load source.

However; the voltage of the battery, it'll return back to it through its positive terminal connection. See figure R-1, the image on the right. Based on an electrician's view during the discharging cycle of a battery, electrons will flow from the positive terminal of the battery to a load source. Then the voltage of battery will return back to it through its negative terminal.

See figure R-1, the image on the left. On the other hand; based on, a chemist's view during the charging cycle of a battery. Its external current flow is flowing from the charging source through the negative terminal of the battery. See figure S-1.

Figure S-1

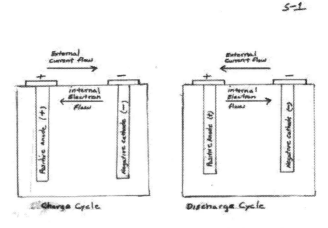

(How current flows in and out of a battery)

Thus; the internal current flow of the battery, it'll be flowing from a more negative to a more positive point within the battery to add electrons back to it. And then, the charging voltage will return back to the charging source from the positive terminal of the battery. In which; it'll be more consistent with, the electromagnetic laws of physics.

Given that; electrons, they'll be flowing from a more negative to a more positive point within the battery as well. However; the internal current flow of the battery, it'll be flowing from its negative electrode toward its positive electrode. In which; it'll be consistent with, the electromagnetic laws of physics as well.

Since electrons, they'll be flowing from a more negative to a more positive point within the battery as well. See figure S-1, the image on the left. On the other hand; in figure S-1, see the image on the right during the discharging cycle of a battery. We'll find the external current flow of the battery will be flowing from its negative terminal to a load source.

The battery voltage will return back to it through its positive terminal. Given that; electrons, they'll be flowing from a more negative to a more positive point within the circuit. But; the internal current of the battery, it'll be flowing from its negative terminal toward a load source to a more positive point; in which, it'll be consistent with the electromagnetic laws of physics.

Since electrons, they'll be flowing from a more negative to a more positive point within the circuit between the battery and the load source. Likewise; based on, an electrician's view during the charging cycle of a battery. Its external current flow will be flowing from a more negative to a more positive point within the circuit between the battery and the load source as well.

Here's why; since current, it'll be flowing from the charging source toward the battery. In which; the charging source, it'll be the most negative point within the circuit between the battery and the load source. Therefore; current, it'll be flowing through the positive terminal of the battery. And then, flow toward its negative terminal.

Wherein; the charging voltage, it'll return back to its source from the negative terminal connection of the battery. Based on, the conventional current flow concept. Given that; the external current flow of the battery, it'll be flowing from a more negative to a more positive point within the battery; in which, it'll be consistent with the electromagnetic laws of physics as well.

However; the conventional current flow concept, it'll be the opposite of the electron current flow concept. See figure S-1, the image on the left. Based on, the conventional current flow concept during the discharging cycle of a battery. Its external current flow will be flowing from its positive terminal to a load source.

In which; the voltage of the battery, it'll return back to it through its negative terminal connection. Then current flow will be flowing from a more negative to a more positive point within the battery. Thus; the current flow, it'll be consistent with the electromagnetic laws of physics as well; but, it'll be the opposite of the electron current flow concept. See figure S-1, the image on the right.

If we really think about the conventional and the electron current flow concepts, we'll find they're the opposite of one another. However; they're the same, if we look at the internal and the external current flow of a battery from two different perspectives. From an electrician's perspective during, the charging cycle of a battery.

The charging current is flowing from a more negative to more positive point within a circuit because current is flowing from outside the battery. Given that; the charging source, it'll be the most negative point within the circuit. Thus; the charging current, it'll able to flow through the positive terminal connection of the battery to charge it.

Then the charging voltage, it'll return back to its source through the negative terminal of the battery. Given that; voltage, it doesn't have to follow the path of least resistance as current does. Therefore; current, it'll be flowing from a more negative to a more positive point within the battery; in which, it'll be consistent with the electromagnetic the laws of physics as well.

From an electrician's perspective during the discharging cycle of a battery, its internal current is flowing from its negative terminal toward its positive terminal and then, toward a load source. Therefore; electrons, they'll be flowing from a more negative to a more positive point; in which, it'll be consistent with the electromagnetic the laws of physics as well.

Thus; the voltage of the battery, it'll return back to it through its negative terminal connection because voltage doesn't have to follow the path of least resistance as current does.

From a chemist's perspective during the charging cycle of a battery, the charging current will be flowing from a more negative to more positive point within the battery.

Given that; the charging voltage, it'll be higher than the normal battery voltage. Then the charging current, it'll be able to flow through the negative terminal of the battery and toward its positive terminal as well. Thus; the charging voltage, it'll return back to its source through positive terminal connection of the battery.

In which; the charging current, it'll be flowing from a more negative to a more positive point within the battery during its charging cycle; thus, consistent with the electromagnetic the laws of physics as well. From a chemist's perspective during the discharging cycle of a battery, its current will be flowing from its negative electrode toward a load source.

In which; it'll be consistent with, the electromagnetic the laws of physics as well. Wherein; the voltage of the battery, it'll return back to it through its positive terminal connection during its discharging cycle because voltage doesn't have to follow the path of least resistance as current does.

Since a chemist and an electrician, they're looking at the current flow of a battery from two different perspectives. Given that; one is looking at it, from the outside of the battery and the other is looking at it from the inside of the battery. Then the idea of current having to enter or exit the positive or the negative terminal of a battery in order to cycle it.

It may be just an illusion created by the internal and the external current flow of a battery. Likewise; our perceptions about why, a battery can't be charged while being discharged because of the laws of physics. It may be an illusion created by the design of the battery because current has to go in it the same way current has to come out as well.

In my book, I used the conventional and electron current flow concept to describe how current will enter or exit a secondary cell battery. Although; some people say that, I'm wrong because current could only enter or exit the positive terminal of a battery. It really doesn't matter which terminal current has to enter or to exit the battery in order to cycle a battery.

We still have to charge a battery on the same terminal connection that, it was discharged on; given that, current has to go in the same way it has to come out in order to cycle the battery; in which, it's my whole point of writing this book. If we're going to charge a battery while discharging it, then it has to have more than one positive or one negative terminal as well.

I found if we could charge our batteries while discharging them, then we could receive the benefits of being able to replenish their chemical energy with electrical energy as well. Thus; the ability to increase, the overall travel ranges of our vehicles on battery power without finding a perfect chemical composition to a longer-lasting battery for them as well.

Here's why; if we're going to rely more heavily on, the electric motor than the gasoline engine to propel our basic form of transportation. Without finding that, perfect chemical composition to a longer-lasting battery. Then we need to be able to charge our automotive batteries while discharging them if we're going to revolutionize the electric vehicle anytime soon.

When there's a conversation about charging our batteries while discharging them, we end up focusing on the wrong science. We're either focusing on the laws of electrochemistry, entropy, inertia or the laws of conservation of energy as well. I find we shouldn't have to focus on those laws of physics; however, we should be focusing on the mechanics of the current; but, we don't.

The laws of electrochemistry, entropy, inertia or the laws of conservation of energy, they don't determine how energy will be added or subtracted from a battery. But; the laws of series, parallel or series-parallel circuitry, they do. They'll determine how current will flow to and from a battery plates. If we think otherwise, then we're focusing on the wrong science.

Here's why; I found that, electrons are simultaneously added and subtracted from a lead acid automotive starting battery plates in one cell; so, they could be added to its plates in the next cell during their charging cycle. The laws of electrochemistry, entropy, inertia or the laws of conservation of energy will have no bearing on whether the battery plates are charging or not.

Energy will be lost to heat during the charging cycle of the battery plates in each cell and they'll be separated by a non-conducting called plastic. The laws of electrochemistry, entropy, inertia or the laws of conservation of energy, they will not be an issue. When energy is taken from the battery plates in one cell, so, it could be added to its plates in the next cell as well.

All this talk about the laws of electrochemistry, entropy, inertia or the laws of conservation of energy will prevent a battery from being charged while being discharged. It all sounds like erroneous beliefs handed down from one generation to the next. I found only the mechanics of the current will prevent a battery from being charged while being discharged.

Here's why; the mechanics of the current, they'll allow electrons to be simultaneously added and subtracted from a battery plates. If they're connected in series-parallel with one another in each cell and each cell, it's connected in series with its terminal connections as well. Although; the plates in each cell, they're separated by a non-conducting material called plastic as well.

During my research on the lead acid automotive starting battery, I found its design is the deciding factor in determining how current will enter or exit it. However; if we're going to charge the battery while discharging it, then it's more to it than changing its design; but, it's about the connection made between it and a load source in their relationship with a charging source as well.

Here's why; a parallel circuit connection made between, a battery and a load source in their relationship with a charging source. It'll make it difficult to charge the battery while it has the load source on it. Since the battery resistance to current, it's different than the load source resistance to current. Then current will flow to the battery and to the load source will be different as well.

Here's another thing; remember, a secondary cell battery plates behave like load sources during their charging cycle. Thus; the battery and the load source, they'll be individual branches of the circuit in their relationship with the charging source. As a result; the charging voltage, it'll be the same across the battery and the load source because of the parallel connection.

However; the charging current, it'll not be the same across the battery and the load source because of the parallel circuit connection made between them. Since the resistance of the battery plates to current, they act like a non-conducting material trying to reflect current away from them during their charging cycle.

Wherein; I found that, a parallel circuit connection made between a battery and a load source in their relationship with a charging source. It doesn't work the same as a parallel circuit connection made between two load sources or two batteries. When it comes to the mechanics of the current; although, a battery plates behave like load sources during their charging cycle.

However; a battery plates have to consume electrons before electrons, are stored on them. In short; electrons have to breakdown the chemical residue build up on the battery plates before electrons, are stored on them. Since the chemical residue on the battery plates, it acts like a non-conducting material to reflect current away from them.

Here's the deal; if there's a load source on, a battery during its charging cycle and they're connected in parallel with one another. Then electrons, they'll not flow into the battery long enough to diminish the chemical residue build up on its plates in order to charge them. Since electrons, they'll have an alternative route to follow; that is, toward the load source.

The chemical residue left on the battery plates from their discharging cycle will reflect electron toward the load source, which is the path of least resistance. Remember like charges will repel one another and they'll follow the path of least resistance as well. If electrons have an alternative route to follow, they'll follow that route.

So; if the resistance at the battery, it's greater than the resistance at the load source. Then the charging electrons will flow toward the load source and not into the battery. Therefore; a parallel circuit connection made between a battery and a load source in their relationship with a charging source, it'll only allow electrons to flow to the load source.

Although; a battery plates, they behave like load sources consuming energy during their charging cycle. Such as; a fan, a compressor or an electric motor; but, remember a parallel circuit connection made between them. It doesn't work the same as a parallel circuit connection made between two load sources; such as, a fan, a compressor or an electric motor.

Remember a battery resistance to current is different than a load source resistance to current. If we're going to understand why, a secondary cell battery can't be charged while being discharged to a load source; such as, a fan, a compressor or an electric motor. We've to understand the different circumstances involved other than the mechanics of the current alone.

Here's the thing; the amount of current flowing to a load source, it'll be based on two factors: the resistance of the load source and the voltage applied by a power source. If we've two load sources connected in parallel with one another and their resistance to current is different, then current will flow to them at a different rate from a power source.

For instance; it's like water flowing downstream in a river, and there're two branches in the river. If one branch in the river, it's larger than the other. Then the volume of water flowing through the larger branch, it'll be greater than the volume of water flowing through the smaller branch of the river; given that, the larger branch in the river has less resistance for water to travel.

However; water, it'll flow through both branches of the river; but, at a different rate. It'll be similar to current flowing to load sources connected in parallel with one another by the same power source. Because more turns that, the wiring has to conduct current inside a load source than greater its resistance to conduct current.

Likewise; less turns that, the wiring has to conduct current inside a load source than less resistance it has to conduct current. Although; two load sources, they could have different resistance and connected in parallel with one another. And yet, they still will receive current at the same time; but, at a different rate based on their individual resistances.

In the case of our secondary cell batteries, a parallel circuit connection made between them. It'll work a little different than a parallel circuit connection made between two load sources; such as, ceiling fans, compressors or electric motors. For instance; if we're charging, two batteries in parallel and one is discharged more than the other.

Then the charging current, it'll flow into the battery with the least resistance first before flowing to the battery with the most resistance. Given that; the charging current, it'll be reflected toward the battery with the least resistance first because current, it'll be following the path of least resistance between the two batteries.

Remember current is the movement of electrons; so, the two batteries resistance to current will act like a non-conducting material to reflect current away from them. Since the charging current, it'll have no other place to go because it can't flow back toward the charging source; thus; current will follow the path of least resistance between the two batteries.

Therefore; current, it'll flow into the battery with the least resistance because it'll take less time to breakdown its resistance. However; when the battery with the least resistance, it can't receive any more electrons. Then its electrical resistance will reverse the current flow back toward the battery had the most resistance at first to breakdown its resistance and charge it as well.

There'll be a twist when it comes to a parallel circuit connection between a battery and a load source, compared to a parallel circuit connection made between two batteries during their charging cycles. The rate at which current will flow into a battery or flow to a load source. It'll not be based on their individual resistance as the laws of parallel circuitry might suggest.

Here's why; unlike, the scenario between the two branches in the river. The twist is that, current will not flow into the battery at all if its resistance is greater than the resistance at the load source. It'll be more like the scenario between the two batteries connected in parallel with one another during their charging cycles.

Given that; current, it'll not flow into the battery at all when there's a charge and a load source on it at the same time. Unlike; the scenario, between the two batteries connected in parallel with one another during their charging cycles. Wherein; current, it'll be reflected back toward one battery or the other when one gets full with electrons.

However; a load source, it'll never get full with electrons in order to reflect them back toward a battery to charge it. Given that; the load source, it'll consume electrons faster than they could break down the resistance of the battery plates in order to charge them. Wherein; it'll be difficult to, charge them while they've a charge and a load source on them at the same time.

Unless; the charging source, it'll create a surplus of electrons at the load source to reverse the current flow back into the battery in order to charge its plates. In other words; electrons, they'll be forced back into the battery to charge its plates with the resistance created at the load source when the charge and the load source is on the battery at the same time.

The key word is force because any additional electrons flowing toward the load source from the charging source. They'll be reflected back into the battery to charge it because the electrical resistance at the load source becomes greater than the resistance of the battery. However; it'll be based on, the amount of electrons that the load source could consume at a given rate.

We must understand a parallel circuit connection made between a battery and a load source. It'll not work the same as a parallel circuit connection made between two load sources or two batteries because of the different circumstances involved. If we don't understand this, then we might get the wrong impression about what we could or couldn't do with a battery as well.

Wherein; voltage, it'll enter or exit a battery different than current will; but, both will enter or exit it based on its design as well. On the other hand; voltage and current, they'll through a battery differently based on which cycling process will be carried out as well. If we don't know this, we might get the wrong impression about what we could or couldn't do with a battery as well.

If we don't understand, how the mechanics of the current will behave flowing in and out of a battery based on its design. Then most likely, we'll blame it on the laws of physics why it can't be charged while being discharged to a load source. All because we haven't acquired enough information to understand what's required to carry out such a process as well?

Likewise; the lack of information, we possess about the mechanics of the voltage and current under the same circumstances. It might give us the wrong impression about what we could or couldn't do with a battery as well. Given that; voltage, it'll flow through a series, a parallel or a series-parallel circuit connection different than current will as well.

We can't begin to understand what's required to charge a battery while discharging it to a load source. If we can't comprehend, a parallel connection made between a battery and a load source. It doesn't work the same as a parallel connection made between two load sources or two batteries when it comes to the mechanics of the current as well. See chapter 5.

If we don't understand those things, then we're inept to assume a battery can't be charged while being discharged to a load source because of the laws of physics. Given that; we'll end up focusing on, the wrong science when it comes to a conversation about charging a battery while discharging it to a load source as well.

Taking an in depth analysis of, the mechanical structure of an automotive starting battery with top and side posts terminal connections. Then it'll give insight into why it can't be charged while being discharged to a load source. Although; the battery, it has two negative terminals in one cell and two positive terminals in another cell as well.

On the other hand; taking an in depth analysis of, the mechanical structure of the battery. It'll give us insight into why it could be charged while being discharged to a load source. However; our erroneous beliefs, they'll not allow us to see beyond our batteries conventional cycling processes because we think, they're set in stone due to their contemporary designs as well.

Our erroneous beliefs about the cycling processes of our secondary cell batteries, they stemmed from the lack of information about the mechanics of the current when it comes to their cycling processes. Likewise; the lack of information about, how current will flow between a battery and a load source in their relationship with a charging source as well.

Focusing on, the mechanics of the current when it comes to the cycling processes of a secondary cell battery. I found our erroneous beliefs only exist because we're focusing on the wrong science. In the next chapter; let's see if, we could find the right science to focus on when it comes to the cycling processes of our secondary cell batteries as well.

CHAPTER 13

FOCUSING ON THE RIGHT SCIENCE

TAKING AN, IN DEPTH analysis of the mechanical structure of an automotive starting battery with top and side posts terminal connections. It'll help us focus on the right science to ascertain; whether, a battery could or couldn't be charged while being discharged without being impeded by the laws of physics without focusing on the vast and complex of them as well.

Our erroneous beliefs about why our secondary cell batteries can't be charged while being discharged due to either the laws of electrochemistry, entropy, inertia or the laws of conservation of energy. They're nothing more than myths stemmed from the lack of information about the mechanics behind our batteries' cycling processes as well.

Let's review figure T-4 from chapter 8 to understand why, we're not focusing the right science. If we're focusing on the laws of physics to ascertain whether our batteries could or couldn't be charged while being discharged. After analyzing, how the negative and positive plates of a lead acid automotive starting battery are interconnected with one another.

In which; I found that, they're interconnected by their straps, tab connectors and electrolyte solution in each cell. And each cell, it's connected in series with its terminal connections as well. I found it'll make it possible for electrons to flow in one direction from its plates in one to its plates in the next cell during its cycling processes as well.

However; this same set-up, it'll make it impossible to charge the battery while discharging it to a load source as well. Since electrons have to go in the battery the same way that, they've to come out in order to charge its plates or for them to be discharged to a load source as well. See figure T-4 again.

Figure T-4

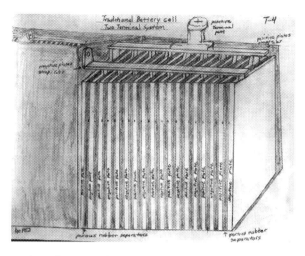

(Inside view of a lead acid automotive battery)

In order to add or subtract electrons from the battery, electrons have to flow toward a negative or a positive terminal in order to enter or to exit it while flowing in one direction at a time; thus, we've to stop its discharging cycle in order to start its charging cycle. Not due to the laws of electrochemistry, entropy, inertia or the laws of conservation of energy as well.

However; I found that, it's due to the contemporary design of the battery because electrons, they've to go in it the same way they've to come out in order to charge its plates or for them to be discharged to a load source as well. But; during my research, I found its design will determine how voltage and current will enter or exit it as well.

On the other hand; I found that, voltage and current will flow to and from the battery plates. The same way voltage and current will flow to and from our load sources; such as, light bulbs, electric motors, ceilings fans and safety switches. Based on, how they're interconnected with one another in their relationship with a power source as well.

For instance; if three load sources, they were connected in series with one another by the same power source. Then voltage will be evenly divided among them equaling to the sum of the voltage flowing from the power source; but, each load source will receive the same amount of current flowing from the power source as well.

However; if the load sources, they were connected in parallel with one another. Then each load source, it'll receive the same amount of voltage flowing from the power source. But; the amount of current flowing to each load source, it'll be based on their individual resistance. Now if the load sources, they were connected in series-parallel with one another.

Then voltage and current will flow to the load sources based on how they were interconnected with one another in their relationship with the power source. For instance; if two of the load sources, they were connected in parallel with one another. And a third load source, it was connected in series with a load source that is connected in parallel with another.

The load sources that are connected in parallel with one another, they'll receive the same amount of voltage flowing from the power source; but, current flowing to them will flow based on their individual resistance. The load source that is connected in series with another, it'll receive the same amount of current as the load source that it's connected in series with.

However; the load source that is connected in series with another, it'll receive a voltage drop from the power source because it's connected in series with another load source. If we're focusing on the right science, then we'll understand why the load sources received the amount of voltage and current they received from the power source as well.

It'll be based on how those load sources were interconnected with one another in their relationship with the power source. Since a lead acid automotive starting battery plates, they behave like load sources consuming energy during their charging cycle. Then the same laws of series, parallel and series-parallel circuitry would apply to them as well.

Here's why; because voltage and current, they'll flow to and from a secondary cell battery plates. The same way voltage and current will flow to and from our load sources as well. Remember if a battery cells, they're connected in series with its terminal connections. Then the charging voltage, it'll be evenly divided amongst the battery plates in each cell.

However; each plate in each cell, it'll receive the same amount of current flowing from the charging source. On the other hand; if a battery cells, they're connected in parallel with its terminal connections. Then the charging voltage, it'll be the same across each plate in each cell; but, each plate in each cell will receive current based on their individual resistance.

Now; if a battery plates in each cell, they're connected in series-parallel with one another. Then the charging voltage and current, they'll flow to those plates. Based on, how they're interconnected with one another in their relationship with the charging source. It'll be no longer how they're interconnected with the charging source; but, with one another as well.

Here's why; the plates connected in parallel with one another, they'll receive the same amount of voltage flowing from the charging source; however; current flowing to them will be based on their individual resistance. Wherein; the plates connected in series with the plates connected in parallel with one another will receive a drop from the charging source.

However; they'll receive, the same amount of current as the plates they're connected in parallel with. If we accurately decipher, how a lead acid automotive starting battery plates are charged during their charging cycle. Then we'll know the laws of series, parallel and series-parallel circuitry, they'll apply to the battery plates as well as they'll apply to our load sources as well.

Understanding, how the cycling processes of the battery will be carried out by the laws of series, parallel and series-parallel circuitry. Then we'll know the laws of electrochemistry, entropy, inertia or the laws of conservation of energy will not be an issue when it comes to charging the battery while discharging it to a load source as well.

Here's why; if we analyze, the mechanics behind the charging cycle of the battery. Then we'll realize, only the electromagnetic laws of physics will be a factor when it comes to a simultaneous cycling process of the battery. Since electrons have to go in it the same way, they've to come out in order to charge its plates or for them to be discharged as well.

Thus; we find ourselves having to, stop one cycling process in order to start the other because of the design of the battery; likewise, energy loss to heat in both of its conversion processes. It'll have little to do with why, it'll be no benefits in a simultaneous cycling process of it. Wherein; it seems like, another erroneous belief because we're focusing on the wrong science.

If we don't understand why, the battery functions the way that it does during its cycling processes. Then we're inept to argue why it can't be charged while being discharged because of the laws of physics as well. Here's why; not knowing that, voltage and current will flow to and from the battery plates based on how they're interconnected with its terminal connections.

Without knowing this type of information, then we're subject mistakes in ascertaining; whether, the battery could or couldn't be charged while being discharged. Not realizing the types of circuit connections; such as, series, parallel or series-parallel circuit connection made between its plates and terminals will determine how voltage and current will flow to and from it.

In which; it'll determine how, voltage and current will flow between the battery and a load source in their relationship with a charging source. See chapter 5. Remember if a battery and a load source, they're connected in parallel with one another in their relationship with a charging source. Then voltage will be the same across the battery and the load source.

However; current, it'll not be the same across the battery and the load source. If the resistance of the battery is greater than the resistance of the load source, then current will not flow into the battery at all because its resistance to current is different than the load source resistance to current due to the different circumstances involved. See chapter 12.

A battery plate's resistance to current will act like a non-conducting material trying to reflect current away from it. But; a load source resistance to current, it'll be based on the number of turns that the wiring has inside of it to conduct current. For these reasons, current will flow to the battery differently than it would flow to the load source. See chapter 12 as well.

Wherein; a parallel connection made between, the battery and the load source in their relationship with the charging source. It's another reason why current will flow to the battery differently than it would flow to the load source as well. Then a parallel connection made between them, it doesn't work the same as a parallel connection made between two load sources.

However; we'll find that, voltage and current will flow through a series, a parallel or a series-parallel circuit connection differently because of the different circumstances involved. For instance; a branch in a circuit means where load sources, they'll interconnected with one another in their relationship with a power source within the circuit.

Thus; within a parallel circuit connection, voltage will be the same at each branch of the circuit between the load sources. But; current, it'll not be the same at each branch of the circuit between the load sources because it'll be based on the individual resistance of each load source at each branch within the circuit flowing from the power source.

Within a series circuit, voltage will drop evenly across each branch of the circuit; but, current will be the same throughout each branch of the circuit regardless of the individual resistance of each load source. Within a series-parallel circuit, voltage and current will flow between each branch of the circuit differently than they would a series or a parallel circuit.

Wherein; it would be based on, how the load sources are interconnected with one another in their relationship with a power source. If we decipher correctly, then the connections made between a battery and a load source in their relationship with a charging source is one of the deciding factors in determining how voltage and current will flow between them as well.

Not only do we've to understand how the mechanical structure of a battery will play a role in its cycling processes. See chapter 3. But; we've to understand, how the connections made between it and a load source in their relationship with a charging source. It'll affect how voltage and current will flow between them as well. See chapter 5.

If we're focusing on the mechanics of the voltage and current alone, then we're still focusing the wrong science to determine why the battery could or couldn't be charged while being discharged because they're not the deciding factor in determining how they'll flow to and from the battery or from it to a load source as well.

Remember we talked about how a lead acid automotive starting battery mechanical structure. It'll form its circuit connections between its plates and cells in their relationship with its terminal connections. The battery mechanical structure revealed that, the mechanics of the voltage and current alone isn't the deciding factors when it comes to its cycling processes.

However; we always start off focusing on, the vast and complex laws of physics when it comes to a conversation about charging the battery while discharging it. Wherein; we should be focusing on, its contemporary design and the type of circuit connections will be made between it and the load source in their relationship with a charging source as well.

Instead of focusing on the simple things, we end up focusing on the vast and complex laws of physics to ascertain; whether, it's possible or not to charge the battery while discharging it to the load source; although, we could charge or discharge the same battery. But; we've to charge it on the same terminal connection that, it's discharged on as well.

And yet, we still focus on the vast and complex laws of physics to determination if the battery could or couldn't be charged while being discharged to the load source as well. In which; it means, we haven't acquired enough knowledge about what's required to charge the battery while discharged it to the load source as well.

When it comes to a conversation about charging a battery while discharging it to a load source, we've to explore the obvious. Since we could charge or discharge the same battery; but, we've to charge it on the same terminal connection it's discharged on. Then it should be obvious why we've to stop its discharging cycle in order to start its charging cycle as well.

It doesn't take a rocket scientist to figure this out since electrons have to go in the battery the same way they've to come out as well. Although; some in the field of automotive battery technology, they shun away from this fact when a conversation arrive about charging the battery while discharging it to a load source as well.

Electrons having to go in the battery the same way they've to come out, it would be a major issue if we attempt to charge the battery while discharging it. On the other hand; some say that, it'll be no benefits in a simultaneous cycling process of it; even if, electrons could enter and exit it at the same time because of the laws of physics as well.

We've to sort through a lot of beliefs to figure out if it's possible or not to charge the battery while discharging it. But; if we don't understand that, voltage will enter or exit a battery different than current will and they'll enter or exit it based on its design as well. However; they'll flow through it differently based on which cycling process will be carried out as well.

If we don't understand these things, then we probably can't sort out what's true or not; thus, we're inept to figure out if it's possible or not to charge the battery while discharging it as well. Since we've to charge it on the same terminal connection that, it's discharged on. Then it's obvious why we've stop the discharging cycle of the battery in order to start its charging cycle as well.

Therefore; if electrons, they've to go in the battery the same way they've to come out in order to charge its plates or for them to be discharged to a load source as well. Then it's due to the contemporary design of the battery why it can't be charged while it's being discharged to the load source; given that, we could charge or discharge the same battery.

Thus; it'll be obvious, we're not focusing on the right science if we're focusing on the laws of physics when it comes to a conversation about charging a lead acid automotive starting battery while discharging it to a load source. Therefore; we probably inept to argue why the battery can't be charged while being discharged due to the laws of physics as well.

Since a lead acid automotive battery chemical structure is two directional, then it shouldn't matter if we're charging it while discharging it to a load source. Given that; the battery chemical structure, it allows electrons to be simultaneously added and subtracted from its plates during their charging cycle.

In which; it should raise questions about assumption, the charge and discharging cycle of the battery is two different processes; therefore, they can't exist at the same time. Wherein; the assumption itself, it contradicts the mechanics behind the charging cycle of the battery because it shows a charge and discharging process is taking place within it during its charging cycle.

All those erroneous beliefs about why the battery charge and discharging cycle, they can't exist at the same time. They all exist because we're focusing on the wrong science when it comes to the cycling processes of our secondary cell batteries. We should be focusing their mechanical structures and the connections made between them and our load sources as well.

Remember a lead acid automotive starting battery is nothing more than a collection of 2-volt batteries. That is connected in series with one another and housed together in a plastic case to make up the whole battery as we know of today. Then the idea its charge and discharging cycle, they can't exist at the same time because of laws of physics is absurd.

When we're charging the battery as a whole as we know of today, we're simultaneously charging and discharging one battery while simultaneously charging and discharging another. Because of how, they're interconnected with one another. Then the question is, are we focusing on the right science when it comes to its energy input and output process.

Also; it raises a red flag, about the assumption if we attempt to charge the battery while discharging it. Its charge and discharging process, they'll cancel themselves out because energy loss to heat, it'll occur in both of its conversion processes; therefore, no gain. This assumption, it contradicts the mechanics behind the charging cycle of the battery as well.

Here's why; remember, any energy loss to heat during the charging cycle of a lead acid automotive starting battery will be made up by the charging source as well. Therefore; neither the laws of entropy, inertia nor the laws of conservation of energy will be an issue when it comes to charging the battery plates in each cell during their charging cycle as well.

It seems like we're not focusing on the right science if we're insinuating that, the laws of entropy, inertia or the laws of conservation of energy will void a simultaneous cycling process of the battery. If we attempt to charge it while discharging it; given that, energy lost to heat will occur in both of its conversion processes during a simultaneous cycling process of it as well.

I find these claims, they're absurd because remember if a lead acid automotive starting battery cells are connected in <u>series</u> with its terminal connections. During its charging cycle, its plates in each cell will receive the same amount of current flowing from the charging source. However; the charging voltage, it'll be evenly divided amongst its plates in each cell.

Therefore; the battery plates in each cell, they'll be able to consume and store energy while supplying it to its plates in the next cell. Thus; the battery plates in one cell, they'll be simultaneously charged and discharged to its plates in the next cell by way of an electrical bus as well. So; all of the plates in each cell, they could be charged during their charging cycle.

However; if the battery cells, they're connected in <u>parallel</u> with its terminal connections. During its charging cycle, its plates in each cell will receive current based on their individual resistance from the charging source. But; each plate in each cell, they'll receive the same amount of voltage flowing from the charging source.

Therefore; the battery plates in each cell, they'll be able to consume and store energy based on their individual resistance. Thus; the battery plates in each cell, they'll <u>not</u> be simultaneously charged and discharged to its plates in the next cell. So; all of its plates in each cell, they could be charged during their charging cycle.

So; how, a battery cells are interconnected with its terminal connections. It'll determine how energy will flow to and from its plates in each cell during its cycling processes. Wherein; it suggests that, neither the laws of entropy, inertia nor the laws of conservation of energy will be an issue; although, energy will be loss to heat during the charging cycle of the battery plates.

In spite of energy being lost to heat, energy is still added back to the battery plates during their charging cycle; although, they're separated by a non-conducting material called plastic as well. Then the idea it'll be no benefits in a simultaneous cycling process of the battery due to the laws of entropy, inertia or the laws of conservation of energy is absurd as well.

Let's reflect back to the electrical aspects of a lead acid automotive starting battery in chapter 5 for a moment. In which; it might help us focus on, the right science when it comes to a conversation about a simultaneous cycling process of the battery. We could reflect back to chapter 3 and 4 to help us focus on the right science as well.

Wherein; it might explain why, the mechanical structure of the battery is the deciding factor in determining how energy is added or subtracted from its plates as well. Remember the mechanical structure of the battery forms, the circuit connections made between its plates and cells in their relationship with its terminal connections as well.

With that said the battery mechanical structure, it determines how energy is added or subtracted from its plates during its cycling processes. Therefore; if we're focusing on, the mechanical structure of the battery. When it comes to a conversation about charging it while discharging it, then we're focusing on the right science as well.

If we're focusing on anything other than the mechanical structure of the battery, then we're <u>not</u> focusing on the right science. Since electrons, they could be simultaneously added and subtracted from its plates in one cell to its plates in the next cell during their charging cycle. Because of how, its plates and cells are interconnected with its terminal connections as well.

Here's why; the battery plates in each cell, they're connected in series-parallel with one another by their straps, tab connectors and the electrolyte solution in each cell. Thus; electrons, they could flow from a strap to a tab connector on one plate. And then, flow across that plate through the electrolyte solution to an adjacent plate without flowing to and from the same tab connector on the same plate.

Therefore; the battery plates in one cell, they're simultaneously charged and discharged amongst themselves in each cell. Since the battery cells, they're connected in series with its terminal connections. Then its plates in one cell, they're simultaneous charged and discharged to its plates in the next cell. So; all of its plates in each cell, they could be charged during their charging cycle as well.

Wherein; we find that, we don't have to stop charging the battery plates in one cell in order to charge them in the next cell. Since electrons, they could flow in one direction from its plates in one cell to its plates in the next cell. Because of how, they're interconnected with one another by their straps, tab connectors and the electrolyte solution in each cell as well.

Focusing on the mechanics behind the charging cycle of a lead acid automotive starting battery, we'll find the right science is already in place for us to charge the battery while discharging it to a load source. If we stop focusing, on the wrong science; such as, the laws of electrochemistry, entropy, inertia or the laws of conservation of energy as well.

Here's why; if we understand that, a parallel circuit connection made between the battery and a load source in their relationship with a charging source. Then we'll know voltage will flow through the parallel circuit connection different than current will. Since the charging voltage, it'll be the same at the battery and at the load source; but, current will not.

Then it's obvious that, the parallel circuit connection made between the battery and the load source in their relationship with the charging source. It's insufficient for the battery to be charged while being discharged to the load source as well. Then it's no need to focus on, the laws of physics to figure out why the battery can't be charged while being discharged to the load source.

Given that; it'll be a host of things, other than the laws of physics will come into play to prevent the battery from being charged while being discharged to the load source. We've to find and eliminate those hosts of things before we could even consider looking at the vast and complex laws of physics; but, that is what we focus on when we begin the conversation.

However; if we're focusing on, the laws of physics to ascertain; whether, a battery could or couldn't be charged while being discharged. Then we're focusing on the wrong science. Given that; we should be focusing on, the laws of mechanics because they determine how electrons are added or subtracted from the battery, not the laws of physics as some believe.

It could be a host of underlining causes, other than the laws of physics will prevent a battery from being charged while being discharged. If we're focusing on the right science, then it'll help us find those underline causes. I believe the right science is the laws of mechanics when it comes to figuring out if a battery could or couldn't be charged while being discharged as well.

For instance; if we know a lead acid automotive starting battery plates in each cell, they're connected in series-parallel with one another by their straps, tab connectors and the electrolyte solution in each cell. Then we'll know, electrons could flow to and from the battery plates in each cell based on the laws governing series and parallel circuitry as well.

Thus; we'll know that, electrons could flow to and from the same plate without flowing to and from the same strap on the same plate. If we understand, how electrons are added and subtracted from the battery plates in each cell. Then we'll know that, it has little to do with the laws of physics why we've to stop its discharging cycle in order to start its charging cycle as well.

We'll know it has more to do with the laws of mechanics than the laws of physics why we've to stop the discharging cycle of the battery in order to start its charging cycle. If we're focusing on, the mechanics behind its charging cycle. Then we'll know that, we're focusing on the right science to ascertain whether it could or couldn't be charged while being discharged as well.

Here's why; we'll know that, the laws of series, parallel or series-parallel circuit connection determine how electrons are added or subtracted from the battery plates. Remember its plates in each cell are connected in series-parallel with one another by their straps, tab connectors and the electrolyte solution in each cell as well.

Thus; electrons, they could flow across the battery plates in each cell based on the laws governing series and parallel circuitry. Since each cell, it's connected in series with one another by way of an electrical bus. Then electrons, they could flow from the battery plates in one cell to its plates in the next cell based on the laws governing series circuitry as well.

Wherein; we don't have to stop charging the battery plates in one cell in order to charge them in the next cell because of the laws of motion and the effect of forces on bodies. We're using the design of the battery to manipulate, how electrons will flow through its body during its charging cycle. Since like charges, they'll repel one another and follow the path of least resistance as well.

We've to ask ourselves, are focusing on the right science when we say, either the laws of electrochemistry, entropy, inertia or the laws of conservation of energy. They'll prevent a secondary cell battery from being charged while being discharged to a load source. If we're adding, energy back to the battery while taking it out to power the load source as well.

Given that; we could add and subtract energy from the same chemical composition of a lead acid automotive starting battery plates in one cell, so, it could be added to its plates in the next cell. Although; its plates, they're separated by a non-conducting material called plastic and energy will be lost to heat during the electrolysis of its plates as well.

On the other hand; we shouldn't expect, the laws of entropy, inertia or the laws of conservation of energy to be an issue. If we're adding, energy back to the battery while taking energy out as well. We can't treat the battery as though it's a closed system under those laws of physics if we're adding energy back to it while taking energy out as well.

Here's why; if we're charging the battery while discharging it, then the laws of entropy, inertia or the laws of conservation of energy will not be an issue. Given that; any energy lost to heat during the process, it'll be made up by a charging source as well. It shouldn't be any different than the battery going through its normal charging cycle when it's just a charge on it as well.

If we focus on the right science, then we'll be focusing less on physics; but, more on mechanics. Since a lead acid automotive starting battery plates in each cell, they're connected in series-parallel with one another by their straps, tab connectors and the electrolyte solution in each cell and each cell is connected in series with its terminal connections as well.

Thus; electrons, they could flow from a strap to a tab connector on one plate. And then, flow across that plate through the electrolyte solution to an adjacent plate. Without flowing to and from, the same tab connector on the same plate. Therefore; the plates in each cell, they're simultaneously charged and discharged amongst themselves and to the plates in the next cell as well.

Although; the battery plates in each cell, they're separated by a non-conducting material called plastic. Then why those in the field of automotive battery technology, they're focusing on the laws of electrochemistry, entropy, inertia or the laws of conservation of energy when it comes to a conversation about the energy input and output process of the battery?

It doesn't make any sense based on the mechanics behind the charging cycle of the battery. So; it seems like, if we're focusing on either the laws of electrochemistry, entropy, inertia or the laws of conservation of energy. Then we're not focusing on the right science when it comes to a conversation about charging a battery while discharging it to a load source as well.

In the next chapter; let's focus on, the laws of mechanics to see if they're the right science to be focusing on when it comes to a conversation about charging a battery while discharging it to a load source as well.

CHAPTER 14

THE LAWS OF MECHANICS

AFTER TAKEN AN IN depth analysis of the mechanics behind, a lead acid automotive starting battery charging cycle. I discovered a simultaneous cycle process of the battery isn't unprecedented after all. It goes through a simultaneous cycling process each time it's charged. Then why all the noise, about it can't be charged while being discharged because of the laws of physics as well.

Looking at Newton's three laws of mechanics, in which, I found this information in the Wikipedia. TNEB TANGEDCO AE Basic Engineering Study Materials states that; the study of rigid body mechanics is based on the following three laws of mechanics. The first Law is that; a particle, it remains in its position (rest or motion) if the resultant force acting on the particle is zero.

The second Law is that; acceleration of the particle, it'll be proportional to the resultant force and in the same direction if the resultant force is not zero. The third Law is that; action and reaction forces between the interacting bodies are in the same line of action, equal in magnitude; but, acts in opposite direction.

If we apply the three laws of mechanics given by Newton and then, equate them to the mechanics behind the charging cycle of a lead acid automotive starting battery. We'll find its charging cycle is carried out by the laws of motion and the effect of forces on bodies. Remember in chapter 9, the question was asked what the laws of physics are any ways.

I said the laws of physics are nothing more than matter and energy behaving differently under different circumstances. And those circumstances, they're bedded in mechanics. Remember a lead acid automotive starting battery plates in each cell, they're connected in series-parallel with one another by their straps, tab connectors and the electrolyte solution in each cell.

Thus; electrons, they could flow from a strap to a tab connector on one plate. And then, flow across that plate through the electrolyte solution to an adjacent plate. Without flowing to and from, the same tab connector on the same plate. Therefore; the battery plates in each cell, they're charged by using the laws of motion and the effect of forces on bodies or the laws of mechanics.

Given how the battery plates are interconnected with one another by their straps, tab connectors and the electrolyte solution in each cell. With the interacting forces of the electrons within the body of the battery, its plates in each cell are charged. Without having to, stop charging one plate in order to charge another; although, they've only one tab connector each.

Here's why; fixed electrons on one plate or its resistance to electrons, it'll reflect free electrons coming from a charging source toward an adjacent plate by way of the electrolyte solution in that cell. Thus; the path of least resistance for the free electrons, it's for them to flow in one direction through the electrolyte solution to an adjacent plate in that cell.

Given that; electrons, they'll repel one another and follow the path of least resistance as well. Since the battery plates in one cell, they're connected in series with its plates in the next cell by way of an electrical bus. Then the resistance of its plates in one cell or fixed electrons on them will reflect free electrons coming from charging source toward the plates in the next cell as well.

Since the acceleration of the particles (electrons), they'll be proportional to the resulting force (voltage) in the same direction if the resulting force is not zero. As a result; the plates in one cell, they're simultaneously charged and discharged amongst themselves by way of the electrolyte solution in each cell and to the plates in the next cell by way of an electrical bus as well.

Given that; the charging electrons, they can't flow back toward the charging source because it'll be the path with the most resistance. Therefore; using, the design of the battery to manipulate the mechanics of the electrons by using the laws of motion and the effect of forces on bodies to charge the battery or the laws of mechanics.

If we're focusing on anything other than the laws of mechanics when it comes to charging the battery while discharging it, then we're focusing on the wrong science. Given that; we don't understand, it's due to the wrong mechanics in play to charge the battery while discharging it because of its contemporary design and the connections involved as well.

In other words; it's due to the design of the battery and the connections made between, it and a load source in their relationship with a charging source. If we think it's due to the laws of physics; such as, the laws of electrochemistry, entropy, inertia or the laws of conservation of energy. Then we're focusing on the wrong science because we don't know better as well.

Since we don't know better, it creates this misconception that we can't charge the battery while discharging it to a load source because of the laws of physics. But; in reality, it's due to the design of the battery and the connections made between it and the load source in their relationship with the charging source is the true reason.

Here's why; if we're charging, the battery while it has the load source. If the charging source, it can't create a surplus of electrons at the load source in order to reverse the current flow back into the battery to charge it. Then it can't be charged, not due to the mechanics of the current alone or the laws of physics; but, it's due to other extenuating circumstances.

For instance; how, a battery is interconnected with a load source in their relationship with a charging source. Then the connection made between them, it determines how electrons will flow between them. In other words; if it's a series, a parallel or a series-parallel circuit connection made between them, then electrons will flow to them accordingly.

It'll have little to do with the laws of electrochemistry, entropy, inertia or the laws of conservation of energy determining how electrons. They'll flow between the battery and the load source in their relationship with the charging source; basically, it's due to the wrong mechanics in play to charge the battery while discharging it to the load source.

If we look closely at the mechanics, behind a lead acid automotive starting battery's charging cycle. We'll find its mechanical structure is used to manipulate the mechanics of the electrons. So; we don't have to, stop charging one plate in order to charge another in each cell or stop charging them in one cell in order to charge them in the next cell as well.

Although; the battery plates have one tab connector each and its plates in each cell, they'll be separated by a non-conducting material called plastic as well. It's not about the laws of physics in and of themselves; but, it's how we go about adding and subtracting electrons from the battery plates is the key. In which; it's all about, the laws of mechanics.

We're using the laws of mechanics to manipulate the laws of physics because like charges, they'll repel one another and follow the path of least resistance; in which, it's a part of the laws of physics as well. We don't have to focus on the laws of electrochemistry, entropy, inertia or the laws of conservation of energy because they're not the reasons why.

Here's why; if we look closely at the mechanics behind, the charging cycle of a lead acid automotive battery plates. Although; energy lost to heat, it'll occur during their charging cycle; but, it has no bearing on whether they're charged or not. Even though; energy, it's taken from the battery plates in one cell; so, it could be added to its plates in the next cell as well.

Wherein; the battery plates in each cell, they're separated by a non-conducting called plastic. And yet, we're still able to charge them by using an electrical bus. This information reveals that, the laws of electrochemistry entropy, inertia or the laws of conservation of energy will not be an issue when it comes to charging the battery plates while discharging them to a load source as well.

Let's forget about the vast and complex laws of physics for a moment and just focus on the mechanics behind the charging cycle of a lead acid automotive starting battery. All we've to do is focus on the technique; that is, used to charge the battery plates in each cell. Without having to, stop charging them in one cell in order to charge them in the next cell.

Then we'll find it's a matter of mechanics and not a matter of physics why the battery can't be charged while being discharged to a load source. Remember in chapter 5, we discussed. If a battery plates in each cell, they're connected in series-parallel with one by their straps, tab connectors and the electrolyte solution in each cell.

Then voltage and current will flow to a battery plates in reverse of how voltage and current will flow from its plates in each cell. Wherein; voltage and current, they'll flow from a strap to a plate different than they'll flow <u>from</u> a plate to a strap connecting them together in parallel. If we understand this, then we know it's about mechanics.

Understanding, how voltage and current will behave flowing to and from a battery plates based on how they're interconnected with one another. Then we'll understand some laws of physics, they're nothing more than matter and energy behaving differently under different circumstances. Thus; the design of the battery, it'll determine how voltage and current will enter or exit it.

Using the conventional current flow concept, I'm going to describe how voltage and current will flow to and from a lead acid automotive starting battery plates during its cycling processes. So; you could see why, voltage and current will behave differently under different circumstances because it's not a matter of physics; but, it's a matter of mechanics.

Here's why; during the charging cycle of a lead acid automotive starting battery, the charging voltage and current. They've to enter the cell with the positive terminal connection it based on the conventional current flow concept. Wherein; voltage and current, they'll behave differently because of the difference in their mechanics as well.

Since the positive terminal connection of the battery, it'll be connected in parallel with its positive plates by a strap and tab connectors. When the charging current enters the positive terminal connection of the battery, then it's no longer about the mechanics of the current alone; but, it's about the design of the battery or the laws of mechanics.

Then the charging current will flow to the battery plates in that cell based on, their individual resistance because they're connected in parallel with the positive terminal connection; however, they'll receive the same amount of voltage flowing from the charging source. As voltage, it flows across each plate <u>from</u> the strap connecting them together in parallel.

Thus; voltage, it'll flow to the battery plates different than current will if they're connected in parallel with one another by a strap. But; the reverse, it'll happen when voltage and current will flow from the battery plates <u>to</u> the strap connecting them together in parallel. The sum of the current flowing across each plate, it'll be combined at the strap connecting them together in parallel.

Although; current, it was divided amongst the battery plates based on their individual resistance when it was flowing from the strap connecting them together in parallel. But; when voltage, it leaves the battery plates and flow to the strap connecting them together in parallel. The sum of the voltage at the strap, it'll be the same as voltage flowing across one plate.

In other words; voltage, it'll not combined at the strap like current will. This is why we've to separate voltage from current if we're going to understand why, it'll be possible to charge a battery while discharging it to a load source. Remember in chapter 5, I used the conventional current flow concept to explain how voltage and current will flow through a battery.

When the charging voltage and current flows from the positive plates to the adjacent negative plates in the cell with the positive terminal connection in it, the negative plates will receive a voltage drop from the positive plates. But; the charging current flowing from the positive plates to the negative plates, it'll be the same as the current flowing from the positive plates.

The reason the negative plates received the amount of voltage and current, they received from the positive plates. It wasn't a matter of physics; but, it was a matter of mechanics. Because of how, the plates were interconnected with one another by way of the electrolyte solution in that cell; in which, it had little to do with the laws of physics as well.

If we sum up, the reasons why voltage and current will flow from and to a strap connecting a battery plates together in parallel. Then it's clear the mechanics of the voltage and current alone, they don't determine how they'll flow through the body of a battery; but, its design will. So; it's no need to focus on, the vast and complex laws of physics as well.

It's clear if we're going to understand why it'll be possible or not to charge a battery while discharging it to a load source. We've to focus on the laws of mechanics because they determine how voltage and current will flow through the body of the battery, not the laws of physics based on the mechanics behind the charging cycle of a lead acid automotive starting battery.

I found we could use the same technique; that is, use to simultaneously charge and discharge. A lead acid automotive starting battery plates amongst themselves in each cell and to its plates in the next cell. We could use it to simultaneously charge and discharge them to a load source without having to stop the discharging cycle of the battery in order to start its charging cycle.

After evaluating, how voltage and current will flow from and to a lead acid automotive battery plates. Given how they're connected in series-parallel with one another by their straps, tab connectors and the electrolyte solution in each cell. It dawn on me if the battery had separate terminals for charging and discharging interconnected by the same plates in the same cell.

Then it would allow us to create a series-parallel circuit connection between a charge and a load source in their relationship with the battery. Thus; allowing, us to use the same technique that is used to simultaneously charge and discharge its plates amongst themselves in each cell; but, using it to simultaneously charge and discharge them to a load source as well.

Given how the charging voltage and current will flow from the charging terminal to a strap connecting it together in parallel with its respective plates. It'll be in reverse of how voltage and current will flow from those respective plates to a strap connecting them together in parallel with the discharging terminal as well.

Since the two terminals, they'll have their own straps and tab connectors leading to the same plates in the same cell. Not only voltage and current will flow in reverse of one another flowing to and from the two terminals. But; I found, how voltage and current will flow through one terminal will not affect how they'll flow out the other terminal connection as well.

Then how voltage and current will flow <u>to</u> the plates connecting the two terminals together in parallel, it'll be independent of how voltage and current will <u>from</u> the plates as well. When the charging current enters a cell through one terminal, it'll be divided amongst the plates based on their individual resistance because they're connected in parallel by a strap.

However; when current, it leaves the plates and flow to the strap connecting them together in parallel with another terminal as well. Then the sum of the current flowing across each plate, it'll be combined at the strap connecting them together in parallel with another terminal; although, current was divided amongst the plates flowing from the strap connecting them together in parallel as well.

When the charging voltage enters a cell through one terminal, it'll be the same flowing across each plate from the strap connecting them together in parallel as well. Now when voltage, it leaves the plates and flow to the strap connecting them together in parallel with another terminal. Then voltage at the strap, it'll be the same as the voltage flowing across one plate.

Although; each plate, it'll receive the same amount of voltage flowing from the strap connecting them together in parallel. When voltage is flowing to a strap from the plates, it'll not combine at the strap connecting them together in parallel as current does. Thus; voltage, it'll flow to and from a strap different than current will.

Then voltage and current will flow <u>to</u> a strap. It'll be in reverse and independent of how they'll flow <u>from</u> it as well. Based on my findings, I found if a battery has separate terminals for charging and discharging in the same cell. And those terminals, they had their own straps and tab connectors leading to the same plates in that cell well.

I found it's the key in charging a battery while discharging it to a load source. We could conjure up all manner of laws of physics to justify why it'll be impossible to a battery while discharging it to a load source. However; it's a waste of time because the laws of physics, they don't determine how electrons are added or subtracted from a battery plates.

It never has been about the laws of physics when it comes to adding or subtracting electrons from a battery plates. We may think so; but, we're wrong because a battery cycling processes are carried out by the laws of mechanics. Maybe; it throws people off, when I'm talking about charging a battery while discharging it because it's not associated with its cycling processes.

Maybe; people think, it's a fictitious concept that I made up. However; I found that, a lead acid automotive starting battery plates. They go through a simultaneous cycling process each time they're charged; given that, they're connected in series-parallel with one another in each cell. And each cell, it's connected in series with its terminal connections as well.

Those in the field of automotive battery technology, they don't define it as a simultaneous cycling process. I find it's no other way to explain it other than a simultaneous cycling process because electrons, they're simultaneously added and subtracted from one plate to another in each cell and to the plates in the next cell during their charging cycle as well.

Maybe; those in the field of automotive battery technology, they don't see it as a simultaneous cycling process. When electrons, are simultaneously added and subtracted amongst its plates in each cell and to its plates in the next cell. So; all of its plates in each cell, they could be charged at once without having to stop charging one plate in order to charge another.

Although; the battery plates in each cell, they're separated by a non-conducting material called plastic. However; electrons, they're taken from the plates in one cell to be added to the plates in the next cell as well. Looking at the battery from the inside of its plastic casing, it's easy to see why it would be possible to charge it while discharging it to a load source as well.

Looking at the battery from a mechanical perspective, then it's easy to see it has little to do with the laws of physics why it can't be charged while being discharged to a load source as well; but, it has more to do with the laws of mechanics. On the other hand; if we know anything about, the mechanics of the voltage and current alone.

Then it'll be easy to assume why a battery could be charged while being discharged to a load source. Given the connections made between the battery and the load source in their relationship with the charging source. It'll determine how voltage and current will flow between the battery and the load source in their relationship with the charging source as well.

We find ourselves having to stop one cycling process of a battery in order to start the other. Not due to the mechanics of the voltage and current alone; but, it's due to the laws of mechanics. And yet, we haven't figured this out because we're focusing on the wrong science. For example; we know that, we could add or subtract water from the same opening of a water jug.

However; we know that, we can't simultaneously add and subtract water from the same opening of the water jug as well. Given it'll violate the laws of hydrodynamics as well as the laws of mechanics as well. Newton's second law of mechanics states that, the acceleration of a particle will be proportional to the resulting force and in the same direction if the resulting force is not zero.

What Newton's second law of mechanics mean to me is that, water can't travel on the same path and flow in the opposite direction with equal force to simultaneously enter and exit the same opening of the water jug? The results will be zero because water will not be able to enter or exit the water jug at all due to the equal force of the water trying to simultaneously enter and exit it.

It's like having two objects trying to occupy the same space; but, going in the opposite direction of one another. If we think, it's due to the laws of hydrodynamics why water will not be able to simultaneously enter and exit the same opening of the water jug. Then we've missed diagnosed why water can't simultaneously enter and exit the water jug as well.

If we make the right diagnosis, then we'll know it's not due to the laws of hydrodynamics; but, it's due to the laws of mechanics because water has to go in the jug the same way it has to come out in order to simultaneously enter and exit the jug as well. In other words; it's due to the design of the jug, why water can't enter and exit it at the same time.

Basically; we should be focusing on, the design of the water jug when it comes to simultaneously adding and subtracting water from it. In other words; we should be focusing on, the mechanics of the water jug when it comes to adding or subtracting water from it. I'm using it as a simple example because it applies to a battery when it comes to adding or subtracting energy from it as well. See figure R-7.

Figure R-7

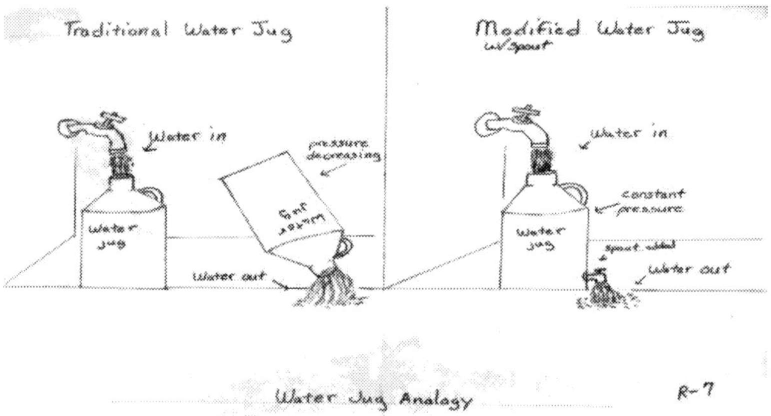

(Water jug analogy)

In figure R-7 on the left, it's an image of a water jug with only one opening for water to enter or to exit it. Wherein; it'll be impossible to simultaneously add and subtract water from the water jug, it's not due to the laws of hydrodynamics; but, it's due to the laws of mechanics because water has to go in the same way it has to come out as well.

In which; water, it could only flow in or out of the water jug flowing in one direction at a time; but, it can't do both because of the design of the water jug. It'll be obvious why water can't be simultaneously added and subtracted from the water jug without the need to bring the laws of hydrodynamics into the conversation if we're making the right diagnosis.

Focusing on the image of a water jug in figure R-7 on the right, it'll be possible to simultaneously add and subtract water from the water jug without violating the laws of hydrodynamics or the laws of mechanics as well. Given that; the jug, it has two openings for water to enter and to exit it at the same time if we attempt to add and subtract water from it as well.

Then the process of simultaneously adding and subtracting water from the water jug, it becomes a matter of mechanics and not a matter of hydrodynamics because we could get around them with the laws of mechanics. So; it's no need to focus on, the vast and complex laws of hydrodynamics to figure out why water can't be simultaneously added and subtracted from the jug.

On the other hand; it'll not be so obvious why, we can't simultaneously add and subtract current from a lead acid automotive starting battery. If we don't know its negative and positive terminal, they only equal to one opening for current to enter or to exit it during its cycling processes. Although; voltage, it could enter one terminal and exit the other to return back to its source.

However; current, it has to enter or exit the same terminal connection of the battery in order to cycle it. If we don't know this, then we might get the wrong impression. It's due to the mechanics of the current alone or the laws of physics why it can't be charged while being discharged; in which, it'll be become a misconception due to the contemporary design of the battery.

Since we've to charge the battery on the same terminal connection that, it was discharged on. Then it brings us to Newton's third law of mechanics, in which, it states action and reaction forces between the interacting bodies in the same line of action; but, flowing in the same direction and not in the opposite direction of one another.

What it means to me is that, we can't charge a battery while discharging it if electrons have to go in it the same way they've to come out as well. Thus; it brings us back to Newton's second law of mechanics, in which, an acceleration of a particle. It'll be proportional to the resulting force and in the same direction if the resulting force is not zero.

Here's the deal; if the resulting force of the electrons, they're equally flowing in the opposite direction of one another in order to enter and exit the battery at the same time. Then we can't charge the battery while discharging it if electrons have to go in it the same way that, they've to come out in order to charge its plates or for them to be discharged as well.

On the other hand; if the resulting force of the electrons, they're not equal. Then electrons, they could only enter or exit the battery; but, they can't do both. Since electrons, they can't travel the same path and flow in the opposite direction of one another in order to enter and exit the battery because they'll repel one another in the opposite direction.

The key words here are "the same opening." If electrons have to go in a battery the same way they've to come out, then we should be focusing on the laws of mechanics and not the laws of physics. When it comes to a conversation about charging the battery while discharging it because they don't determine, how electrons will be added or subtracted from it.

If we're going to charge a battery while discharging it to a load source, then the type of connections made between them. It'll be crucial because voltage and current will flow through a series, a parallel or a series-parallel circuit connection differently. Wherein; these types of circuit connections, they'll determine how voltage and current will flow between devices.

In other words; the connections made between, a battery and a load source in their relationship with a charging source. It's one of the things will prevent the battery from being charged while being discharged to the load source, beside the battery having only one opening for current to enter or to exit it; thus, we've to look beyond the contemporary design of the battery as well.

However; most people, they'll use their conventional wisdom about the cycling processes of a secondary cell battery to conclude why it can't be charged while being discharged. Wherein; I find that, most people will come to the wrong conclusion because their conclusions will be based on the contemporary design of the battery as well.

First of all; our secondary cell batteries, they're not built for charging while discharging. Since we've to charge them on the same terminal connection that, they were discharged on. Secondly; we've to look at our secondary cell batteries from, a different perspective because we can't use our conventional wisdom about their cycling processes to figure this out.

If we don't venture outside the status quo, then we could only assume what affects the laws of physics will have on charging our batteries while discharging them; however, it'll be based on their contemporary designs as well. If we don't change their contemporary designs, so, energy could enter and exit them at the same time. Then we'll be speculating about the effects.

Remember if a battery cells, they're connected in <u>series</u> with its terminal connections. Then a load source, it'll receive the current potential of the battery plates in one cell and the voltage potential of all of its plates in each cell during its discharging cycle. But; during its charging cycle, the charging voltage will be evenly divided amongst its plates in each cell.

However; the plates in each cell, they will receive the same amount of current flowing from the charging source. As a result; the battery plates in each cell, they could consume and store energy while supplying it to its plates in the next cell. So; all of its plates in each cell, they could be charged during their charging process as well.

Another example; if a battery cells, they're connected in <u>parallel</u> with its terminal connections. Then a load source, it'll receive the current potential of all of the battery plates in each cell and only receive the voltage potential of its plates in one cell during its charging cycle. But; during its charging cycle, the charging current will flow to its plates in each cell based on their resistance.

However; the battery plates in each cell, they'll receive the same amount of voltage flowing from the charging source. Thus; the design of a battery, it'll determine how voltage and current will flow to and from its plates during its cycling process. We don't have to speculate about what affects the laws of physics will have on charging our batteries while discharging them.

Given that; it's not about the laws of physics when it comes to charging our batteries while discharging them, it's about the laws of mechanics or how we go about designing our batteries. In other words; how, a battery cells are interconnected with its terminal connections; whether, they're connected in series, parallel or series-parallel with its terminal connections.

I found it'll determine how voltage and current will flow to and from the battery plates accordingly. Thus; the fault lies with, the contemporary design of our secondary cell batteries why they can't be charged while being discharged and not the laws of physics as most people believe; however, most people will not agree with me as well.

We could all agree on one thing; that is, we could charge or discharge the same battery. But; we've to charge it on the same terminal connection that, it was discharged on. Then it's a matter of mechanics and not a matter of physics; given that, electrons have to go in the same way they've to come out in order to charge the battery plates or for them to be discharged as well.

Using either the laws of electrochemistry, entropy, inertia or the laws of conservation of energy to, discredit a simultaneous cycling process of a lead acid automotive starting battery. It shows we're focusing on the wrong science and the wrong mechanics when it comes to the cycling processes of our secondary cell batteries as well.

Here's why; remember each cell in a lead acid automotive starting battery, it's nothing more than a battery in and of itself. That is connected in series with one another and housed in a plastic case to make up the whole battery as we know of today. When it's charged as a whole, we're simultaneously charging and discharging one battery to another as well.

Based on the mechanical structure of the battery and the mechanics behind, its charging cycle. It's no doubt neither the laws of electrochemistry, entropy, inertia nor the laws of conservation of energy will be an issue. When it comes to a simultaneous cycling process of it because we could already, add or subtract energy from its chemical composition first of all.

Secondly; if we're adding energy back to the battery while taking it out, then the laws of entropy, inertia or the laws of conservation of energy will be an issue. Because the battery, it'll not be considered a closed or perpetual system under those laws of physics as well. If we think the battery is, then we're focusing on the wrong science and the wrong mechanics as well.

Here's why; since a lead acid automotive starting battery plates in each cell, they're connected in series-parallel with one another by their straps, tab connectors and the electrolyte solution in each cell. And each cell, it's connected in series with its terminal connections as well.

Then the battery plates in each cell, they could be simultaneously charged and discharged amongst themselves and to its plates in the next cell. Without having to, stop charging them in one cell in order to charge them in the next cell as well. All because of the connections made between them; although, they're separated by non-conducting materials as well.

If we're focusing on the laws of physics to ascertain; whether, it's necessary to stop the discharging cycle of the battery in order to start its charging cycle. Then we're focusing on the wrong science to ascertain this information. We should be focusing on the laws of mechanics because they allow its plates in one cell to be charged and discharged to its plates in the next cell.

Without having to, stop charging the battery plates in one cell in order to charge them in the next cell; although, they're separated by a non-conducting material called plastic as well. If we're focusing on, the mechanical structure of the battery and the mechanics behind its charging cycle. Then we're focusing on the right science and the right mechanics to figure this out.

Whether; it's necessary to stop, the discharging cycle of the battery in order to start its charging cycle. Given how a battery plates and cells, they're interconnected with its terminal connections. It'll determine how voltage and current will flow to and from its plates. Thus; it'll determine if, it's necessary to stop its discharging cycle in order to start its charging cycle as well.

For example; if the charging voltage and current, they could flow into one cell in the battery before flowing to a load source. Then the charging current, it'll flow into the battery based on its resistance and the voltage applied by the charging source. But; current, it'll flow to the other cells in the battery based on how they're connected to the cell current has to enter in first.

However; current, it'll flow to the load source based on its resistance and the voltage applied by the charging source as well. Not based on the resistance between the load source and the battery in their relationship with the charging source. As it's when current, it has to flow to the load source first before it's reversed back into the battery to charge it.

This is why we can't use, our conventional wisdom about the cycling processes of our secondary cell batteries. To speculate, what affects the laws of physics will have on a simultaneous cycling process of them because it'll be based on their contemporary designs. Given that; voltage and current, they'll behave differently under different circumstances as well.

Who's to say that, we can't charge our batteries while discharging them if energy doesn't have to go in them the same way it has to come out? Since voltage and current, they'll behave differently under different circumstances. In the next chapter; let's see why, we don't have the right mechanics in play to charge our batteries while discharging them as well.

CHAPTER 15

THE RIGHT MECHANICS IN PLAY

We could conjure up all manner of laws of physics to justify why a battery can't be charged while being discharged. But; the question is, are we focusing on the right mechanics to make this determination? If we take an in depth analysis of the mechanics behind the charging cycle of a lead acid automotive starting battery, then we could make this determination as well.

Those who use the laws of physics to explain why, a lead acid automotive starting battery can't be charged while being discharged. They might want to re-evaluate their assumptions because they don't add up with, the mechanics behind the charging cycle of the battery. In which; it shows, the laws of physics will not be an issue concerning the matter.

On the other hand; it doesn't make any sense to use, the laws of physics to discredit a simultaneous cycling process of the battery. Given its mechanical make up is nothing more than a collection of 2-volts batteries. That is connected in series with one another and housed together in a plastic case to make up the whole battery as we know of today.

When the battery is charged as a whole, we're simultaneously charging and discharging one battery while simultaneously charging and discharging another. Because of how, they're interconnected with one another; wherein, it raises a red flag about the assumption. It's due to the laws of physics why the battery can't be charge while being discharged to a load source as well.

Those who are skeptical of charging the battery while discharging it to a load source, they might want to re-think their assumptions because they don't add up. That it's due to the laws of physics why the battery can't be charged while being discharged to the load source as well. Given that; the battery, it's nothing more than a collection of 2-volts batteries as well.

What it seems like, we don't have the right mechanics in play to charge the battery while discharging it the load source without being impeded by the laws of physics; therefore, we blame it on the laws of physics. But; in reality, it's due to the laws of mechanics. If we focus on the mechanics of the voltage and current alone, then we'll find it's true in its entirety.

Here's why; the wrong mechanics are in play, in which, it makes it impossible to charge the battery while discharging it the load source. Let's review figure B-7 from chapter 10 again; but, this time we're going to focus on water tank-B scenario. So; you could understand, we're not focusing on the right mechanics in order to charge a secondary cell battery while discharging it.

Figure B-7

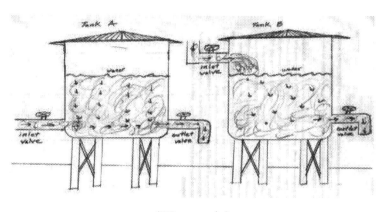

(Water tanks)

In water tank-B scenario, I've manipulated the path which the incoming water will travel in order to enter or to exit the tank by changing its design. Therefore; the incoming water, it has to travel the depth of the tank before it has a chance to flow out of tank; unlike, water tank-A. Using a different design concept by changing, the location of the inlet and outlet valves of the tank.

We could actually use the weight of the incoming water and gravity to subtract water from the tank while adding water back to it; thus, simultaneously adding and subtracting water from it. Unlike; tank-A, tank-B will not end up empty if we attempt to simultaneously add and subtract water from it, all because we changed its mechanical structure.

With the new design concept of the water tank, then the incoming water doesn't have a direct path to the open outlet valve if both valves are open at the same time; unlike, water tank-A. Therefore; the incoming water, it has to go in the tank before it has a chance to come out; thus, using the weight of the incoming water and gravity to add and subtract water from it.

Since water is like electrical current, it'll follow the path of least resistance. Then water tank-B, it'll represent what'll happen if a battery is built to the right specification for charging while discharging. Although; an automotive starting battery with top and side posts terminal connections, it has two positive terminals in one cell and two negative terminals in another as well.

However; the battery, it's not built to the right specification for charging while discharging; so, it can't be charged while being discharged because the right mechanics aren't in play. What the battery, top and side posts terminal connections are lacking is that. They don't have their own strap and tab connectors leading to their respective plates as well.

Therefore; if there's a charge on, one terminal and a load source on the other respective terminal of the battery. Then the charging current, it doesn't have to come in contact with the battery plates before flowing to the load source. Given that; the charging current, it'll have a direct path to the load source similar to water tank-A scenario in figure B-7.

It has little to do with the mechanics of the current alone or the laws of physics; but, it has more to do with the connections made between the two respective terminal connections of the battery why current will not come in contact with its plates. Given that; its two respective terminal connections, they're interconnected by a strap and not by their respective plates.

Therefore; the battery, it's not design to the right specifications for charging while discharging. Since one of its respective terminal, it's connected to its respective plates by a strap and tab connectors; but, the other one isn't because it's only connected to its respective terminal connection by a strap; thus, they only equal to one terminal connection.

Here's why; the two respective terminals, they're joined by a strap and not by their respective plates; so, they'll be connected in parallel with one another; wherein, it's like having a charge and a load source on the same terminal connection of the battery. Thus; current, it has to flow to the load source before current is reversed back into the battery to charge it.

If the battery top and side posts terminal connections, they had their own straps and tab connectors leading to their respective plates. Then the charging current, it has to come in contact with those respective plates before flowing to a load source. If a charge is on, one respective terminal and the load source is on the other respective terminal as well.

Wherein; the plates interconnecting, the two respective terminals together in parallel will be energized. Then they'll transfer energy to the adjacent plates in that cell and to the load source that is connected to the other respective terminal connection as well; thus, allowing the battery to be charged while being discharged to the load source as well.

If the battery top and side posts terminal connections, they had their own straps and tab connectors leading to their respective plates. Then we could manipulate which path the charging current will travel within the battery, similar to water tank-B scenario. Because we could use the mechanics of the electrons to, add and subtract them from the battery plates as well.

Since like charges, they'll repel one another and follow path of least resistance. Thus; we could use, the design of the battery to manipulate the mechanics of the electrons to charge it while discharging it to a load source. I know it's hard for some people to believe this. But; it'll be similar to, a lead acid automotive starting battery plates being charged in each cell as well.

Here's why; the battery plates in each cell, they're connected in series-parallel with one another by their straps, tab connectors and the electrolyte solution in each cell. Since like charges, they'll repel one another and follow path of least resistance as well. Then electrons, they could flow from a strap to a tab connector on one plate.

And then, electrons could flow across that plate and through the electrolyte solution to an adjacent plate. Thus; charging and discharging each plate without electrons flowing to and from, the same tab connector on the same plate. In which; it's called mechanics because we're manipulating, the mechanics of the electron with the design of the battery as well.

If we understand, voltage will flow to and from a battery plates different than current will. Then we'll know, we're focusing on the right mechanics when it comes to a conversation about charging a battery while discharging it. Given we'll be focusing on the mechanics of the voltage and current and how they'll flow through the body of the battery as well.

If we separate voltage from current, then we'll find current consists of electrons and they'll repel one another and follow the path of least resistance as well. However; voltage, it doesn't consist of electrons and it doesn't follow the path of least resistance as well. Since voltage, it's nothing more than an electromotive force use to move electrons.

Therefore; if there's a charge and a load source on the same terminal of a battery at the same time, then the charging voltage will behave differently than the charging current will. This is why it's important to focus on the right mechanics because voltage, it'll be the same across the battery and the load source as well. See chapter 12.

However; current, it'll not be the same across the battery and the load source because they'll be connected in parallel with one another in their relationship with the charging source. Remember if a battery cells, they're connected in <u>series</u> with its terminal connections. Then a load source, it'll only receive the current potential of the battery plates in one cell.

But; the load source, it'll receive the voltage potential of all of the battery plates in each cell during its discharging cycle. However; during its charging cycle, its plates in each cell will receive the same amount of current flowing from the charging source. But; the charging voltage, it'll be evenly divided amongst the battery plates in each cell during its charging cycle.

If we're focusing on the mechanics of the voltage and current, then we'll find voltage will flow to and from a battery plates different than current will. Based on, the type of connections; such as, series, parallel or series-parallel circuitry connections made between the battery plates and cells in their relationship with its terminal connections as well.

In which; those connections, they'll determine how voltage and current will flow between the battery plates and cells in their relationship with its terminal connections as well. Knowing voltage and current, they'll flow through a series, a parallel or a series-parallel circuitry connection differently. Then we don't need to focus on the vast and complex laws of physics as well.

Given that; we'll know, it's all about mechanics and not about physics when it comes to the cycling processes of a secondary cell battery. Remember a lead acid automotive starting battery plates in each cell, they're connected in series-parallel with one another by their straps, tab connectors and the electrolyte solution in each cell as well.

Thus; having the right mechanics in play, it'll allow electrons to flow from a strap to a tab connector on one plate. And then, flow across that plate through the electrolyte solution to an adjacent plate. Therefore; free electrons flowing from the charging source, they'll be reflected toward an adjacent plate by way of the electrolyte solution in each cell as well.

Without electrons flowing to and from, the same tab connector on the same plate in each cell. Since each cell in the battery, it'll be connected in series with its terminal connections by way of an electrical bus. Then fixed electrons on its plates in one cell, they'll reflect free electrons coming from charging source toward its plates in the next cell as well.

Having the right mechanics in play, we don't have to stop charging the battery plates in one cell in order to charge them in the next cell; although, they're separated by a non-conducting material called plastic. With this revelation, it seems like we've been focusing on the wrong science when it comes to a conversation about charging the battery while discharging it as well.

Focusing on the right mechanics, we'll know it's no need to focus on the laws of electrochemistry, entropy, inertia or the laws of conservation of energy. Given they're not an issue when electrons, are simultaneously added and subtracted from the battery plates in one cell to its plates in the next cell; although, they're separated by a non-conducting material as well.

In view of my finding, it's not necessary to stop the discharging cycle of the battery in order to start its charging cycle because of the laws of physics. We could use the same technique that is used to simultaneously charge and discharge its plates in one cell to its plates in the next cell. It could be use to simultaneously charge and discharge them to a load source as well.

Without having to, stop the discharging cycle of the battery in order to start its charging cycle as well. Those who are skeptical of these ideas, it seems like they're focusing on the wrong science and the wrong mechanics as well. Remember in chapter 10 in figure B-1, it showed the design concept of an automotive starting battery with top and side posts terminal connections.

Also; in chapter 10 in figure B-2, it showed the design concept of a modern day cell phone battery with its multi-terminal connections as well. I'm going to show those same design concepts in this chapter as well. Hoping it'll shed some light on the difference between a battery having a charge and a load source on the same terminal, compared to them being on separate terminals of it.

I found those differences, they're important for us to know because they'll let us know if we're focusing on the right mechanics or not when it comes to a conversation about charging a secondary cell battery while discharging it to a load source as well. But; the design concepts of those batteries in figure B-1 and B-2, they'll help clarify those differences for us as well.

Wherein; we'll find that, the design of a battery is the deciding factor in determining how its cycling processes will be carried out. Let's review figure B-1 and figure B-2 from chapter 10 again.

Figure B-1

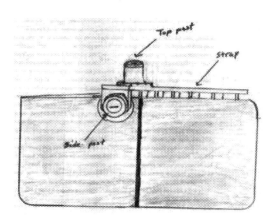

(Auto battery w/ top and side posts connections)

An automotive starting battery with its top and side posts terminal connections, it's similar to a cell phone battery with its multi-terminal connections as well. Given that; their respective terminal connections, they're interconnected by a strap and not by their respective plates. Their design concepts show they're not designed for charging while discharging as well.

On the other hand; their design concepts show why, they can't be charged while being discharged to a load source as well. Given that; the right mechanics, they're not in place to carry out such a process as well.

Figure B-2

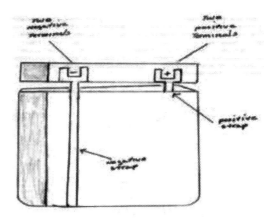

(Cell phone battery w/ multi-terminals)

Looking at an automotive starting battery with top and side posts terminal connections and a cell phone battery with multi-terminal connections, what they've in common is that. All of their negative terminals, they share the same straps connecting them in parallel. Likewise; all of their positive terminal connections, they share the same straps connecting them in parallel as well.

In which; we'll find that, the two respective terminal connections. They only equal to one terminal connection. For example; only one of the negative terminal connection is connected to, its respective negative plates by a strap and tab connectors; however, the other respective negative terminal connection is only connected to its respective terminal by a strap.

Thus; the strap, it joins the two respective negative terminals together in parallel; likewise, the same for the two respective positive terminal connections as well. Therefore; putting, a charge on one terminal and a load source on the other respective terminal. Then it's like having a charge and a load source on the same terminal connection of the battery as well.

Since the two respective terminal connections, they only equal to one terminal connection because they're joined by a strap and not by their respective plates. Then we'll find that, the battery and the load source will be in parallel with one another in their relationship with the charging source because the two terminals, they only equal to one terminal.

Therefore; the charging voltage and current, they'll be subject to the laws of parallel circuitry if there's a charge on one terminal and a load source on the other respective terminal connection as well. Given that; the two respective terminals, they're connected in parallel with one another by way of a strap and not by their respective plates.

Thus; we can't charge the battery while, discharging it to the load source because wrong mechanics are in play to carry out such a process. It has little to do with the laws of physics in and of themselves; but, more to do with the connection made between the two respective terminal connections. Although; voltage, it'll be the same across the battery and the load source.

However; current, it'll not be the same across the battery and the load source because of the parallel circuit connection made between them. In which; it'll make, it easier for current to flow to the load source instead of flowing into the battery. Given that; current, it'll follow the path of least resistance between them because of the parallel connection made between them.

I found it's not only due to the parallel connection made between the battery and the load source why it would be easier for current to flow to the load source instead of flowing into the battery. It's also due to the battery resistance to current because it's different than the load source resistance to current, which it makes it easier for current to flow to the load source as well.

Since the battery respective terminals, they're joined by a strap and not by their respective plates. Then the path of least resistance for the current is to flow across the strap connecting the two terminals together and not flow to their respective plates. Given that; the path of least resistance for the current, it's to flow directly to the load source.

Thus; current, it'll flow through one respective terminal and out the other without coming in contact with the battery plates. This is one reason why, some people will assume it'll make no difference if electrons could enter and exit a battery at the same time. We still couldn't charge it while discharging it to a load source because of the laws of physics as well.

Wherein; this line of thinking, it's only guided by the lack of information about the design concept of a battery and how it'll play a role in its cycling processes. Given that; the laws of physics, they don't determine how electrons will be added or subtracted from the battery. However; the laws of mechanics, they determine how electrons will be added or subtracted from it.

Bottom line is that, we haven't taken an in depth of analysis of the mechanical structure of our batteries to learn otherwise. For example; a cell phone's battery with its multiple terminal connections, it only uses one set of terminals for adding or subtracting electrons from it; however, the other sets will be used for thermistors for overcharge protection for it.

The problem with the cell phone's battery having multi-terminal connections is that, the respective terminals share the same strap. It only joins them together in parallel; so, the battery will have more than one negative or one positive terminal connection. I promised to talk about why it'll have multi-terminal connections and how they'll play a role in its cycling processes as well.

The multi-terminal connections of the battery, they only allow us to receive or make calls while the phone is in its charging mode; however, it's a down side for this as well. The respective terminals of the battery sharing the same strap, it'll allow the charging cycle of the battery to be interrupted when the phone is receiving or making calls as well.

This is especially true, if the cell phone is consuming electrons faster than they could break down the internal resistance of the battery to charge it. Since its multi-terminal connections, they'll only allow electrons to flow to and from one set of terminals to charge or discharge it. But; its other set of terminals, are used for overcharge protection for it.

Since one set of terminals, they only allow electrons to flow from a charging apparatus to the battery plates and from its plates to a voltage and current regulatory system. However; the other set of terminals, they only allow electrons to flow from the battery plates to a thermistor for overcharge protection for it.

Then the mechanical make up of the battery will not allow its plates to be charged while they're being discharged. Given that; electrons, they could only travel the same path in order to flow to its plates from the charging apparatus or flow from its plates to the voltage and current regulatory system of the cell phone; therefore, the battery isn't built for charging while discharging.

Taking an, in depth analysis of the mechanical structure of an automotive starting battery with top and side posts terminal connections. We'll find it's not built for charging while discharging; although, it has two negative terminals in one cell and two positive terminals in another cell; but, it still can't be charged while being discharged as well.

Taking an in depth analysis of the mechanical structure of the batteries, it's self evident they're not built to the right specification for charging while discharging. Because electrons, they still have to go in the batteries the same way they've to come out as well. On the other hand; do we actually know, what it means to charge a battery while discharging it?

Remember if a battery plates in each cell, they're connected in series-parallel with one another by their straps, tab connectors and the electrolyte solution in each cell. And each cell, it's connected in series with its terminal connections as well. During its charging cycle, its plates in one cell will be charged and discharged amongst themselves and to the plates in the next cell as well.

Here's why; electrons, they'll be able to flow from a strap to a tab connector on one plate. And then, flow across that plate through the electrolyte solution to an adjacent plate. Without flowing to and from, the same tab connector on the same plate; wherein, one plate will be charged and discharged to an adjacent plate by way of the electrolyte solution in that cell.

Since each cell is connected in series with one another by way of an electrical bus, then the plates in one cell. They'll be simultaneously charged and discharged to the plates in the next cell. So; all of the plates in each cell, they could be charged during their charging cycle without having to stop charging them in one cell in order to charge them in the next cell as well.

Those in the field of automotive battery technology, they don't define it as a simultaneous cycling process during a battery charging cycle. If its plates in each cell, are connected in series-parallel with one another and each cell is connected in series with its terminal connections as well. It's no other way to explain this other than a simultaneous process.

A lead acid automotive starting battery goes through a simultaneous cycling process each time, it's charged. Wherein; it'll not be out of the realm of physic to, charge it while discharging it to a load source. Since the process, it'll be nothing more than the extension of the battery charging cycle; providing that, the right mechanics are in play as well.

In other words; electrons, they don't have to go in the battery the same way they've to come out in order to charge its plates or for them to be discharged to the load source as well. Then it becomes a matter of mechanics and not a matter of physics when it comes to charging the battery while discharging it to the load source as well.

Here's why; if electrons, they could enter and exit the battery at the same time. Then it becomes a matter of mechanics if the charging voltage is greater than the normal battery voltage. Wherein; the charge and discharging cycle of the battery, they could exist at the same time. Given that; electrons, they could go in one way and come out another.

Then we'll find charging a battery while discharging it to a load source, it's not about physics; but, it's about mechanics. In other words; it's about, how we go about adding or subtracting electrons from the battery. Here's why; if a battery cells are connected in series with its terminal connections, then it'll be a voltage drop throughout each cell in it.

However; current, it'll be the same throughout each cell in the battery during its charging cycle. Basically; we'll be manipulating, the mechanics of the voltage and current with the laws of mechanics or with the design of the battery. Given how we connect the battery cells to its terminal connections will determine how voltage and current will flow through it.

Thus; when it comes to charging the battery, it's not about the laws of physics; but, it's about the laws of mechanics or the laws of motion and the effect of forces on bodies. The real question is that, do we actually understand the mechanics that is taking place within our secondary cell batteries during their cycling processes. If we do, then it's not apparent.

If we think it's inconceivable or farfetched to charge our batteries while discharging them because of the laws of electrochemistry, entropy, inertia or the laws of conservation of energy as well. Wherein; it seems like, it's due to connections; such as, series, parallel or series-parallel will determine how voltage and current will be added or subtracted from our batteries.

Here's why; I found a parallel circuit connection made between, a battery and a load source in their relationship with a charging source. It'll be an insufficient connection to charge the battery while discharging it to the load source. The parallel connection made between them will allow current to flow to the load source first because the battery resistance to current is different.

It has little to do with the mechanics of the current alone or the laws of physics. Why current, it'll flow to the load source first; but, it's mainly due to the parallel connection made between the battery and the load source in their relationship with the charging source. A host of things will come into play, other than the mechanics of the current or the laws of physics as well.

We just can't focus on one thing; but, we've to focus on a host of other things. Here's why; remember a secondary cell battery plates, they behave like load sources consuming energy during their charging cycle. Wherein; current, it'll not flow into the battery and to a load source at the same time if they're connected in parallel with one another. See chapter 12.

Although; the battery plates, they may behave like load sources consuming energy during their charging cycle. There's a difference between a battery plates consuming energy than a load source consume energy. When a battery consumes energy, it's due to the electrolysis process of its plates. When a load source consumes energy, it's due to its functions.

On the other hand; a battery resistance to current, it'll reflect current away from it. But; a load source resistance to current, it'll be based on the number of turns that the wiring has inside of it to conduct current. Therefore; a parallel circuit connection made between a battery and a load source, it doesn't work the same as a parallel circuit connection made between two load sources.

We've to understand what mechanics, are in play when it comes to charging the battery while discharging it to the load source. A host of things will come into play, other than the mechanics of the current alone or the laws of physics as well. Wherein; these hosts of other things, they'll prevent the battery from being charged while being discharged to the load source as well.

If a battery has only one opening for voltage and current to enter or to exit it, then it and a load source will be in parallel with one another in the relationship with a charging source as well. Then we know that, we've to change the relationship between them knowing voltage will flow through a parallel circuit connection different than current will. See chapter 12 as well.

If a battery and a load source, they're in parallel with one another in their relationship with a charging source. Then the wrong mechanics are in play to charge the battery while it has the load source on it. Unless; the charging source, it'll create a surplus of electrons at the load source in order to reverse the current flow back into the battery to charge it.

Wherein; the parallel circuit connection made between the battery and the load source, it'll allow the battery to be charged while it has the load source on it; although, it has only one opening for electrons to enter or to exit it. In which; it'll be the same reason, it can't be charged while being discharged to the load source because the same mechanics are in play as well.

If we're going to focus on charging a battery while discharging it to a load source, then we've to ask ourselves. Are we focusing on the right mechanics to carry out such a process? Since we could charge or discharge the same battery; but, we've to charge it on the same terminal connection that it was discharged on.

Therefore; we've to change how, we carry out the cycling processes of a battery since we've to stop one of its cycling processes in order to start the other. Given that; energy has to go in it the same way that, energy has to come out as well. Thus; we've to implement or put into play, the right mechanics to charge it while discharging it to a load source as well.

While understanding that, voltage and current will flow from a strap <u>to</u> a battery plate different than they'll flow <u>from</u> its plate to a strap connecting them together in parallel. Wherein; voltage and current, they'll flow from a strap <u>to</u> a battery plate in reverse and independent of how they'll flow <u>from</u> its plate to a strap connecting them together in parallel.

It dawn on me if the battery had separate terminals for charging and discharging, like a capacitor. And those terminals, they had their own straps and tab connectors leading to the same plates. Then we could implement or put into play the right mechanics to charge the battery while discharging it to a load source, just like we do with our capacitors as well.

I found voltage and current will flow <u>to</u> a strap connecting, the plates and the charging terminal together in parallel. It'll be in reverse and independent of voltage and current flowing <u>from</u> those plates to a strap connecting them together in parallel with the discharging terminal as well. Then energy, it doesn't have to go in the same way it has to come out as well.

With the right mechanics in play, we could get around the problems that the mechanics of the current poses when there's a charge and a load source on a battery at the same time. The idea of a battery having separate terminals for charging and discharging, it came from a run capacitor because I understood it the most; although, it's other types of capacitors out there as well.

Focusing on the mechanical structure of a run capacitor, it shed light on why. It'll be possible to charge a battery while discharging it to a load source without being impeded by the mechanics of the current. In the next chapter; let's see why, it'll be possible to get around the problems current poses when there's a charge a load source on a battery at the same time.

CHAPTER 16

A RUN CAPACITOR

IN THE LATTER PART of 1745, the capacitor or the Leyden jar was competitively invented by Ewald Jurgen Von Kleist and Pieter van Musschenbroek. But; in the early part of 1746, Musschenbroek product the first working example of a capacitor based on the WIKIKPEDIA. The capacitor is an innovated device and its modern day design will give us insight into charging a battery while discharging it as well.

I found the interior design of a lead acid automotive starting battery. It'll allow its plates in one cell to be simultaneously charged and discharged to its plates in the next cell. Without being impeded by the mechanics of the current. It has never been about the mechanics of the current preventing us from charging a battery while discharging it to a load source.

However; it always has been about, the laws of motion and the effect of forces on bodies preventing us from charging a battery while discharging it to a load source. What I discovered is that, the design concept of a lead acid automotive starting battery is the reason it can't be charged while being discharged to a load source. See chapter 3.

Given that; the battery, it has to be charged on the same terminal connection it was discharged on. In which; its design concept, it'll create an insufficient cycling process for it since we've to stop its discharging cycle in order to start its charging cycle. In which; it's due to its exterior design since electrons, they've to go in it the same way they've to come out as well.

We either don't understand or we don't care that, the design concept of the battery poses a problem when it comes to its cycling processes. Wherein; it seems like, we haven't taking an in depth analysis of its mechanical structure to see what affects. It'll have on its cycling processes since we've to charge it on the same terminal that, it was discharged on.

It raises a series of questions, is it due to the mechanics of the current or is it due to the mechanical structure of the battery why it can't be charged while being discharged? For instance; induction motors, they operate very efficiently when they're powered by direct current; however, they've an efficiency problem when they're powered by alternating current.

Since most homes and businesses, they use alternating current to power their electrical devices. It'll be a very small market for induction motors if the companies only sold them to businesses using direct current to power their operations. Those companies who manufacture induction motors, they had to figure out how to make their product work for everyone.

If the manufactures going to be successful, they had to figure out why their induction motors had efficiency problems when they're powered by alternating current. Subsequently; they found out that, the motors efficiency problems were due to the current alternating within the circuit. They couldn't stop the current from alternating within the circuit because of its source.

However; those companies who manufacture induction motors, they didn't abandon the idea of the motors. But; they figured out, how to make the alternating current work in their favor by using run capacitors within the circuit along with the motors; so, they could operate efficiently like they do in a Dc circuit. See figure 2B.

Figure 2B

(Pictorial diagram of a run capacitor)

A run capacitor, it has multi-terminal connections and they've their own plates; wherein, they're separated by a non-conductive material called dielectric. As a result; a run capacitor, it could be charged or discharged on either terminal connection back into an Ac circuit. Without having to, stop its discharging cycle in order to start its charging cycle.

Therefore; I found that, a run capacitor is <u>simultaneously</u> charged and discharged within an Ac circuit because of the behavior of the current within the circuit. So; adding, a run capacitor within an Ac circuit along with an induction motor. It'll prevent the current from alternating within the circuit; thus, the current will no longer affect the efficiency of the motor.

In short; each time, current alternates within a circuit. It changes the direction that the electrons, are flowing within the circuit. In which; this action, it affects the performance of the induction motor. So; having, a run capacitor within the circuit along with the motor. It'll allow electrons to flow in one direction within the circuit between their source and the motor.

For instance; when the current alternates within the circuit, it'll simultaneously charge and discharge a run capacitor within the circuit. Thus; filling, the void or gap created by the alternating current with additional electrons. Therefore; the capacitor, it'll allow a constant flow of electrons to flow in one direction within the circuit without alternating their direction.

As a result; the induction motor, it'll operate efficiently within an Ac circuit like it does in a Dc circuit as well. Once the engineers figured out the alternating current, it caused the efficiency problem for the induction motor. They took advantage of the alternating current allowing it to work in their favor after identifying the efficiency problem with the induction motor.

The actions of the engineers made it possible for everyone to use induction motors in their homes and businesses. My point is that, they didn't abandon the idea of induction motor. However; to a certain degree, we've abandon the idea of the electric vehicle becoming our primary source of transportation for more than a century as well.

Given that; we've efficiency and capacity problems with our secondary cell batteries because as more and more electrons, are released from them. Then less and less electrons could be released from them; wherein, it's due to their ever increasing internal resistance because we're taking electrons from them as well.

It doesn't make any sense to put our hope in finding a perfect chemical composition to a longer-lasting battery before the electric vehicle could become our primary source of transportation. Given that; it has been more than a century since the first electric vehicle was made and yet, we haven't found a perfect chemical composition to a longer-lasting battery for it.

Although; we could, replenish our batteries chemical energy with electrical energy. And yet, we still search for a perfect chemical composition for them as well. Wherein; it seems like, all we've to do is figure out how to charge them while using them to power our vehicles. Since we could, replenish our batteries chemical energy with electrical energy as well.

Focusing on the mechanics behind a run capacitor, it helped me figure out how to replenish our batteries' chemical energy with electrical energy while using them. It has little to do with the laws of physics; but, more to do with the laws of mechanics when it comes to charging our batteries while discharging them to a load source as well.

Here's why; the mechanical structure of a run capacitor, it gives us insight into why it'll not be necessary to stop the discharging cycles of our batteries in order to start their charging cycles. Since the run capacitor doesn't have to be charged on the same terminal connection that, it was discharged on; unlike our secondary cell batteries.

However; the general consensus is that, it's no other way to carry out the cycling processes of a secondary cell battery; but, to stop its discharging cycle in order to start its charging cycle because of the laws of physics. The design concept of a run capacitor proves otherwise if the right mechanics are in play to charge the battery while discharging it.

If we take an in depth analysis of the mechanics behind, a lead acid automotive starting battery's charging cycle. It'll prove it has little to do with the mechanics of the current alone or the laws of physics; but, it has more to do with the exterior design of the battery why we've to stop its discharging cycle in order to start its charging cycle.

Let's forget about the laws of physics for a moment and focus on the mechanics involved in carrying out the charging cycle of the battery. It'll help debunk the theory why we've to stop its discharging cycle in order to start its charging cycle because of the mechanics of the current alone or the laws of physics as well.

Remember the battery plates in each cell are connected in series-parallel with one another by their straps, tab connectors and the electrolyte solution in each cell. Thus; electrons, they could flow from a strap to a tab connector on one plate. And then, flow across that plate through the electrolyte solution to an adjacent plate without flowing to and from the tab connector.

Therefore; one plate, it's simultaneously charged and discharged to an adjacent plate by way of the electrolyte solution in each cell. All because electrons, they could flow in one direction from one plate to the next; although, they've one tab connector each. Since the plates in one cell, they're connected in series with the plates in the next cell by way of an electrical bus.

Then the plates in one cell, they're simultaneously charged and discharged to the plates in the next cell without having to stop charging them in one cell in order to charge them in the next cell; although, they're separated by a non-conducting material called plastic as well. Given that; electrons, they could flow in one direction from the plates in one cell to the plates in the next cell.

The theory we've to stop the discharging cycle of a lead acid automotive starting battery in order to start its charging cycle because of the mechanics of the current or the laws of physics. It'll be debunked if electrons don't have to go in the battery the same they've to come out in order to charge its plates or for them to be discharged to a load source as well.

Then we would need to stop the discharging cycle of the battery in order to start its charging cycle because its interior structure proves this. Since we don't have to stop charging its plates in one cell in order to charge them in the next; although, they're separated by a non-conducting material; so, theory is based on energy has to go in the same way it has to come out.

The mechanics behind a run capacitor, they verify this because they allow it to be charged while being discharged to a load source since energy don't have to go in it the same way energy has to come out. Given that; a secondary cell battery, it's something similar to a capacitor because you've to add energy to it before you could take energy out as well.

The exterior mechanical structure of a capacitor, it proves it's not necessary to stop the discharging cycle of the battery in order to start its charging cycle due to the mechanics of the current or the laws of physics. If energy, it doesn't have to go in the battery the same way it has to come out in order to charge its plates or for them to be discharged to a load source as well.

The idea that it'll defy the laws of physics to simultaneously charge and discharge a lead acid automotive starting battery sounds ridiculous. It's very much within the realm of physics; however, our perceptions won't allow us to consider a simultaneous cycling process of it; although, it goes through a simultaneous cycling process each time it's charged as well.

I found the only thing will prevent the battery from being simultaneously charged and discharged to a load source, like a capacitor. It's the exterior design of the battery because energy has to go in it the same way energy has to come out as well. Remember earlier we talked about why we couldn't charge a cell phone's battery while using it to power the phone.

Given that; it would be difficult due to the parallel connection made between the phone's voltage and current regulatory systems and its battery in their relationship with its charging apparatus; wherein, it wasn't entirely due to the parallel connection made between them. It was mainly due to its battery having only one opening for current to enter or to exit it.

Let me expand on this for a moment because a host of things will come into play to prevent a battery from being charged while being discharged, other than the mechanics of the current alone. Have you ever noticed when you're playing a certain app on your cell phone or tablet, the charging apparatus was able to charge the battery while the app was running.

I found while running of the app, it wasn't consuming electrons faster than the charging apparatus could add electrons back to the battery; wherein, I realized that. There're other reasons why it'll be difficult to charge a cell phone or a tablet battery while using it. Sometimes, it's about the speed at which electrons will be added or subtracted from the battery as well.

Although; a parallel connection made between, a cell phone voltage and current regulatory system and its battery. It'll allow the resistance at its voltage and current regulatory system to dictate the amount of electrons flowing into its battery to charge it. However; it depends on, the amount of electrons that the cell phone will consume at a given rate as well.

In which; the given rate, it'll determine how difficult it'll be to charge the battery while using it to power the cell phone. In short; running, certain apps on a cell phone or a tablet may cause it to consume electrons faster than the charging apparatus could create a surplus of electrons at the voltage and current regulatory system to reverse the current flow back into the battery to charge it.

Thus; a host of things could come into play to prevent a battery from being charged while being used, other than the mechanics of the current alone. I find the primary reason is that, a battery has only one opening for electrons to enter or to exit it in order to charge its plates or from them to be discharged to a load source as well.

The design concept of the battery, it'll create a scenario where a charge and a load source will be on the same terminal connection of it. In which; it'll create, a parallel circuit connection between them; thus, the laws of parallel circuitry will come into play to determine how voltage and current will flow between them as well.

This is why we need to know, voltage will flow through a parallel circuit connection different than current will. Although; voltage, it'll be the same across a battery and a load source; however, current will not because of the parallel circuit connection made between them; given that, it'll allow the mechanics of the current to dictate which path they'll follow.

Whichever path has the least resistance for the current to follow, it'll follow that path. If the resistance at the battery is greater than the resistance at the load source, then current will flow to the load source and not into the battery at all since its resistance to current. It'll act like a non-conducting material to reflect current away from it.

The load source resistance to current will not reflect current away from it; but, current will flow to it based on the number of turns that the wiring has inside of the load source to conduct current. A battery resistance to current is different than a load source resistance to current; so, it doesn't have to be the same across them as voltage will since it reacts differently to resistance.

If a battery has only one opening for current to enter or to exit and it'll be in parallel with a load source as well. These host of things coming together, they'll prevent the battery from being charged while it has the load source on it. Unless; a charging source, it'll create a surplus of electrons at the load source to reverse current back into the battery to charge it.

The parallel connection that exist between a battery and a load source in their relationship with a charging source, it'll assist certain laws of physics; such as, like charges will repel one another and follow the path of least resistance as well. These certain aspects of the laws of physics, they'll prevent a battery from being charged while it has a load source on it as well.

I found the laws of mechanics will prevent a battery from being charged while it has a load source on it as well. If electrons, they've to go in it the same way they've to come out in order to carry out its cycling processes as well. So; both, the laws of physics and the laws of mechanics will prevent a battery from being charged while being discharged as well.

If we go back to, where I was talking about it was possible to charge a cell phone or a tablet battery while playing an app. It seems like there're other ways to charge a battery while using it without a charging source creating a surplus of electrons at a voltage and current regulatory system to reverse the current flow back into a battery to charge it as well.

Providing a cell phone or a tablet, they didn't consume electrons faster than electrons could break down the internal resistance of a battery to charge it. I found those unusual events taking place with a cell phone or a tablet. Where electrons flowed into the battery and to the voltage and current regulatory systems at the same time while the apps, were running as well.

Those unusual events, they occurred because of the rate at which the apps consumed electrons; wherein, it's like the scenario with the two branches in a river. However; those unusual events with a cell phone or a tablet battery, they give insight into why it could be charged while it has a load source on it if it was a charge on one terminal and a load source on anther.

This revelation, it brought a capacitor to mind. If a surplus of electrons could be created within a battery instead at a load source, then we could create the same effect with a battery as we do with a capacitor as well. For instance; since a surplus of electrons, they'll be created within a capacitor before they're discharged back into a circuit as well.

We could use this technique to charge a battery while discharging it to a load source if we allow the charging electrons to flow into the battery before flowing to the load source. Then the charging electrons will be able to flow into the battery and to the load source at the same time. Like the scenario where water will flow through both branches of a river at the same time. See chapter 12.

Wherein; the charging electrons, they'll have two paths to follow. One path toward the other cells in the battery and the other toward the load source, it'll be similar to the two branches in a river because current will flow to the battery based on its resistance and to the load source based on its resistance; in which, it'll be similar to a run capacitor as well.

Since the surplus of electrons, they'll be created within the capacitor before they're discharged back into a circuit. Therefore; if the charging electrons, they could flow into a battery before flowing to a load source. Then the surplus of electrons would be created within the battery first in order to be discharged back toward the load source as well.

The design concept of a run capacitor reveals if a battery had separate terminals for charging and discharging and those terminals had, their own straps and tab connectors leading to the same plates. Then we could get around the mechanics of the current that poses a problem when a charge and a load source will be on the battery at the same time as well.

For instance; a cell phone's battery, it could have a charging apparatus on one terminal and a voltage and current regulatory system on another. Then the battery will not be charged on the same terminal connection that, it was discharged on. I'm not talking about a strap that leads to a battery plates; but, I'm talking about a terminal connection.

In which; a charging apparatus or a voltage and current regulatory system will be connected to. See figure B-2 in chapter 15. With that in mind; although, a cell phone battery will work a little different than a run capacitor will. Given that; the battery, it can't be simultaneously charged and discharged to a voltage and current regulatory system.

Since the battery manufactures, they haven't taken the extra step needed to simultaneously charged and discharged the cell phone battery to a voltage and current regulatory system. It seems like the battery manufactures intentions wasn't to charge the battery while discharging it to a voltage and current regulatory system as well.

The battery manufactures, they only wanted to design the battery; so, the cell phone could receive calls during its charging mode and have a thermistor for overcharge protection for its battery as well. If the battery manufactures had taken, the extra step by allowing the battery multi-terminals to have their own straps and tab connectors leading to their respective plates.

Then the battery, it could be charged while being discharged to a voltage and current regulatory system. Although; it seems like, electrons could enter one terminal flowing to the battery plates and exit another terminal flowing to a voltage and current regulatory system; but, it's not the case. Let's not be fooled by the multiple terminal connections of the battery.

Given one set of terminals share the same strap leading to their respective plates, which it allows electrons to flow to the battery plates and to the voltage and current regulatory system; but, the other set of terminals. They share the same strap leading to their respective plates, which it allows electrons to flow from the battery plates to a thermistor for overcharge protection.

In short; the cell phone battery, it has two positive terminals and two negative terminals. The two positive terminals share the same strap leading to their respective plates; likewise, the two negative terminals share the same strap leading to their respective plates. Only one positive and one negative terminal will be used to add or subtract electrons from the battery.

Likewise; only, one positive and one negative terminal will be used for electrons to flow from the battery to a thermistor for overcharge protection for it. Remember I'm not talking about a strap that leads to the battery plates; but, I'm talking about the terminal connections that a voltage and current regulatory system or a thermistor will be connected to.

Nevertheless; the battery, it'll have two sets of terminals for current to enter or to exit it at the same time. However; current, it can't flow to the battery plates from a terminal; that is, joined to its respective terminal by a strap. If a charge is on, one and a load source is on the other since one terminal isn't connected to its respective plates by a tab connector as the other.

Thus; current, it could only flow back and forth across the strap that joins the two respective terminals together. So; current, it'll flow through one respective terminal and out the other without coming in contact with their respective plates. Since the charging current, it'll be reflected toward one of the respective terminal if there's a charge on one and a load source on the other.

All because the respective terminals, they're interconnected by a strap and not by their respective plates. This is why a cell phone battery or an automotive starting battery with multi-terminal connections, they can't be charged while being discharged because the wrong mechanics is in play to charge them while discharging them as well.

Unlike; a cell phone battery with its multi-terminal connections or an automotive starting battery with its top and side posts terminal connections, a run capacitor's multi-terminal connections. They have their own straps and tab connectors leading to their respective plates; wherein, it makes it possible to charge the capacitor while discharging it as well.

This is where, the concept of a run capacitor gives us insight into charging a battery while discharging it; although, it doesn't need separate plates for charging while discharging like a run capacitor does. However; the battery, it'll need separate terminals for charging while discharging like a run capacitor does since we'll be using direct current and not alternating current.

Thus; the charge and the discharging terminal of the battery, they could use the same plates to charge it while discharging it as well. If we focus on certain aspects of the mechanical structure of a run capacitor, we could eliminate some of the problems that the mechanics of the current poses when there's a charge and a load source on a battery at the same time as well.

If we've a charge on one terminal and a load source on another terminal, then current could flow to the battery on one terminal from the charging source. And then, current could flow from the battery to the load source on another terminal connection at the same time. Carrying out my research on a run capacitor, I found it's all about mechanics and not physics.

When it comes to charging a battery while discharging it to a load source; given that, the mechanics of the current alone or the laws of physics will not prevent the battery from being charged while being discharged the load source; but, the design of the battery will. We could use some of the mechanics that is applied to our capacitors and apply them to our batteries as well.

Then we could simultaneously charge and discharge our batteries to a load source, like we do our capacitors as well. I know it sounds crazy to some folks when I say it's possible; but, it's only impossible. If we don't change, the contemporary design of our secondary cell batteries; so, energy doesn't have to go in them the same way it has to come out as well.

If we focus on, the design concept of a run capacitor on the outside of it; wherein, it has multi-terminal connections; but, on the inside its multi-terminals have their own straps and tab connectors leading to their respective plates and they're separated by a non-conductive material called dielectric; thus; its design concept allow it to be simultaneously charged and discharged.

Think about how our iPods, laptops and computers because their functions is based on their mechanical, chemical and electrical aspects coming together to carry out their functions. It's all about manipulating the mechanics of the current with their mechanical structures. Thus; our secondary cell batteries, they're no different when it comes to carrying out their functions as well.

If we want to charge our batteries while discharging them, then all we've to do is manipulate the mechanics of the current with the design of our batteries to achieve this goal as well. While focusing on the innovated design of a run capacitor, it brought my attention to the design concept of a power transformer because it's simultaneously charged and discharged as well.

The innovated design of the power transformer will give insight into why it'll be possible to simultaneously charge and discharge a battery. In the next chapter; let's explore, the design concept of the power transformer to see why it'll be possible to simultaneously charge and discharge a battery if it has separate terminals for charging while discharging, like a capacitor does.

CHAPTER 17

A POWER TRANSFORMER

An English scientist named; Michael Faraday, he made the first transformer in 1831. A power transformer is an electrical device that produces electrical current in a second circuit through by electromagnetic induction. Thus; the power transformer, it has a primary and a secondary winding; that is, interconnected by an iron core as well.

The iron core is wrapped with copper wiring to create a magnetic field when it's energized by the primary winding. In which; the iron core, it'll transfer energy from the primary winding to the secondary winding; so, it could power a load source; such as, an electric motor, a fan or a safety switch by way of an electrical bus.

A power transformer is another innovated device that gives insight into why it'll be possible to simultaneously charge and discharge a secondary cell battery to a load source. If the battery has separate terminals for charging and discharging, like a run capacitor does and those terminals have their own straps and tab connectors leading to the same plates as well.

What I learned in college while earning, my Associate Degree in Science in the field of heating, cooling and refrigeration. A power transformer is used to help operate our electrical devices such as; washers, dryers, refrigerators, furnaces and so forth. The power transformer is the one that caught my attention because energy could enter and exit it at the same time.

In figure B-4, it shows the type of power transformer; that is, used in our electrical devices; such as, washers, dryers, refrigerators and furnaces to increase or to decrease line voltage for a particular application for the devices.

Figure B-4

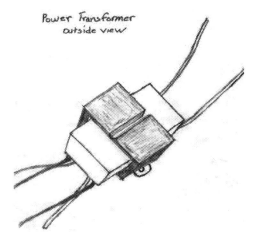

(Pictorial diagram of a Power transformer)

What a power transformer reveals is that, it'll be possible to simultaneously charge and discharge a secondary cell battery if the right mechanics are in play. Then we don't have to charge it on the same terminal connection that, it was discharged on, similar to a run capacitor. The power transformer is the blue-print for charging a battery while discharging it to a load source.

Here's why; unlike, a run capacitor where energy has to flow to one plate while flowing from another in order to simultaneously charge and discharge it. Given that; its plates, they're separated by a non-conductive material called dielectric. A power transformer simultaneous cycling process is carried out a little different than a run capacitor simultaneous cycling process.

Given that; the power transformer, it has a primary and a secondary winding that is interconnected by an iron core. This is how, it's simultaneously charged and discharged to a load source. Unlike; a run capacitor using, separate plates to simultaneously charge and discharging it to a load source; so, the power transformer is the blue print to follow in this case.

For instance; the amount of current will flow to a load source that is connected to a secondary winding of a power transformer, it'll be based on the number of turns the wiring. The secondary winding has to conduct current, opposed to the number of turns the wiring the primary winding has to conduct current wrapped around the iron core of the power transformer.

The number of turns the wiring, the primary and the secondary winding has wrapped around the iron core of the power transformer. It'll tell us if we're stepping up or stepping down the current flow to the load source. In short; if we step-down the transformer, then more turns the wiring the primary has wrapped around the iron core to conduct current.

Compared to, less turns the wiring the secondary winding has wrapped around the iron core to conduct current of the power transformer. Thus; the voltage inside the secondary winding, it'll be half the voltage inside the primary winding because it'll be directly proportional to the number of turns, the wiring inside the secondary winding will have to conduct current.

If we step-up the power transformer, then the opposite will happen because the primary winding will have less turns of the wiring wrapped around the iron core to conduct current. Compared to, more turns the wiring the secondary winding will have wrapped around iron core to conduct current within the power transformer as well.

Also; I found a power transformer, it could be a load source and a power source at the same time. Since its primary winding, it could consume energy from a power source while its secondary winding is supplying energy to a load source at the same time. Thus; it makes, the power transformer a load source and a power source at the same time.

Since the secondary winding, its voltage could be increased or decreased to supply the right amount of current to the load source for a particular application. Then the amount of current flowing to the primary winding, it'll be different than the amount of current flowing to the load source from the secondary winding as well.

In which; it makes, the secondary winding the power source for the load source and not the primary winding; but, the primary winding will be the power source for the secondary winding. In which, it's a load source consuming, energy from the iron core to supply energy to the load source; that is, connected to the secondary winding by way of an electrical bus.

Therefore; energy, it has to pass through one load source in order to flow to another; but, powering both load sources at the same time. In which; it's similar to, energy flowing through a lead acid automotive starting battery. So; all of its plates in each cell, they could be charged at once during their charging cycle; although, they're separated by plastic as well.

The similarity between the battery and the power transformer during their energy transfer process. It'll give insight into why the power transformer is the blue-print for charging the battery while discharging it to a load source; wherein, a run capacitor. It was one step in understanding how to charge the battery while discharging it by changing its exterior design.

The power transformer is going to show why the battery having separate terminals for charging while discharging in the same cell and have their own straps and tab connectors leading to the same plates in that cell. It'll be possible to charging the battery while discharging it because voltage and current will behave differently under different circumstances as well.

Here's why; the mechanics behind a power transformer energy transfer process, it reveals line voltage will return back to its source from the primary winding as well as from a load source; that is, connected to the secondary winding by way of an electrical bus as well. In which; this is a clue, why the battery will need separate terminals for charging while discharging as well.

Given that, the charging voltage, it must be able to return back to its source from the battery as well as from a load source; that is, on the battery at same time as well. A run capacitor and a power transformer, they'll help us understand how to design our batteries; so, they could be charged while being used to power our load source as well.

Since a power transformer, it's nothing more than two load sources connected in parallel with one another in some form or fashion. Wherein; its primary and secondary winding, they're interconnected by an iron core to manage how energy will flow to a third load source; that is, connected to the secondary winding by way of an electrical bus as well.

The power transformer energy transfer process is similar to the lead acid automotive starting battery energy transfer process as well. Given that; its plates in each cell, they're connected in series-parallel with one another. And each cell, it's connected in series with its terminal connections to manage how energy will flow to and from it as well.

Thus; how, the battery is constructed allows its plates in each cell to be simultaneously charged and discharged amongst themselves and to its plates in the next cell as well. So; all of its plates in each cell, they could be charged at once during their charging cycle without having to stop charging them in one cell in order to charge them in the next cell as well.

The internal energy transfer process of the battery is similar to the internal energy transfer process of the power transformer. Unlike; the external energy transfer process of the power transformer, the external energy transfer process of the battery will not allow energy to enter it from a power source and exit it to a load source at the same time.

However; the similarities between the battery and the power transformer during their energy transfer process, it suggests it'll be possible to charge the battery while using it to power a load source as well. The similarities between the internal energy transfer process of the battery and the power transformer brought my attention back to the run capacitor as well.

Given that; if the battery, it had separate terminals for charging and discharging like the run capacitor. Then the battery, it could consume energy while supplying it to a load source like the power transformer as well. Bases on, the mechanics behind the battery and the power transformer internal energy transfer processes as well.

The battery having separate terminals for charging and discharging, then its exterior design will allow energy to enter it from a charging source and energy to exit it to a load source at the same time. Wherein; it'll be similar to, the exterior design of the power transformer allowing energy to enter and exit it at the same time as well.

We know a lead acid automotive starting battery plates in one cell, they're simultaneously charged and discharged to its plates in the next cell. The million dollar question is that, can we simultaneously charged and discharged them to a load source as well? I say yes; but, some in the field of automotive battery technology says no.

They're skeptical of charging the battery while discharging it to a load source because they think it'll be inconceivable or farfetched due to the laws of electrochemistry, entropy, inertia and the laws of conservation of energy as well. During my research, I found those laws of physics will not be an issue when it comes to charging it while discharging it to a load source as well.

Here's why; based on, the similarities between the internal energy transfer process of the battery and the power transformer. It seems like those skeptics, they're focusing on the wrong science because they haven't compared the internal energy transfer process of the battery with the internal energy transfer process of the power transformer as well.

When the battery plates in one cell, they're simultaneously charged and discharged to its plates in the next cell. The laws of electrochemistry, entropy, inertia and the laws of conservation of energy, they're not an issue. Although; its plates in each cell, they're separated by a non-conducting material and energy will be lost to heat during the process as well.

Likewise; the iron core inside the power transformer, it'll be simultaneously charged and discharged to the secondary winding in order to power a load source; that is, connected to it by an electrical bus. The laws of entropy, inertia and the laws of conservation of energy will not be an issue; although, energy will be lost to heat during the process as well.

It seems ridiculous to assume it'll be inconceivable or farfetched to simultaneously charge and discharge the battery plates to a load source. Given that; they're simultaneously charged and discharged amongst themselves in each cell and to the plates in the next cell. Without the laws of entropy, inertia or the laws of conservation of energy being an issue as well.

What's ironic, we're overlooking the battery has to be charged on the same terminal connection it was discharged on; thus, having to stop its discharging cycle in order to start its charging cycle. However; we don't have to, stop charging its plates in one cell in order to charge them in the next cell; although, they're separated by a non-conducting material called plastic as well.

From a mechanical respective, it seems plausible to charge the battery plates while discharging them to a load source. Without having to stop, the discharging cycle of the battery in order to start its charging cycle if we eliminate the process of having to charge it on the same terminal connection that it was discharged on as well.

Since the battery plates in one cell, they're simultaneously charged and discharged to its plates in the next cell by way of an electrical bus. In which; it's similar to, an iron core being simultaneously charged and discharged to a secondary winding inside a power transformer to power a load source. That is, connected to the secondary winding by way of an electrical bus as well.

If we understand, the mechanics behind the energy transfer process of a run capacitor because it has separate terminals for charging and discharging. And understand, the energy transfer process of a power transformer because its primary and secondary winding, they're interconnected by way of its iron core to transfer energy between them.

Then it wouldn't seem so inconceivable or farfetched to charge the battery while discharging it to a load source. If the battery had separate terminals for charging while discharging, like a run capacitor. And those terminals, they were interconnected by the same plates. Like a primary and secondary winding, are interconnected by an iron core inside a power transformer.

If we look at the mechanics behind a run capacitor and a power transformer, they reveal charging the battery while discharging it to a load source. It's nothing more than an extension of the battery charging cycle. If it had separate terminals for charging while discharging in the same cell and those terminals, they had their own strap and tab connectors leading to the same plates as well.

This is why I'm talking about the power transformer in this chapter because it's all about mechanics or procedures in how, we go about adding and subtracting electrons from the battery. It's not about laws of physics or the mechanics of the current alone as some people want us to believe when it comes to charging it while discharging it a load source as well.

If we're focusing on the wrong science or have the wrong mechanics in place, then it'll seem inconceivable or farfetched to charge a secondary cell battery while discharging it to a load source. This is how, all the erroneous beliefs comes into existence about what we could or couldn't do with our secondary cell batteries because of the laws of physics as well.

However; if we observe, the energy transfer process of a run capacitor or a power transformer. Then it'll clear up some of those erroneous beliefs about what we could or couldn't do with our secondary cell batteries because of the laws of physics. We'll begin to see the laws of physics will have little to do with, how we go about adding and subtracting energy from a battery.

What I mean by little is that, like charges will repel one another and follow the path of least resistance as well. Other than that, the laws of physics will have little to do with how electrons will be added or subtracted from a battery plates. Here's why; based on, the conventional current flow concept when charging a lead acid automotive starting battery.

The charging current has to flow through the positive terminal of the battery. And then, flow across the strap and tab connectors connecting the positive plates together in parallel with the terminal connection. Thus; current, it'll flow across the positive plates and then it'll be transferred to the adjacent negative plates by way of the electrolyte solution in that cell as well.

Wherein; current, it'll flow from the negative plates in the cell with the positive terminal in it to the positive plates in the next cell by way of an electrical bus. If we're familiar with the energy transfer process of a power transformer, then it'll seem similar to its energy transfer process. Because of how, energy is transferred from its primary to its secondary winding by way of its iron core as well.

Understanding the mechanics of the voltage and current, then we'll realize their mechanics alone will have little to do with the energy input and output process of our batteries or our power transformers as well. Wherein; it's how, we go about adding and subtracting energy from them will determine their energy input and output process.

This is why, we must take an in depth analysis of the mechanics behind the charging cycle of the lead acid automotive starting battery. We'll find it goes through a simultaneous cycling process each time, it's charged. Thus; we could use, the design concept of a run capacitor and a power transformer to understand what's required to charge the battery while discharging it.

The design concept of a run capacitor shows if the battery has separate terminals for charging and discharging. And those terminals, they'll have their own straps and tab connectors leading to the same plates. Then its energy input and output process, it'll be similar to the run capacitor because it could be charged on one terminal and discharged on another as well.

Likewise; the design concept of a power transformer, it shows if the battery charge and discharging terminal. They were interconnected by the same plates as a primary and a secondary winding, are interconnected by an iron core inside the power transformer. Then those terminals, they'll determine how energy will flow to and from the battery as well.

I discovered the connections made between a lead acid automotive starting battery plates and cells in their relationship with its terminal connections. The connections will determine how voltage and current will enter or exit it as well as they determine how voltage and current will enter or exit a run capacitor or a power transformer as well.

On the other hand; based on, the design of a battery, voltage will enter or exit it differently than current will. However; based on, which cycling is carried out. It'll determine how voltage and current will flow through the battery as well. This is why it's important to, separate voltage from current because they behave differently under different circumstances.

Although; current, it's a combination of voltage and electrons flowing in one direction; but, current will act different than voltage will because of the electrons within the current; given that, they'll repel one another and follow the path of least resistance as well. However; voltage, it doesn't repel itself and it doesn't follow the path of least resistance as well.

Wherein; voltage, it's nothing more than an electromotive force use to move electrons. If we don't understand the differences between voltage and current, then we can't begin to comprehend. Why we can't charge a battery while discharging it to a load source, other than blaming it on the mechanics of the current alone or the laws of physics as well.

Likewise; if we don't understand, voltage will flow through a circuit connection different than current will. Then it'll seem inconceivable or farfetched to charge a battery while discharging it to a load source. If we don't understand, the connections made between them will determine how voltage and current will flow between them as well.

For more than a century, we've become so accustom to stopping the discharging cycles of our batteries in order to start their charging cycles. Given that; we think, it's no other way to carry out their cycling processes; but, to stop one cycling process in order to start the other because of the laws of physics; however, this assumption doesn't hold true based on my research.

Remember a lead acid automotive starting battery plates in each cell, they're connected in series-parallel with one another by their straps, tab connectors and the electrolyte solution in each cell. Thus; electrons, they could flow from a strap to a tab connector on one plate. And then, flow across that plate through the electrolyte solution to an adjacent plate as well.

I found electrons don't have to flow to and from the same tab connector on the same plates in order to flow to another plate; although, each plate has only one tab connector each. Also; I found the battery plates in each cell, they're simultaneously charged and discharged amongst themselves by way of the electrolyte solution in each cell as well.

Since the battery cells, they're connected in series with one another by way of an electrical bus. Then its plates in one cell, they're simultaneously charged and discharged to its plates in the next cell; so, they all could be charged at once during their charging cycle without having to stop charging them in one cell in order to charge them in the next cell as well.

Although; the battery plates in each cell, they're separated by a non-conducting material called plastic. They're still charged in one cell while they're charging in another. Without the laws of electrochemistry, entropy, inertia or the laws of conservation of energy being an issue; even though, energy will be lost to heat during the process as well.

It seems like we shouldn't be talking about the laws of physics when it comes to a conversation about charging our batteries while discharging them; given that, we've to stop their discharging cycles in order to start their charging cycles. Since we've to charge them on the same terminal connection that, they were discharged on as well.

Since we could charge and discharge the same battery; but, we can't charge it while discharging it because we've to charge it on the same terminal connection that, it was discharged on. Then we should be talking about, the laws of motion and the effect of forces bodies or the laws of mechanics when it comes to charging it while discharging it.

If energy could enter the battery on one terminal and exit it on another, then its mechanics will be similar to a run capacitor since we could charge it or discharge it. But; we don't have to charge it on the same terminal connection that, it was discharged on. I found if a battery had separate terminals for charging and discharging, like a run capacitor.

It'll separate the battery's discharging cycle from its charging cycle if its charge and discharging terminal had their own straps and tab connectors leading to the same plates. Those straps connecting the plates and the charge and the discharging terminal together in parallel, they'll act like a medium between a charge and a load source in their relationship with the battery.

Since a lead acid automotive starting battery and a power transformer, they use a medium of some kind to transfer electrons from one point to the next without the two points coming in contact with one another. The battery uses a process called ionization to carry out its functions; however, the power transformer uses a process called induction to carry out its functions.

Although; the power transformer and the lead acid automotive starting battery, they use two different methods to achieve their goal of moving electrons from one point to the next without the two points coming in contact with one another. The battery uses an electrolyte solution and the power transformer uses an iron core to transfer electrons between two points.

The electrolyte solution in the battery, it transfers electrons between the battery's positive and negative plates in each cell without them coming in contact with one another. The iron core inside the power transformer, it transfers electrons between the primary and secondary winding without them coming in contact with one another as well.

The electrolyte solution in the battery, it functions is similar to the iron core inside the power transformer. Thus; the relationship between the positive and negative plates in each cell in the battery, it's similar to the relationship between the primary and the secondary windings inside the power transformer as well.

No matter how we slice it, there're similarities between the energy transfer process of a lead acid automotive starting battery and a power transformer. If we understood those similarities, then it wouldn't seem so inconceivable or farfetched to charge the battery while using it to power a load source as well.

Here's why; if the battery had separate terminals for charging and discharging, then the load source would have access to the simultaneous cycling process. That is taking place within the battery during its charging cycle. As a result; its plates in one cell, they could consume and store energy while supplying it to the load source as well

Given that; charging the battery while discharging it to a load source, it'll be nothing more than an extension of the battery charging cycle. Since its plates in each cell, they're already simultaneously charged and discharged amongst themselves in each cell and to the plates in the next cell as well.

All because electrons, they could flow in one direction from the battery plates in one cell to its plates in the next cell. Therefore; the wrong mechanics, are in place to charge it while discharging it to a load source because electrons have to go in the same way they've to come out in order to charge its plates or for them to be discharged to the load source as well.

On the other hand; some skeptics think, it's due to either the laws of entropy, inertia or the laws of conservation of energy as well. Why the battery, it can't be charged while being discharged because energy is lost to heat in both conversion processes of the battery. If we attempt to carry them out at the same time, they'll cancel themselves; therefore, no gain.

The skeptics conclusion, aren't consistent with the mechanics behind the charging cycle of the battery. Given that; energy will be lost to heat during its charging cycle and yet, its plates in each cell will still be charged. If we focusing on the energy transfer process of a power transformer, energy will be lost to heat between its primary and secondary winding as well.

And yet, a load source that is connected to the secondary winding by way of an electrical bus will still be powered; although, energy will be lost to heat during the process. Wherein; any energy lost to heat during the process, it'll be made up by the line voltage. A similar process happens during the charging cycle of the battery; but, it'll be made up by the charging source.

Then the skeptics, they can't use the laws of entropy, inertia or the laws of conservation of energy to discredit a simultaneous cycling process of the battery if we're charging it while discharging it as well. Here's why; those laws of physics, they'll not an issue if we're adding energy back to the battery while taking it out as well.

If we're charging the battery while discharging it to a load source, the same principles that apply to the power transformer will apply to the battery as well. Thus; when the charging current, it'll flow into the cell with the charge and discharging terminal in it first before flowing to the load source and the other cells in the battery as well.

In which; current, it'll be deployed to the other cells in the battery and to the load source. Based on, how they're interconnected with the cell with the charge and discharging terminal in it. Wherein; the straps connecting, the plates together and the plates themselves will act like a medium between the charge and the load source in their relationship with the battery.

Therefore; the medium, it'll determine how current will be deployed to the other cells in the battery and to the load source when the charge source is on one terminal and the load source is on the other terminal. Then current will flow <u>to</u> the strap connecting the plates together in parallel with the charging terminal.

It'll be in reverse of current flowing <u>from</u> those plates to the strap connecting them together in parallel with the discharging terminal as well. Also; how, current will flow <u>to</u> the strap connecting the plates together in parallel with the charging terminal. It'll be independent of current flowing <u>from</u> those plates to the strap connecting them together in parallel with the discharging terminal as well.

I found it'll be similar to current flowing to and from an iron core inside a power transformer as well. The iron core is the medium between a primary and a secondary winding inside the power transformer. The amount of current will flow from the primary winding to the iron core. It'll be based on, the number of turns the wiring the primary winding has wrapped around the iron core to conduct current.

Likewise; the amount of current flowing from the iron core to the secondary winding will be based on, the number of turns the wiring the secondary winding has wrapped around the iron core to conduct current as well. In which; it'll determine, the amount of current flowing to a load source; that is, connected to the secondary winding by way of an electrical bus as well.

Understanding, the mechanics behind the energy transfer process of a power transformer. It'll enlighten us on why it'll be possible to charge the battery while discharging it if it had separate terminals for charging and discharging in the same cell. And those terminals, they had their own straps and tab connectors leading the same plates in that cell as well.

Then the similarities between the energy transfer process of the battery and the power transformer becomes manifested. Given that; the plates joining, the charge and the discharging terminal together in parallel. The charging source becomes the power for them; wherein, they become the power source for the other cells in the battery and for a load source as well.

Although; a power transformer; it's not a device that could store energy, like a lead acid automotive starting battery could. But; the power transformer, it could consume energy while supplying energy to a load source. Wherein; the lead acid automotive starting battery plates in one cell, they could consume and store energy while supplying it to the plates in the next cell as well.

With those similarities between the battery and the power transformer in their energy transfer process, it gives credence if the battery had separate terminals for charging and discharging. Then a load source will have access to the process; that is, taking place within the battery during its charging cycle; thus, charging it while discharging it to the load source as well.

Here's why; based on the conventional current flow concept, energy has to enter the positive terminal connection of the battery in order to charge it. Then the positive plates in the cell with the positive terminal connection in it. They'll transfer energy to the adjacent negative plates by way of the electrolyte solution in that cell.

It'll be similar to a primary winding transferring energy to a secondary winding by way of an iron core inside a power transformer. Wherein; the battery negative plates in the cell with the positive terminal connection in it, they'll transfer energy to the positive plates in the next cell by way of an electrical bus.

It'll be similar to a secondary winding inside a power transformer transferring energy to a load source; this is, connected it by way of an electrical bus as well. The similarities between the power transformer and the battery energy transfer process shows we're only restricted by their designs when it comes to their energy transfer process as well.

Here's the deal; we can't get hung up on, the meaning of induction or ionization because they're result of a medium of some kind being used to transfer electrons from one point to the next without the two points coming in contact with one another. Since the battery, it could store energy from a power source or supply energy to a load source as well.

However; the battery, it can't store energy from a power source while supplying it to a load source. Given that; energy, it has to go in the battery the same way it has to come out. This is why we've to focus on, the design concept of a run capacitor and a power transformer. To understand why, it'll be possible for a battery to store energy while supplying it to a load source as well.

In the next chapter; I'm going to share with you some important facts about, the similarities between the design concept of a lead acid automotive starting battery and a power transformer. So you could decide for yourself; whether, we're restricted by the laws of physics or by the design of the battery when it comes to its energy input and output process as well.

CHAPTER 18

RESTRICTED BY ITS DESIGN

THERE'RE OTHER SIMILARITIES BETWEEN a lead acid automotive starting battery and a power transformer that relates to their design concepts. Their similarities reveal the battery is only restricted by its design when it comes to its energy input and output as well. Wherein; it's proven by the mechanics, behind its charging cycle.

On the other hand; the charging cycle of the battery, it proves its exterior design is the reason why it can't be charged while being discharged to a load source as well. Here's the thing; the battery, it has only one opening for electrons to enter or to exit it; in which, it prevents it from being charged while being discharged as well.

Wherein; it's safe to assume that, the battery can't be charged while being discharged because of the laws of physics; especially, if like charges will repel one another and follow the path of least resistance as well. Then the mystery is solved why, we've to stop the discharging cycle of the battery in order to start its charging cycle because of the laws of physics as well.

If we're focusing on, the mechanical structure of the battery. Then we'll find its mechanical structure, it consists of individual parts like a power transformer as well. You would assume we could manipulate the design of the battery, so that, it could be charged while being used to a power load source like the power transformer as well.

Since a lead acid automotive starting battery and a power transformer, they consist of individual parts that are interconnected with one another in some form or fashion to make up the devices as we know of today. For instance; the power transformer, it consists of a primary and a secondary winding and they're interconnected by an iron core as well.

The number of turns in the wiring of the primary winding will be different than the number of turns in the wiring for the secondary winding. In which; the number of turns in the wiring for the secondary winding, it'll determine if current will be step-up or step-down to flow to the load source; that is, connected to it by an electrical bus.

On the other hand; a lead acid automotive starting battery, it consists of a collection of 2-volt batteries connected in series with one another and housed in a plastic case to make up the whole battery as we know of today. So; its voltage and current potential, it could be increase for a particular application when one battery isn't enough.

Wherein; those collections of batteries in a lead acid automotive starting battery, they're called cells by those in the field of automotive battery technology. In which; those cells, they're nothing more than batteries in and of themselves. Wherein; each cell in the battery, it has an equal number of negative and positive plates in it.

Those negative and positive plates, they're connected in series-parallel with one another by their straps, tab connectors and the electrolyte solution in each cell. Remember a single strap joins all the negative plates together in parallel. And a single strap, it joins all the positive plates together in parallel in each cell as well.

The electrolyte solution in each cell within the lead acid automotive starting battery, it joins the negative and the positive plates together in series. This is how the battery plates in each cell, they're constructed; in which, it makes each cell nothing more than a battery in and of itself. That is housed together in a plastic case to make up the whole battery as we know of today.

Thus; during the discharging cycle of the battery, its voltage potential is the sum of all of its plates in each cell and its current potential will only equal to the sum of its plates in one cell. However; during its charging cycle, voltage will be evenly divided amongst its cells and the charging current will be the same throughout in each cell in it.

So; how, the battery is constructed will determine how voltage and current will flow in and of it during its cycling processes. In which; its design, it's the deciding factor in determining how its cycling processes will be carried out as well. If we understanding, how the battery is constructed and how it'll play a role in its cycling processes as well.

Then we'll know that, we're not restricted by the laws of physics when it comes to the cycling processes of the battery. Given that; the positive and the negative plates in each cell, they're separated by a non-conducting porous material; so, they'll not touch one another; in which, it allows electrons to flow between the plates by way of the electrolyte solution in each cell.

Since the battery plates in each cell, they're connected in series-parallel with one another by their straps and tab connectors and the electrolyte solution in each cell. Then during its charging cycle, electrons could flow from a strap to a tab connector on one plate. And then, flow across that plate through the electrolyte solution to an adjacent plate.

Without electrons flowing to and from, the same tab connectors on the same plate. In which; electrons, they'll flow amongst the battery plates in each cell. And then, flow to the plates in the next cell by way of an electrical bus. A non-porous material called plastic, it'll separate the plates in each cell; in which, creating each cell in the battery as well.

Thus; the battery plates in each cell, they're simultaneously charged and discharged amongst themselves in each cell and to the plates in the next cell as well. So; all the plates in each cell, they could be charged at once during their charging cycle. It's no other way to explain this, other than the battery goes through a simultaneous cycling process when it's charged.

Therefore; we're not restricted by the laws of physics when it comes to simultaneously charging and discharging the battery to a load source as well. In which; we're only restricted by, the design of the battery when it comes to its cycling processes; in other words, we don't have to stop its discharging cycles in order to start its charging cycles as well.

If we're not sure about this, then all we've to do is take a lead acid automotive starting battery apart piece by piece and see why it functions the way that it does. See chapter 3. We don't have to conjure up all manner of laws of physics to see if they'll restrict its cycling processes because they're governing by the laws of motion and the effect of forces on bodies.

If we're focusing on the laws of physics, then we're focusing on the wrong science. See chapter 8. Remember a lead acid automotive starting battery as a whole is nothing more than a collection of 2-volt batteries. That is connected in series with one another and housed together in a plastic to make up the whole battery as we know of today.

When the battery is charged as a whole, we're simultaneously charging and discharging one battery. While simultaneously charging and discharging another battery because of how, they're interconnected with one another; wherein, it has little to do with the laws of physics; but, more to do with the laws of motion and the effect of forces on bodies.

We don't have to be a physicist, electrochemist or someone who works in the field of automotive battery technology to figure this out. If we take the battery apart piece by piece and observe its inner workings, then we'll know it has little to do with the laws of physics why its plates in one cell are simultaneously charged and discharged to its plates in the next cell as well.

If we think about, how the battery plates are simultaneously charged and discharged amongst themselves in each cell and to the plates in the next cell. It's similar to the energy transfer process between a primary winding, an iron core and a secondary winding inside a power transformer as well.

Once we focus on, the inner workings of a lead acid automotive starting battery and a power transformer during their energy transfer processes. Then we'll know that, we're not restricted by the laws of physics when it comes to their energy input and output processes; but, we only restricted by their designs when it comes to their energy input and output processes.

Once we see beyond, the exterior structure of a lead acid automotive starting battery and see its inner works. Then our abilities to charge it while discharging it to a load source, it'll be up to us. Based on, how we design the battery. We've to figure out what's needed to change its exterior structure; so, it'll not impede it from being charged while being discharged as well.

In figure B-5, it's an inside view of a power transformer showing if the battery had separate terminals for charging and discharging like a run capacitor. And those terminals, they had their own straps and tab connectors leading to the same plates. Like a primary and a secondary winding are interconnected by an iron core inside a power transformer.

We could charge the battery while using it to power a load source. If the battery has separate terminals for charging and discharging in the same cell and those terminals, they'll have their own straps and tab connectors leading to the same plates in that cell. Then energy could enter and exit the battery at the same time, like a run capacitor or a power transformer as well.

Not only could energy enter and exit the battery at the same time; but, it could store energy while it's supplying it to a load source. Since the battery plates in one cell, they could store energy while supplying it to the plates in the next cell; so, they could store energy as well; although, they're separated by a non-conducting material called plastic as well.

If we understand the mechanics behind the battery, then we'll know it's not out of the realm of physic for it to store energy while supplying it to a load source by way of electrical bus as well. We're only restricted by our imaginations in figuring out, how to design our batteries; so, they could be charged while being discharged to our load sources as well.

My imagination is that, a battery has to have separate terminals for charging and discharging like a capacitor; so, it could be charged while being discharged. Wherein; they must be in the same cell and interconnected by the same plates, like a primary and a secondary winding are interconnected by an iron core inside a power transformer as well.

Then I find it's not out of the realm of physic to simultaneously charge and discharge a battery to a load source. Given that; a capacitor and a power transformer, they're simultaneously charged and discharged to a load source; in which, it does violate the laws of physics since we've the right mechanics in play to carry out such a process as well.

Here's why; if a lead acid automotive starting battery has separate terminals for charging and discharging in the same cell. And those terminals, they'll have their own straps and tab connectors leading to the same plates in that cell. Not only does the relationship between the positive and the negative plates remain the same within the battery.

However; adding, an additional positive terminal to the battery in a cell with a terminal already in it and they'll be interconnected by the same plates with their own straps and tab connectors as well. Then the relationship between the two terminals become similar to the relationship between a primary and a secondary winding wrapped around an iron core inside a power transformer as well.

Adding, another terminal to the battery plates with a terminal already connected to them. They'll take on two roles; one role will be similar to a primary winding and the other role will be similar to an iron core inside a power transformer as well. See figure B-5, it'll show what I'm alluding to.

Figure B-5

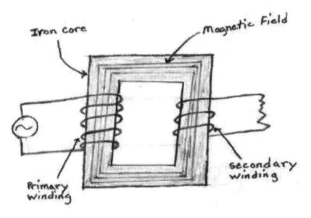

(Schematic diagram of an inside view of a power transformer)

In figure B-5; imagine that, the primary and the secondary winding inside a power transformer. They'll represent two positive terminals in the same cell within the battery and the positive plates connecting the two terminals together, they'll represent an iron core inside the power transformer as well.

Thus; energy, it could be transferred to and from the positive plates connecting the two positive terminals together in parallel by those two positive terminal connections in the same cell within the battery. It'll be similar to energy being transferred to and from an iron core by way of a primary and a secondary winding inside a power transformer as well.

The positive and the negative plates in the cell with the two positive terminal connections in it within the battery, they'll represent a primary and a secondary winding inside a power transformer as well. The electrolyte solution in the cell with the two positive terminal connections in it within the battery, it'll represent an iron core inside a power transformer as well.

Thus; energy, it could be transferred from the battery positive plates to its adjacent negative plates by way of the electrolyte solution in the cell with two positive terminal connections in it. It'll be similar to energy being transferred from a primary to a secondary winding by way of an iron core inside a power transformer as well.

So; energy, it'll be transferred from one positive terminal to the other by way of the same plates within the battery. Since energy, it has to flow directly from the charging source to those plates joining the two positive terminals together. It'll energize the electrolyte solution in that cell and then, transfer energy to the adjacent negative plates in that cell as well.

It'll be similar to the role of a primary winding energizing an iron core to transfer energy to the secondary winding inside a power transformer as well. Since the battery positive plates in the cell with the two positive terminals in it, they'll be taking on two roles. Then they'll be able to store energy while supplying it to the plates in the next cell and to a load source as well.

Here's why; if there's a charge on, one positive terminal and a load source on the other as well. Since the positive plates will join the two positive terminals together, it'll allow energy to be transfer from one terminal to the next. Since energy has to flow into the battery before flowing to the load source, it'll be similar to a power transformer energy transfer process.

For instance; the positive plates joining the two positive terminal connections together in parallel within the battery, they'll act like a medium between a charge and a load source in their relationship with the battery. Like an iron core being a medium between a primary and a secondary winding inside a power transformer as well.

Since current has to flow into the battery before flowing to the load source, it'll allow us to charge the battery while discharging it to the load source because relationship between them has changed. In other words; current, it no longer has to flow to the load source before its reversed back into the battery to charge it as usually.

Although; energy, it'll be lost to heat during the simultaneous cycling process of the battery; but, it'll become irrelevant if the charging source makes up any energy lost to heat during the process. Similar to a power transformer energy transfer process, it loses energy to heat; but, it'll become irrelevant because it'll be made up by the power source.

Thus; if the charging source is constantly adding energy back to the battery during its simultaneous cycling process, then any energy lost to heat becomes irrelevant. Therefore; the idea that, the battery can't be charged while being discharged due to either the laws of entropy, inertia or the laws of conservation of energy is an erroneous belief because it doesn't pan out.

Here's why; if we're adding energy back to the battery while taking it out, then those laws of physics will not apply as well. Wherein; it seems like, we're focusing on the wrong science or the wrong mechanics or both if we assume those laws of physics. They'll be an issue when it comes to a simultaneous cycling process of the battery as well.

We don't have to focus on the vast and complex laws of physics to figure out if energy lost to heat in both conversion processes of the battery, it'll render a simultaneous cycling process of it void. All we've to do is look at the energy transfer process of a power transformer to figure this out since energy will be lost to heat when its iron core is energized by its primary winding.

Although; energy lost to heat, it does occur between a primary winding and an iron core inside a power transformer; but, it doesn't stop its secondary winding from powering a load source; that is, connected to it by way of an electrical bus as well. So; charging, the battery while discharging it will not render the process void because we're charging it as well.

Here's why; if a battery, it has separate terminals for charging and discharging in the same cell. And those terminals, they'll have their own straps and tab connectors leading to the same plates in that cell. Then how those terminals, they're interconnected with those plates. It'll determine how energy will flow to and from the battery and to a load source as well.

It'll be similar to energy being added or subtracted from an iron core inside a power transformer. Wherein; the number of turns the wiring, a primary winding has wrapped around the iron core to deliver current to it. Compared to the number of turns the wiring, a secondary winding has wrapped around the iron core to conduct current from it as well.

The wiring wrapped around the iron core by the primary and secondary winding, it'll determine how energy will flow to and from the iron core inside the power transformer to a load source; that is, connected to the secondary winding by way of an electrical bus. The energy transfer process of a power transformer has little to do with the laws of physics.

Likewise; the energy transfer process of a lead acid automotive starting battery, it has little to do with the laws of physics as well. But; just like, it's with the power transformer. It has more to do with the laws of motion and the effect of forces on bodies. The laws of entropy, inertia or the laws of conservation of energy will have no bearing on the process as well.

For example; a primary winding, it'll lose energy to heat when it transfers energy to an iron core inside a power transformer. Likewise; energy, it'll be lost to heat when it's transferred from an iron core to a secondary winding inside a power transformer as well. Wherein; energy, it'll be lost to heat between them before it's delivered to a load source by way of an electrical bus.

However; we'll find that, any energy lost to heat will be made up by a power source; but, the connections made between a primary winding, an iron core and a secondary winding inside a power transformer. The connections made between them will determine the amount of voltage and current a load source will receive from the power source.

Here's another thing; a secondary winding inside a power transformer, it'll act like a medium between an iron core and a load source; that is, connected to the secondary winding by way of an electrical bus. The number of turns the wiring, the secondary winding has wrapped around the iron core. It'll determine the amount of energy the load source will receive.

Likewise; if the charging current, it has to flow into one cell in the battery before flowing to the other cells in it and to a load source. Then the amount of energy flowing to the other cells in the battery and to the load source, it'll have little to do with either the laws of entropy, inertia or the laws of conservation of energy as well.

However; the charging source and the connections made between, the battery and the load source in their relationship with one another will determine if they'll receive energy from the charging source. Not the laws of entropy, inertia or the laws of conservation of energy; but, the laws of motion and the effect of forces on bodies will determine this.

Here's why; remember a lead acid automotive starting battery plates in each cell, they're connected in series-parallel with one another by their straps, tab connectors and the electrolyte solution in each cell. And each cell, it's connected in series with its terminal connections as well. So; all we've to do is add, an additional terminal to the battery in the same cell.

Therefore; if the charging voltage and current, they've to enter the cell with the two terminal connections in it first. Then the laws of parallel circuitry will begin when the charging voltage and current reaches the strap and tab connectors connecting the plates together in parallel. The straps and tab connectors will create the parallel connections between them.

Thus; the charging current, it'll be divided amongst the plates connecting the two terminals together in parallel based on the plates individual resistance. Then the charging current, it'll energize the electrolyte solution in that cell. The same amount of current flowing across the previous plates will be transferred to the adjacent plates by way of the electrolyte solution as well.

This is where the laws of series circuitry will begin and a voltage drop will occur within the battery as well. Given that; the battery positive and negative plates in each cell, they'll be connected in series with one another by way of the electrolyte solution in each cell. Then the adjacent negative plates will receive the same amount of current as the adjacent positive plates in that cell.

However; the adjacent negative plates in the cell with the two terminal connections in it. They'll receive a voltage drop from the adjacent positive plates because they're connected in series with them by way of the electrolyte solution in that cell. Therefore; a voltage drop, it'll continue throughout each cell within the battery because they're connected in series with one another.

While voltage and current is flowing through each cell in the battery, voltage and current will be flowing toward a load source; that is, connected to one of the two terminal connections as well. Then the reverse will happen when current leaves the plates and flow to the strap connecting them together in parallel as well.

The current potential of each plate, it'll be combined at the strap connecting them together in parallel with the terminal connection that the load source is connected to. However; the voltage potential at the strap, it'll be the same as the voltage flowing across one plate connecting them together in parallel as well.

In which; it makes, the two terminals in reverse and independent of one another because of how, voltage and current will flow from the strap <u>to</u> the plates joining one terminal connection together in parallel. It'll be in reverse and independent of, how voltage and current will flow <u>from</u> the plates to strap joining them together in parallel with the other terminal as well.

We'll not know this if we don't separate voltage from current; although, current is a combination of voltage and electrons. However; current, it'll flow through a parallel circuit connection different than voltage will. We must understand the difference between them if we're going to understand, why it'll possible to charge a battery while discharging it. See chapter 12.

If we don't know the difference, then we'll continue to blame it on the laws of physics why we can't charge a battery while discharging it as well. If we understand how voltage and current will through a battery based on its design, then we'll know it's not due to the laws of physics; but, it's due to the laws of motion and the effect of forces on bodies.

Understanding the mechanics of the voltage and current alone, we'll know a battery will need more than one terminal connection for voltage and current to enter and exit it at the same time if we're going to charge it while discharging it as well. It'll be no need to focus on the laws of electrochemistry, entropy, inertia or the laws of conservation of energy as well.

We just need to focus on the simple mechanics of the voltage and current. For instance; if a battery has separate terminals for charging and discharging and they're interconnected by the same plates, then the terminals will be in reverse and independent of one another; thus, it becomes possible to charge it while discharging it because it'll be no different than a capacitor or a transformer.

Here's why; if energy goes in one way and comes out another, it's the same thing as charging while discharging. But; most people, they don't see it my way because of semantics when it comes to charging a battery while discharging it. With a capacitor, it's easy to make a comparison between it and a secondary cell battery when it comes to charging while discharging.

However; we can't see, the comparison between a secondary cell battery and a transformer when it comes to charging while discharging because the words ionization and induction throws people off. Ionization and induction, they're used as a medium to transfer electrons from one point to the next without the two points coming in contact with one another.

We can't get hung up on semantics if we're going to understand the similarities between a secondary cell battery and a transformer. Although; a transformer, it can't store energy like the battery could; but, both could consume energy from a power source and supply it to a load source. However; the battery, it can't do both like the transformer could.

If we're going to charge the battery while discharging it like a capacitor or charge the battery while using like a transformer, energy has to be able to enter and exit the battery at the same time like a capacitor or a transformer as well. Wherein; a charging source, it becomes a power source for the battery and it becomes a power source for a load source.

In which; the battery, it becomes like a capacitor or a transformer because energy could enter and exit it at the same time as well. Here's why; if the battery, it has separate terminals for charging and discharging in same cell. And those terminals, they'll have their own straps and tab connectors leading to the same plates as I've mention over and over as well.

Then the charge and discharging terminal of the battery, they'll be able to behave like a primary and a secondary winding inside a power transformer. In which; the primary winding, it'll be a power source for an iron core and it'll be the power source for the secondary winding inside the power transformer to power a load source connected to the secondary winding by an electrical bus.

I'm going to use the conventional current follow concept to show, how a simultaneous cycling of a lead acid automotive starting battery will work. If it had separate terminals for charging and discharging, like a run capacitor does. While comparing, the battery energy transfer process to that of a power transformer based on the conventional current flow concept as well.

Here's why; current, it has to flow in or out of the positive terminal of the battery in order to cycle it. Therefore; I'm going to explain charging the battery while discharging it to a load source in that fashion, so, it'll be clearer to the average person as well. For instance; if the battery's charge and discharging terminal, they're interconnected by its positive plates.

Then the battery's positive plates, they'll be a medium between the charge and the discharging terminal; in which, they'll allow energy to flow to and from the positive plates connecting them together in parallel. Wherein; the positive plates, they'll act like an iron core since it's the medium between a primary and a secondary winding inside a power transformer.

The electrolyte solution in the cell with the charge and the discharging terminal, it'll be a medium between the positive and negative plates in that cell; wherein, it'll allow energy to be transferred between the plates in that cell. In which; it'll be similar to an iron core allowing, energy to flow between a primary and a secondary winding inside a power transformer as well.

When placing a charge on, the charging terminal of the battery. It'll energize the positive plates in that cell like a primary winding will energize an iron core inside a power transformer as well. The battery's positive plates will ionize the electrolyte solution in the cell with the charge and the discharging terminals in it to transfer energy to the adjacent negative plates in that cell.

Then energy will be transferred from the negative plates in the cell with the charge and the discharging terminal it in to the positive plates in the next cell by way of an electrical bus. In which; It'll be similar to, a secondary winding inside a power transformer transferring energy to a load source; that is, connected to it by an electrical bus as well.

Since the charge and the discharging terminal of the battery, they'll be interconnected by the positive plates in the same cell. If a load source is connected to the discharging terminal, then the positive plates in the cell with the charge and the discharging terminal in it. They'll allow energy to be transferred to the load source as well.

In which, it'll be similar to, an iron core inside a power transformer being energized by a primary winding to transfer energy to a load source; that is, connected to a secondary winding by way of an electrical bus as well. Thus; energy will flow from the discharging terminal of the battery to the load source based on how, it's connected to the positive plates in that cell.

It'll be similar to energy flowing from a secondary winding to a load source by way of an electrical bus. In which; it'll be based on, the number of turns the wiring the secondary winding has wrapped around the iron core to conduct current inside the power transformer as well.

After making a comparison between, the energy transfer process of a lead acid automotive starting battery and a power transformer. I found the laws of physics will not restrict the battery from being charged while being used to power a load source; but, only the design of the battery will. It's all about mechanics or procedure in how energy is added or subtracted from it.

We don't have to focus on the vast and complex laws of physics to figure out, is it possible or not to charge a battery while discharging it. All we've to do is focus on the mechanical structure of a capacitor or a transformer to figure this out. They give credence we're not restricted by the laws of physics when it comes to a battery's energy transfer process as well.

After my research on the mechanical structure of a lead acid automotive starting battery, a run capacitor and a power transformer, it was one thing left for me to do; that is, to go from theorization to actualization. In the next chapter; let me tell you, how I made a prototype of a lead acid automotive starting battery; so, it could be charged while being discharged as well.

CHAPTER 19

A PROTOTYPE OF AN AUTO BATTERY

To PROVE MY HYPOTHESIS, it's not due to the laws of physics; but, it's due to the contemporary design of our secondary cell batteries why they can't be charged while being discharged. Wherein; I had to venture outside the status quo because our secondary cell batteries, they're only built with one opening for energy to enter or to exit them.

Thus; I had to figure out, how to build a prototype so energy could enter and exit it at the same time. At first it seen impossible to, build one on my own because I didn't have the proper tools on hand. I couldn't afford to have a prototype built to test my hypothesis because after eleven years of working in the automotive industry.

I had lost my job two years prior to completing my research on the mechanics behind the cycling processes of a lead acid automotive starting battery. After a couple weeks of rumors that, the plant was going to shut down suddenly it shut down in March of 2009. We didn't get our accumulated vacation pay or our severance pay because the company didn't have it.

The company was in business for more than fifty years because of the down turn in the economy in 2008, the company closed its doors on a very cold day in March of 2009. General Motors and Chrysler, they had to be bail out by the government or close their doors as well; so, things had gotten pretty bad between 2008 and 2009.

After a few months of being jobless, it dawn on me one day how to build my own prototype of a lead acid automotive starting battery; so, electrons could enter and exit it at the same time. In this chapter; I am going to share with you, how I built my own prototype; so, electrons could enter and exit it at the same time.

Plus; I am going to share with you, the results of my experiments on charging the battery while discharging it to a load source as well. In which; I found that, it's neither complex nor impossible if we change the design of the battery; so, it could be charged while being discharged to a load source a

Let me be earnest with you because on many occasions when I attempted to change the design of a lead acid automotive starting battery; so, it could be charged while being discharged to a load source. My attempts had failed because I wasn't paying attention to the design concept of the battery.

Here's why; at first, I had two positive terminals at one end and two negative terminals at the other end of the battery cells when I carried out my experiments. I thought I was charging it while discharging it; but, it was nothing more than an illusion. Given that; electrons, they can't flow from one positive terminal to one negative terminal while flowing in the opposite direction from the other positive and negative terminal. See chapter 3 and 24.

Since the positive and the negative plates in each cell in the battery, they're setup in sequence because it'll be a positive and then a negative and then a positive plate setup in this sequence; until, each cell has an equal number of positive and negative plates in it. Thus; electrons, they've to flow across one plate through the electrolyte solution to an adjacent plate.

Therefore; the cells <u>without</u> a terminal connection in it, electrons could only flow in or out of that cell flowing in one direction at a time. Because of the battery plates and cells, they're connected in series with one another. Thus; electrons, they can't simultaneously flow in the opposite direction of one another while flowing on the same path as well.

I didn't carry out the proper analysis on the lead acid automotive starting battery at first because if I had, then I would have realized I couldn't charge all the plates in each cell while discharging it to a load source as well. Because the battery plates and cells, they were connected in series with its terminal connections as well.

I had to figure out how to modify the battery; so, I could charge all the plates in each cell while discharging it to a load source as well. My first prototype was called; "Miracle Auto Battery", a deep cycle battery for the twenty-first century; maybe, you've heard of it. I built it; so, electrons could flow across every plate in it; but, it wasn't the most cost efficient way to build it.

However; I was able to, charge my prototype while discharging it. Given that; the two positive terminals, they were interconnected by the positive plates in each cell as well as the two negative terminals were interconnected by the negative plates in each cell as well. After publishing my first book, I found a more efficient way to design a lead acid automotive starting battery.

A couple of months after publishing my first book, I re-visited my notes and I realized I had overlooked some important facts about the discharging cycle of a lead acid automotive starting battery. Its total amp draw will only equal to the plates in one cell during its discharging cycle because its cells, they're connected in series with its terminal connections as well.

I said to myself what does this mean, its total amp draw only equal to the plates in one cell during its discharging cycle. We're getting amp draw from all of the plates in each cell; but, it'll only equal to the amp potential of its plates in one cell during its discharging cycle this is what I thought while reviewing my notes.

After further investigation, it became apparent what it meant by the total amp draw of the battery will only equal to one cell in it. What I found was that; when a battery cells, they're connected in series with its terminal connections. Then its amp draw, it'll only come from the cell with a terminal connection in it. See chapter 5.

Given that; the other cells in the battery, they'll be connected in series with the cell with the terminal connection in it to increase its voltage and current potential. On the other hand; if its cells, they're connected in parallel with its terminal connections. Then its total amp draw, it'll come from all of its plates in each cell.

However; the battery's voltage potential, it'll only come from its plates in one cell to increase its voltage and current potential during its discharging cycle. Since a lead acid automotive starting battery cells, they're connected in series with its terminal connections; thus, its total amp draw will only come from a cell with a terminal connection in it.

My notes also revealed that, I didn't need to build my prototype; so, electrons could flow through every cell in it to charge it before electrons could reach a load source as well. In which; it seen more feasible to, build my own prototype with the tools I had on hand as well. I realized something else while, reviewing my notes as well.

I found a voltage drop will only occur within the battery after voltage, it leaves the strap connecting the plates together in parallel with a terminal connection. Upon reviewing my notes, I realized it was a more cost efficient way to design a lead acid automotive starting battery; so, it could be charged while being discharged to a load source as well.

I found electrons didn't have to flow across every plate or through every cell in the battery before flowing to a load source in order to charge the battery while discharging it to the load source. It was clear to me that electrons, only needed to simultaneously enter and exit a cell within the battery with a terminal connection in it.

Based on, the electron current flow concept. It'll be the negative terminal connection of the battery or based on the conventional current flow concept, it'll be the positive terminal connection of the battery. However; either way it goes, it seen more plausible to build my own prototype of a lead acid automotive starting battery.

Given that; I only needed to, change the cell with a terminal connection in it; in which, it'll be either the cell with the negative or the positive terminal connection in it. Since the terminal connection, it'll be connected in parallel with its respective plates in that cell. Then those plates, they'll be connected in parallel with a charging source during the battery charging cycle.

Thus; I thought, if we add another terminal to a cell with a terminal connection already in it and they're interconnected by the same plates as the other terminal. Then only the plates connected in series with the plates connecting, the two terminals together will receive a voltage drop coming from the charging source.

Then we don't have to worry about, a voltage drop occurring before voltage reaches a load source. If the two terminals, they were in the same cell and interconnected by the same plates as well. Since current, it could only enter or exit the same terminal connection of a battery during its cycling processes.

Then it was clear to me that, the battery must have two negative or two positive terminal connections in the same cell instead of one in order to charge it while discharging it. After gaining additional information from my notes, I thought it was a lot easier to build my own prototype with the tools I had on hand compared to changing the whole design of a battery.

Also; other data revealed from my notes showed that, voltage and current could enter or exit a battery based on how its plates and cells were interconnected with its terminal connections as well. Then I realized that, the two terminal connections in the same cell must have their own straps and tab connectors leading to the same plates in that cell as well.

Thus; electron, they could simultaneously flow to and from the same plates. Without having to flow to and from, the same strap and tab connectors on the same plates. All of sudden, it was becoming clearer and clearer how to build my own prototype. And yet, I knew it wasn't going to be easy as well.

Here's why; a lead acid automotive starting battery plates, they're made of lead-peroxide and sponge-lead. Wherein; they're so fragile, they'll fall apart at a touch. See chapter 4. Nevertheless; I was determine to, add another terminal connection to a cell with a terminal connection already in it without damaging the battery or bringing harm to myself.

Here's the deal; an automotive starting battery, it was easier to modify than a deep cycle automotive battery. Given that; an automotive starting battery, its case isn't as thick as a deep-cycle automotive battery case is; plus, an automotive starting battery is less expensive to buy as well.

The two batteries are practically the same; except, the deep-cycle automotive battery plates and straps, they're much thicker than an automotive starting battery plates and straps. Plus; the deep cycle automotive battery plates, they've more surface space than an automotive starting battery plates as well.

Since I was going to, use a tire inflation device as a scale down version of an electric motor for an electric vehicle to carry out my experiments. Then I didn't need to, use a deep-cycle automotive battery to perform my experiments; given that, an automotive starting battery is just a scale down version of a deep-cycle automotive battery as well.

The sleeves housing the negative plates inside a lead acid automotive starting battery, they're porous rubber used to separate them from the positive plates in each cell. I decided to insert lead strips between the rubber sleeves and the positive plates to make additional tab connectors for them based on the conventional current flow concept.

So; I could attach another positive terminal connection to the positive plates, already had a positive terminal connected to them. Thus; the two positive terminals, they would have their own straps and tab connectors leading to the same plates in that cell as well. Since the battery plates in each cell, they're connected in series-parallel with one another as well.

Then having, two terminals in the same cell and interconnected by the same plates. I discovered when there's a charge on one terminal and a load source on the other terminal. It would create a series-parallel circuit connection between the charge and the load source in their relationship with the battery as well.

I found the series-parallel circuit connection made between the charge and the load source in their relationship with the battery. It created two paths for the charging current to travel during the charging cycle of the battery. I found one path was toward the other cells in the battery based on the laws governing series circuitry.

The other path for current to follow, it was toward the load source based on the laws governing parallel circuitry. Thus; the charging source, it no longer had to create a surplus of electrons at the load source in order to reverse the current flow back into the battery to charge it. Given that; the charging current, it had to flow into the battery first before flowing to the load source.

I found it would be in reverse of the traditional cycling process of the battery when it's a charge and a load source on it at the same time. Wherein; the surplus of electrons, they would be created in the cell with the charging current has to enter in first and then reversed back toward the other cells in the battery and toward the load source as well.

To simulate a simultaneous cycling process of a deep-cycle automotive battery, I used a 12 volt-750 amp capacity lead acid automotive starting battery instead of a deep-cycle automotive battery. I used a 12 to 24 volt tire inflation device with a maximum amp draw up to 15 amps to represent a scale down version of an electric motor for an electric vehicle.

I also used an 18 volt-40 amp charging system as the charging source to carry out my experiments. For safety reasons, I attached a twenty-five foot extension cord to the charging device. So; I could plug it in, from a safe distance in case the battery exploded when I turn on the charger. I stood outside my garage door about twenty-five feet away from the battery.

Then I plugged in the charger as I've done many occasions. After plugging in the charger, it heard no explosion and I saw no sparks; although, I heard the tire inflation device and the charging system running at the same time. I waited for a couple of minutes before I went back into the garage to see what was going on.

I had already cut off the top of the plastic casing of the battery and I left its terminals, plates and cells intact. So; I could observe, how electrons would flow through it based on the bubbles they created in the electrolyte solution during the cycling processes of the battery. When I finally went back into the garage to, see what was going on.

I saw the electrolyte solution bubbling around the positive terminal connection that the tire inflation device was connected to. Thus; it seen like, the battery was powering the tire inflation device. Also; the electrolyte solution, it was bubbling around the positive terminal connection that the charging source was connected to as well.

Thus; it seen like, the charger was charging the battery. Therefore; it seen like, the battery was charging while it was discharging to the tire inflation device as well. However; I thought that, I had to do more testing other than observing the bubbles created in the electrolyte solution around the two positive terminal connections to be sure my theory was correct.

I performed a series of additional tests using ammeters and voltmeters. Wherein; I could measure, the amount of voltage and current flowing in and out of the battery during each experiment. The beginning voltage of the battery, it was around 13.5 volts; but, I discharged it until it reached around 10.5 volts before I put a charge on it.

Then I connected an ammeter in series with the charging source and the charging terminal of the battery. So; I could, read the intensity of the current between the charger and the charging terminal. Likewise; I connected an ammeter in series with, the tire inflation device and the discharging terminal; so, I could read the intensity of the current between them as well.

Then I wouldn't have any doubt about, the charging source was charging battery while it was powering the tire inflation device as well. After one hour of simultaneously charging and discharging the battery to the tire inflation device, I turned off the charger and the tire inflation device to place a voltmeter across the positive and negative terminals of the battery.

I wanted to see if the battery had gained or lost voltage during the experiment; but, it actually gained voltage during the experiment because its voltage level, it had risen around 12.5 volts after I discharged it to 10.5 volts at the beginning of the experiment. In which; the experiment, it verified my data I had compiled during my research as well.

Wherein; my data suggested that, a lead acid automotive starting battery plates in each cell. They could consume and store energy while supplying it to its plates in the next cell as well. Not only the test I performed verified my data; but, I found the battery plates in each cell could consume and store electrons while they're flowing to the tire inflation device as well.

Based on my experiments, I found it was not necessary to stop the discharging cycle of the battery in order to start its charging cycle because of the laws of physics. However; it's due to the contemporary design of the battery, why we've to stop its discharging cycle in order to start its charging cycle because electrons have to go in the same way they've to come out as well.

The information I gather from the mechanical structure of a run capacitor and a power transformer during their energy input and output process. It suggests we don't have to stop the discharging cycle of the battery in order to start its charging cycle because of the laws of physics. Given that; its energy input and output process, it's a product of its design as well.

Charging the battery while discharging it to the tire inflation device, the battery didn't get hot; however, it seen strange at the time because it usually gets hot during its charging cycle when it's just a charge on it and not when it's just a load source on it. I wondered about this for awhile and then, I went back through my notes.

What I found during the charging cycle of a lead acid automotive starting battery, the chemical reactions taking place within it during its cycling processes. One is exothermic or it releases and the other is endothermic or it absorbs heat. The endothermic chemical reaction taking place during its discharging cycle will balance out the heat released by its electrical resistance. See chapter 4.

Thus; the battery, it doesn't appear to get hot during its discharging cycle; but nevertheless, energy lost to heat does occur during both of its conversion processes. While carrying out my experiments, I was wondering was this what some in the field of automotive battery technology was talking about.

When they say if we attempt to carry out both conversion processes of the battery at the same time, they'll cancel themselves out; therefore, no gain because energy lost to heat does occur in both of its conversion processes. In other words; we'll not be able to, charge the battery while discharging it because of its endothermic and exothermic chemical reactions.

What I discovered during my experiments on, the lead acid automotive starting battery is that. The amount of energy lost to heat during its charging cycle becomes irrelevant because it only determine if it's getting hot or not; but, it doesn't determine if it's charging or not. Plus; its endothermic and exothermic chemical reaction, they'll not take place at the same time.

If we're charging the battery while discharging it, then the charging voltage has to be higher than the normal battery voltage. Then the electromotive force created by the charging source, it'll be able to add electrons back to the battery; in other words, only the exothermic chemical reaction will take place within the battery during a simultaneous cycling process of it.

Then the next question is that, why the battery didn't get hot when it was charging while discharging it? I concluded subtracting electrons from it while adding electrons back to it created an endothermic affect. In which; it'll balance out, the heat released by the exothermic chemical reaction because electrons could escape from the battery as well.

Wherein; charging the battery while discharging it to a load source, it allows electrons to escape from the battery; therefore, less heat will accumulate within the battery as well. In other words; electrons, they'll not only be jolting for space on the battery plates while creating heat; but, they could flow to a load source during a simultaneous cycling process of the battery as well.

It really doesn't matter if the endothermic and the exothermic chemical reaction of the battery will take place at the same time. Given that; they only determine, the intensity of the heat being created by energy lost to heat. The idea energy lost to heat in both conversion processes of the battery will cancel themselves out if we attempt to carry them out at the same time is absurd.

If we're charging the battery while discharging it, then we should expect energy to be lost to heat because we're charging it; however, we should expect to charge it because we're charging it as well. If we're assuming, we can't add energy back to it while taking energy out. Then we're not accounting for the value of the energy being added back to it as well.

Given that; energy, it'll be accounted for during the charging cycle of the battery because any energy lost to heat during its charging cycle, it'll be made up by the charging source as well. It'll be unusual not to account for the value of the energy being added to the battery if we're charging it while discharging it as well.

We either don't understand the concept of charging a lead acid automotive starting battery while discharging it or we don't understand the mechanics behind its charging cycle as well. Then the question is that, are we focusing on the right science or the right mechanics when we assume that? I'll be no benefits in a simultaneous cycling process of it as well

I decided to carry out a second series of tests while intermittently charging the battery for five hours to see what effects. It'll have on its temperature and its voltage levels over a long period of time. My experiments consisted of five minutes of charging it while it had the tire inflation device on it and five minutes of not charging it while it had the tire inflation device on it as well.

I discharged the battery until its voltage level had dropped around 10.5 volts before I started my five hour experiment. I was taking random readings from time to time to see if the battery had any significant voltage drop during the five hour experiment. After five hours, the battery temperature seen to be lukewarm and its voltage level had risen around 12.5 volts as well.

Thus; the results of the five hour experiment, it was similar to the results of the one hour experiment. Wherein; I found that, the intermittent charge on the battery. It didn't allow it to become almost empty or almost full; therefore, it didn't get hot and its voltage level rose around 12.5 volts as well.

At first; I thought that, the battery voltage level rose because the charging system. It could have had produced up to 18 volt and 40 amps. Thus; it was capable of producing enough amps to, power the tire inflation device and recharge the battery at the same time. I was thinking this at first; but, after further investigations. I found it wasn't creditable enough.

Here's why; the charging current, it had to flow into one cell in the battery before flowing to the tire inflation device. I found the electrical resistance at the tire inflation device didn't have any bearing on the amount of electrons flowing into the battery. Thus; I was surprised, about the outcome; although, my data was telling me it was possible as well.

Actually; I really didn't know, what to expect before I carried my experiments because of all the opposition. I had received from some in the field of automotive battery technology. It was impossible to charge a battery while discharging it because of the laws of physics. Upon carrying out a few more experiments, I concluded they're focusing on the wrong science.

Although; energy, it'll be lost to heat in both conversion processes of a lead acid automotive starting battery. It doesn't mean it'll be no gain if we attempt to charge it while discharging it. I found the process of adding energy back to the battery while taking it out will not be a zero gain because it's proven by the mechanics behind its charging cycle.

Here's why; energy, it's simultaneously added and subtracted from the battery plates in one cell; so, it could be added to its plates in the next cell. Although; they're separated by a non-conducting material called plastic and yet, they're still charged even though energy lost to heat will occur during the process as well.

It seems like it'll be no different than adding energy back to the battery while taking it out to power a load source as well. Likewise; the idea both chemical reactions of the battery will occur at the same time, it seems absurd. Since its plates in one cell, they're charged and discharged to its plates in the next cell without both chemical reactions occurring at the same time.

My final conclusion after carrying out my experiments on a lead acid automotive starting battery was that. I found it was possible to simultaneously charge and discharge the battery to a load source and it'll be no benefits in it as well. I found the assumptions about why it'll be impossible to charge the battery while discharging it and it'll be no benefits in it as well.

Apparently; all those assumptions, they're based on energy havening to in the battery the same way energy has to come out in order to cycle it as well. Having energy to enter the battery on one terminal and exit the battery on another terminal connection, its charge and discharging cycle could exist at the same time and it'll be a benefit in it as well.

Here's why; in order to charge the battery, the charging voltage must be higher than the normal battery voltage; thus, its charge and discharging cycle could exist at the same time. If energy could enter and exit it at the same time, then it'll be no conflict between its charge and discharging cycle; wherein, it has little to do with the laws of physics as many believe it does.

I found it has more to do with the laws of motion and the effect of forces on bodies. What I discovered was that, it boils down to how we go about adding and subtracting energy from the battery. In which; our thought process about, how we go about adding and subtracting energy from it will be created by its contemporary design as well.

Therefore; we think, it's impossible to charge the battery while discharging it and it'll be no benefits in it because of our perceptions about, how we go about carrying out its cycling processes; wherein, it's established by its contemporary design. Thus we see no other way to carry out its cycling processes; but, to stop one cycling process in order to start the other as well.

We blame it on the laws of physics why we've to stop the discharging cycle of the battery in order to start its charging cycle. Given that; we're overlooking, the fundamentals of how it's charged in the first place. It seems like we don't understand the mechanics behind its charging cycle to figure out, how to charge it while discharging it to a load source as well.

We're too busy focusing on the wrong science because it's not about, the laws of electrochemistry, entropy, inertia or the laws of conservation of energy; but, we like to think they are. However; it's about them, it's about the laws of motion and the effect of forces on bodies or it's about the laws of mechanics.

If electrons have to go in the battery the same way they've to come out in order to charge its plates or for them to be discharged to a load source as well. Then the laws of physics should be the last thing, we should focusing on. The mystery behind why we could charge and discharge the same battery; but, we can't charge it while discharging it because of the laws of physics.

If we've to charge the battery on the same terminal connection that, it was discharged on. Then why it can't be charged while being discharged because of the laws of physics is no mystery at all. Each cell in a lead acid automotive starting battery is nothing more than a battery in and of itself, connected in series with another and housed together in a plastic.

When the battery is charged, we're simultaneously charging and discharging one battery while simultaneously charging and discharging another because of how, they're interconnected with one another. It has little to do with the laws of physics; but, more to do with the laws of mechanics why it can't be charged as a whole while being discharged to a load source as well.

In the next chapter; let me give you, an in depth analysis of the mechanics behind my prototype of a lead acid automotive starting battery. So; you could understand why, I was able to charge every cell in it while discharging it to the tire inflation device as well. Although; the battery plates and cells, they're connected in series with its terminal connections as well.

CHAPTER 20

ANALYSIS OF A PROTOTYPE

During my research on a lead acid automotive starting battery, I've heard many different opinions why it can't be charged while being discharged. Most of those opinions were pointed toward the laws of physics. I compiled and compared them for future reference while carrying out a number of experiments. I found those opinions were nothing more than erroneous beliefs. See chapter 11.

When it comes to a conversation about charging a secondary cell battery while discharging it as another option to, increase its efficiency and overall capacity. Some in the field of automotive battery technology say it's impossible due to the laws of physics; thus, finding a perfect chemical composition is the only way to increase the overall travel range of a vehicle on battery power.

Most people immediately pivot toward the laws of physics when I bring up a conversation about charging a secondary cell battery while discharging it. Basically; the skeptics, they think the conversation is about the laws of physics; but, the conversation is about the laws of mechanics because it's about how we go about adding or subtracting energy from the battery.

I discovered having two terminals in the same cell and those terminals having their own strap and tab connectors leading to the same plates in that cell. It'll create a series-parallel connection between a charge and a load source in their relationship with the battery. At first; I didn't know, how this series-parallel circuit connection would work as well.

Since the two terminals were interconnected by the same plates, I assumed the two terminals would be in parallel with one another. Wherein; the charge and the load source, they would be in parallel with one another as well. Therefore; the amount of current flowing into the battery, it would be dictated by the amount of resistance at the load source as well.

However; all I was looking for at first was, a way for electrons to enter and exit the battery at the same time. After carrying out a number of experiments with my prototype, it didn't appear it and the tire inflation device was in parallel with one another as I thought they were. Given how, electrons were flowing to the tire inflation device as well.

It didn't affect how electrons were flowing into the battery during my experiments when charging it while discharging it to tire inflation device. In which; electrons, they should have been flowing into the battery based on the amount of resistance at the tire inflation device. Given the parallel circuit connection made between them as well.

After a careful analysis of the mechanical structure of my prototype, I realized having two terminals in the same cell. And those terminals, they had their own straps and tab connectors leading to the same plates in that cell. It made the two terminals in reverse and independent of one another as well.

Thus; how, electrons will flow <u>to</u> those plates from one terminal. It would be independent of how electrons will flow <u>from</u> those plates to the other terminal connection. Although; the two terminals, they were connected in parallel with one another by way of the same plates; but, the terminals didn't act like they were in parallel with one another.

I found out later on, why the two terminals didn't act like they were connected in parallel with one another. Given that; the charging current, it had to flow to and from those plates joining the two terminals together in parallel. Based on, how those terminals were connected to those plates in that cell as well.

Likewise; current, it had to flow to the plates in the other cells in the battery based on how they were connected to the plates joining the two terminals together in parallel as well. What was normal about the two terminals being connected in parallel with one another by the same plates? The voltage flowing from charging source into the battery and to the tire inflation device was the same.

In which; I thought that, it was consistent with the laws governing parallel circuitry. However; the amount of current flowing into the battery and toward the tire inflation device, it was based on the laws governing parallel circuitry as well; wherein, I thought it wasn't consistent with the laws of parallel circuitry as well.

Given that; the battery resistance to current, it'll reflect current away from it. But; a load source resistance to current, it'll base on the number of turns the wiring has inside of it to conduct current. Since the battery and the tire inflation device, they were in parallel with one another by way of the same plates.

Then I thought the amount of resistance at the tire inflation device, it should have dictated the amount of current flowing into the battery; however, it didn't dictate the amount of current flowing into the battery because current had to flow into it based on its resistance and not the resistance at the tire inflation device.

At first; I didn't understand why, the amount of resistance at the tire inflation device didn't dictated the amount of current flowing into the battery. Given that; I didn't understand, the different circumstances involved since the battery and the tire inflation device resistance to current is very different as well.

What I found during my investigation is that, the plates in each cell in a lead acid automotive starting battery. They're connected in series-parallel with one another by their straps, tab connectors and the electrolyte solution in each cell. In which; I found that, the electrolyte solution in each cell connected the positive and the negative plates together in series.

What I learned while venturing into the field of heating, cooling and refrigeration is that. In order to have, a series-parallel circuit connection between load sources. It must be three or more load sources interconnected with one another by the same power source. At least two of the load sources, they must be connected in parallel with one another.

However; one or more load sources, they must be connected in series with one of the load sources that is connected in parallel with another load source. Since the battery plates, they behave like load sources during their charging process. Then the plates joining the two terminals together, they're load sources connected in parallel with the tire inflation device.

Therefore; we have, two or more load sources that is connected in parallel with one another. But; at the same time, we've other load sources connected in series with load sources that is connected in parallel with one another as well. Since the battery plates, they behave like load sources during their charging cycle as well.

Thus; the plates in the other cells in the battery, they're connected in series with the plates joining the tire inflation device together in parallel with a terminal connection. Wherein; the positive and the negative plates in the cell with the two terminal connections in it, they're connected in series-parallel with one another. And each cell, it's connected in series with one another as well.

In which; the negative plates in the cell with the two terminals in it, they're connected in series with the positive plates in the next cell by way of an electrical bus; thus, we've load sources connected in series-parallel with one another and other load sources connected in series with one another as well.

Since the positive plates in the cell with the two terminals in it, they're connected in series-parallel with the negative plates in the same cell. Therefore; the charging source and the tire inflation device, they'll be connected in series-parallel with one another in their relationship with the battery as well. Thus; understanding, how these connections will regulate voltage and current.

It was a game changer for me because I was able to understand having two terminals in the same cell and they've their own straps and tab connectors leading to the same plates in that cell. It was the key in charging the battery without having to stop its discharging cycle because current has to flow into the battery before flowing to the tire inflation device.

Wherein; the amount of resistance at the tire inflation device, it didn't dictate the amount of electrons flowing into the battery. Although; the two terminals, they were connected in parallel with one another by way of the same plates. However; the two terminals, they were independent of one another; thus, I didn't have to stop the discharging cycle in order to start the charging cycle.

I found how current flowed <u>to</u> those plates joining the two terminals together in parallel. It was independent of how current flowed <u>from</u> those plates. Likewise; how, current flowed <u>to</u> those plates joining the two terminals together in parallel. It was in reverse of current flowing <u>from</u> those plates as well.

It made the two terminals in reverse and independent of one another; although, they were interconnected by the same plates. What I didn't understand at first was that, current flowed into the battery and to the tire inflation device. Based on, the laws governing parallel circuitry as well; in which, I wasn't expecting because of the parallel connection made.

What I learned was that, the charging current had to flow directly into the battery on one terminal from the charging source. On the other hand; current, it had to flow directly from the battery to the tire inflation device. In which; voltage and current, they were flowing into the battery and to the tire inflation device based on the laws governing parallel circuitry as well.

What came into light for me was that, it defined what a parallel circuit connection meant; in which, voltage and current will flow directly from a power source to a load source. In other words; voltage and current, they don't have to flow through one load source in order to flow to another load source.

Then I understood that, voltage and current flowed directly into the battery from charging source and directly from the battery to tire inflation device. Wherein; the charging source, it was the power source for the battery and it was the power source for the tire inflation device because of the connections made between them.

It gave me a new perspective on, how a series-parallel circuit connection will work between a charge and a load source in their relationship with a battery. I was able to figure this out because of the training I had in the field of heating, cooling and refrigeration. However; I couldn't figure it out, until I realized a battery plates behave like load sources as well.

Given that; the battery plates, they'll consume and store energy during their charging cycle. Then it became clear to me if the charging current had to flow into the cell with the two terminal connections in it. Before flowing to the tire inflation device, then the path of least resistance for the current was to flow into the battery before flowing to the tire inflation device.

I found once current entered the battery, it had to flow to the other cells in the battery and to the tire inflation device. Based on, how they were connected to the plates in the cell with the two terminal connections in it. Then the amount of current flowed to the other cells in the battery. It was the same as the current flowed from the charging source as well.

Given that; the other cells in the battery, they were connected in series with the cell with the two terminals in it. However; those cells, they received a voltage drop from the charging source because they were connected in series with the cell with the two terminals in it as well.

Since the tire inflation device was connected in parallel with the plates in the cell with the two terminals, it received the same amount of voltage flowing from the charging source before the voltage drop occurred within the battery; but; current flowed to the tire inflation device based on its resistance because current flowed directly to it as well.

After understanding, how voltage and current will flow through a series, a parallel or a series-parallel circuit connection. Then I understood why the battery and the tire inflation device received the amount of voltage and current, they received; although, they were connected in parallel with one another by the same plates as well.

Wherein; the parallel connection between the battery and tire inflation device, it didn't act like a parallel circuit connection as I knew of because of the different circumstances involved. Looking at my prototype of a lead acid automotive starting battery, it revealed a battery with its plates and cells connected in series with its terminal connections.

We could charge all the plates in each cell in the battery while discharging it to a load source as well. If the battery, it had separate terminals for charging and discharging in the same cell and they had their own straps and tab connectors leading to the same plates as well. I found it created a medium between a charge and a load source in their relationship with the battery.

It reminded me of an iron core being a medium between, a primary and a secondary winding inside a power transformer. Wherein; current, it has to flow through the primary winding and to the iron core and then to the secondary winding to a power a load source; that is, connected the secondary winding by an electrical bus.

Then the line voltage is the power source for the primary winding and it's the power source for the iron core, in which, it's the power source for the secondary winding and it's power for a load source connected to it by way of an electrical bus. Wherein; it revealed, what happens if the charging current has to flow into one cell in the battery before flowing to the tire inflation device.

In which; the charging source, it became the power source for the plates connecting the two terminals together inside the battery. They became the power source for the tire inflation device and the other cells in the battery as well. The plates connecting the two terminals, they created a medium between the charge and the tire inflation device in their relationship with the battery.

This is how I came to understand, why it was possible to charge my prototype of a lead acid automotive starting battery while discharging it to the tire inflation device. It didn't seem so complex after I understood, the straps connecting the two terminals together in parallel with the plates. Those straps, they created a series-parallel circuit in their relationship with the plates.

At first; I was wrong to assume, the battery and the tire inflation device were in parallel with one another. Given that; the two terminals, they were interconnected by the same plates. Since the two terminals, they had their own straps and tab connectors leading to the same plates. It created a series-parallel connection between the two terminals and the plates instead.

Here's why; the plates connecting, the two terminals together in parallel. Those plates, they were connected in series-parallel with the other plates in that cell by their straps, tab connectors and the electrolyte solution in that cell. Since the charging current, it had to flow into one cell in the battery before flowing to the tire inflation device.

Then it created a series-parallel circuit connection between the charging source and the tire inflation device in their relationship with the battery. Understanding, how current flowed through a series, a parallel or a series-parallel circuit connection. I realized current had to flow to the battery and the tire inflation device based on the laws governing parallel circuitry.

In other words; current, it had to flow directly into the battery from the charging source and flowed directly from the battery to the tire inflation device; so, they both received current based on the laws governing parallel circuitry. While current flowed to the other cells in the battery based on, the laws governing series circuitry.

I found it had little to do with the mechanics of the current alone; but, it had more to do with the type of connections made between the battery and the tire inflation device in the relationship with the charging source. Whether; they were connected in series, parallel or series-parallel with one another, current flowed to them accordingly.

For instance; if load sources, they were connected in series with one another. Then each load source will receive a drop voltage from the power source; but, they'll receive the same amount of current flowing from it. Likewise; if the same load sources, they were connected in parallel with one another as well.

Then the load sources, they would receive the same amount of voltage flowing from the power source; however, current will flow to them based on their individual resistance. Now if the same load sources, they were connected in series-parallel with one another. Not only would they receive voltage and current based on, how they were interconnected with a power source.

However; the load source, they would receive voltage and current based on how they were interconnected with one another as well. For example; if two load sources, they were connected in parallel with one another and a third load source, it was connected in series with one of the load sources that is connected in parallel with another load source.

Then the two load sources connected in parallel with one another, they would receive the same amount of voltage flowing from the power source; but, current flowing to them would be based on their individual resistance. The third load source that is connected in series with another, it would receive a voltage drop from the power source.

However; the third load source that is connected in series with another, it would receive the same amount of current as the load source that it's connected in series with. Understanding how voltage and current will flow through a series, a parallel or a series-parallel circuit connection, I didn't have to focus on the vast and complex laws of physics.

All I had to do was focus on, the connections made between my prototype and the tire inflation device in their relationship with the charging source. To determine, how voltage and current would flow between them. It's a host of other things, other than the laws of physics will prevent a battery from being charged while being discharged to a load source as well.

We can't just focus on one thing and say it's the reason why the battery can't be charged while being discharged to the load source. Then we become inept to use the laws of physics to justify the reasons why, the battery can't be charged while being discharged to the load source; given that, they allow us to charge or discharge the same battery as well.

Therefore; I find that, the laws of physics won't allow a battery to be charged while being discharged to a load source because of other extenuating circumstances involved. For instance; we've to charge the battery on the same terminal connection that, it was discharged on because electrons have to go in it the same way they've to come out as well.

Then it has little to do with the laws of physics, why the battery can't be charged while being discharged the load source as well. If we didn't have to charge the battery on the same terminal connection that, it was discharged on would our assumptions hold true. It's due to the laws of physics why it can't be charged while being discharged the load source as well.

I don't think so because every sign indicates, the laws of physics will not prevent a battery from being charged while being discharged to a load source as well. Given that; we don't have to stop charging, a lead acid automotive starting battery plates in one cell in order to charge them in the next cell; although, they're separated by a non-conducting material called plastic.

In which; it indicates that, we don't have to stop the discharging cycle of the battery in order to start its charging cycle because of the laws of physics as well. Given the only difference between charging the battery plates while discharging them to a load source, compared to not having to stop charging them in one cell in order to charge them in the next cell as well.

It's the direction that electrons have to flow in order to charge the battery plates or for them to be discharged to the load source as well. Then it has little to do with the laws of physics, why the battery can't be charged while being discharged to the load source; but, it has more to do with the laws of motion and the effect of forces on bodies or the laws of mechanics.

Having a charge on one terminal and a load source on another terminal connection of a battery, it's not the same as having a charge and a load source on the same terminal connection of the battery; given that, different circumstances will be involved since matter and energy will behave differently under different circumstances as well.

The reason we could charge or discharge the same battery; but, we can't charge it while discharging it because of the laws of physics. It's due to the different circumstances involved because of them. The laws of physics will not allow us to charge the battery while discharging it because it would violate them if we don't change the circumstances involved to carry out such a process.

The laws of physics will allow us to charge or discharge a secondary cell battery; but, they'll not allow us to charge it while discharging it because of the circumstances due to its contemporary design. Given that; electrons, they've to go in it the same way they've to come out in order to charge its plates or for them to be discharged as well.

Wherein; the laws of physics will prevent us from charging the battery while discharging it, only to the extent electrons have to go in it the same way they've to come out as well. Then the question is that, is it due to the laws of physics or is it due to the laws of motion and the effect of forces on bodies why the battery can't be charged while being discharged?

Those are the fundamental questions we've to ask ourselves when it comes to a conversation about charging our secondary cell batteries while discharging them to a load source. After carrying out my experiments with my prototype of a lead acid automotive starting battery, I believe the answer is the laws of motion and the effect of forces on bodies.

Understanding our batteries, they're electrical devices just like our capacitors and transformers as well. Therefore; how, voltage and current will flow to and from them based on their designs. It's the key in understanding how voltage and current will flow to and from our batteries when they're connected in series, parallel or in series-parallel with our load sources as well.

If a battery and a load source, they're connected in parallel with one another in their relationship with a charging source. Then voltage flowing to the battery and to the load source, it'll be the same; however, current flowing to them will not be the same because of the different circumstance involved. We've to understand what different circumstances, are involved. See chapter 12.

Wherein; voltage and current, they'll behave differently under different circumstances because they're different; given that, one is made of electrons and the other is not. They'll flow through a series, a parallel or a series-parallel circuit connection differently. So; we've to look at the connections made to the battery and the load source in their relationship with the charging source as well.

Having two terminals in the same cell within the battery and those terminals, they had their own straps and tab connectors leading to the same plates in that cell as well. It's not the same as having a charge and a load source on the same terminal of a battery because of the different circumstances involved.

Without my training in the field of heating, cooling and refrigeration, I wouldn't have been able to figure out those difference circumstances involved as well. I thought the two terminals would be in parallel with one another since they were interconnected by the same plates; thus, the charge and the tire inflation device would be in parallel with one another as well.

However; the outcome, it was different than I thought it would be because of the different circumstances involved. It caused the charging source and the tire inflation device to be in series-parallel with one another in their relationship with the battery because the two terminals, they had their own straps and tab connectors leading to the same plates as well.

Let me explain this in a different way and just focus on the mechanics of the voltage and current alone. And then, you might understand why I was able to charge the battery while discharging it to the tire inflation device. Wherein; it might help you understand, how the charging voltage and current will enter the cell with the two terminal connections in it as well.

In which; it'll be based on, how the two terminals are interconnected with the same plates. Let's use the conventional current flow to plain this because sometime I might use the electron current flow concept as well. Think about the charging cycle of a lead acid automotive starting battery when it's just a charge on it as well.

The charging voltage and current have to enter the positive terminal of the battery connection based on the laws governing parallel circuitry. Since its positive terminal, it'll be connected in parallel with its positive plates in that cell. Each plate will get the same amount of voltage flowing from the charging source because of the mechanics of the voltage.

However; the charging current, it'll flow to those positive plates based on their individual resistance because of the mechanics of the current. When the charging voltage and current, they're transferred from the positive plates to the adjacent negative plates in that cell by way of the electrolyte solution.

The electrolyte solution in the cell with the positive terminal will create a series circuit connection between the positive and negative plates in that cell. Now because different circumstances, they're involved since current will be flowing across the positive plates in the cell with the positive terminal connection in it based on their individual resistance.

Then the same amount of current flowing across each positive plate, it'll be transferred to an adjacent negative plate by way of the electrolyte solution in that cell. It'll be a twist when the charging current begins to flow from the negative plates in that cell to the strap joining them together in parallel; wherein, current will regroup or combined at the strap.

As a result; the sum of the current flowing to the positive plates in the next cell, it'll be the same as the current flowing from charging source; although, it was divided amongst the negative plates in the previous cell based on their individual resistance as well. Thus; current, it'll behave differently flowing <u>to</u> a strap than flowing from a strap connecting the plates together in parallel.

In other words; current, it'll be divided amongst the plates based on their individual resistance as it flows them from the strap. However; when current, it flows <u>from</u> the plates to the strap. It'll regroup at the strap that joins the plates together in parallel. Thus; current will flow to a strap in reverse of how, it'll flow from a strap.

When the charging voltage flows across the positive plates in the cell with the positive terminal in it, voltage will be the same across each positive plate as it flows from the charging source. Given that; those plates, they're connected parallel with the positive terminal connection in that cell.

However; the adjacent negative plates in that cell with the positive terminal in it, they'll receive a voltage drop from the adjacent positive plates. Given that; the negative plates in that cell, they're connected in series with the adjacent positive plates by way of the electrolyte solution in that cell with the positive terminal connection in it.

It'll be a twist when voltage begins to flow from the negative plates in the cell with the positive terminal connection in it to the positive plates in the next cell. Voltage will not regroup at the strap joining the negative plates together in parallel; thus, voltage will be the same as it's flowing across one negative in the cell with the positive terminal connection in it.

Therefore; the positive plates in the next cell, they'll receive a voltage drop from the charging source; consequently, voltage will continue to drop evenly throughout each cell within the battery. Thus; the mechanics of the voltage will change because flowing <u>from</u> a strap, it'll be the same across each plate; but, flowing <u>to</u> a strap. It'll be the same at the strap as it's across one plate.

As a result; voltage, it'll flow from a strap different than it'll flow to a strap. Looking at the mechanics of the voltage and current, they'll behave differently flowing from a strap to a plate than flowing from a plate to a strap because of the different circumstances involved. It's more to it than just due to the laws of physics why a battery can't be charged while being discharged.

Here's why; when there's a charge and a load source on, the same terminal connection of a battery. Then the reason why voltage, it'll be the same across the battery and the load source; but, current will not. It's more to it than the mechanics of the voltage and current alone; but, it has something to do with the connections made between the battery and the load source as well.

We just can't blame it on the mechanics of the voltage and current alone or the laws of physics. To justify why a secondary cell battery can't be charged while being discharged to a load source, then we become inept to use them because it's more to it than the mechanics of the voltage and current alone or the laws of physics as well.

Here's another point; the amount of <u>voltage</u>, each cell in a lead acid automotive starting battery will receive from a charging source. It'll be based on, how the battery cells are interconnected with its terminal connections as well. Since its cells, they're connected in <u>series</u> with its terminal connections.

Then the charging voltage, it'll drop evenly throughout each cell in the battery during its charging cycle; but, <u>current</u> will be the same throughout each cell in the battery. During its discharging cycle, its voltage potential will equal to the voltage potential of all of its plates in each cell and its current potential will only equal to the plates in one cell.

Now if a battery cells, they're connected in <u>parallel</u> with its terminal connections. During its charging cycle, voltage will be the same throughout each cell; but, current will flow to each cell in it based on their individual resistance. During its discharging cycle, its voltage potential will equal to the plates in one cell; but, its current potential will equal to all of its plates in each cell.

Upon evaluating, the mechanical structure of a battery. We'll find voltage and current will flow in and out of it based on its design. We don't have to focus on the vast and complex laws of physics to figure out why, it can't be charged while being discharged to a load source. If we focus on, the details on how voltage and current will behave under different circumstances.

However; we've to separate voltage from current to understand, the difference between them. If we're going to understand, voltage will enter or exit a battery different than current will based on its design; but; voltage will flow through it different than current will based on which cycling process is carried out as well.

On the other hand; when I was carrying out, my experiments with my prototype. I found it wasn't using, its chemical electromotive force to discharge it to the tire inflation device; however, the electromotive force created by the charging source. It was discharging my prototype to the tire inflation device.

Wherein; the lead-sulfate build up on, the battery plates was decreasing while minimizing their electrical resistance; thus, increasing their efficiency and overall capacity for their next chemical discharging cycles as well. In short; when I stop simultaneously charging and discharging my prototype and then, started using its chemical electromotive force to discharge it.

I found my prototype had less internal resistance when I started its next chemical discharging cycle to the tire inflation device. Thus; I realized that, a simultaneous cycling process of our secondary cell batteries. It'll be a mechanical solution to increase their efficiency and overall capacity without finding a perfect chemical composition for them.

On the other hand; we think, it's impossible to carry out a simultaneous cycling process of our batteries and it'll be no benefits in it as well. Wherein; the experiments, I carried out with my prototype of a lead acid automotive starting battery says otherwise. If we design our batteries, so, energy could enter and exit them at the same time.

I believe it comes down to facts versus perceptions. Why we haven't thought about adding, energy back to our secondary cell batteries while taking it out. As an option to, increase their efficiency and overall capacity without finding a perfect chemical composition for them. After all, we could replenish their chemical energy with electrical energy as well.

In the next chapter; let's talk about facts versus perceptions because after my analysis of the mechanics behind, a lead acid automotive starting battery's cycling processes. I found our perceptions about what we could or couldn't do with the battery because of the laws of physics, it doesn't add up with the mechanics behind its cycling processes as well.

CHAPTER 21

FACTS VERSUS PERCEPTIONS

AFTER MY CAREFUL ANALYSIS of the mechanics behind, a lead acid automotive starting battery's cycling processes. I found facts versus perceptions get very murky when it comes to what we could or couldn't do with it because of the laws of physics. It seems like we haven't carried out the experiment for ourselves. To see if we changed its design, we could charge it while discharging it.

Although; I found, energy lost to heat occurs in both conversion processes of the battery; but, it'll be irrelevant during a simultaneous cycling process of it because we'll be adding energy back to it while taking energy out as well. Using the laws of entropy, inertia or the laws of conservation of energy to, discredit a simultaneous cycling process of the battery.

I find the lack of information will create a false perception about what we could or couldn't do with our batteries because of the laws of physics as well. For more than a century those in the field of automotive battery technology, they've been designing our automotive batteries with only one opening for energy to enter or to exit them.

On the other hand; some will turn around and say that, it'll be impossible to charge our batteries while discharging them because of the laws of physics as well. However; some will say, even if energy could enter and exit our batteries at the same time. It'll be no benefits in it because of the laws of entropy, inertia or the laws of conservation of energy as well.

Some in the field of automotive battery technology, they'll not consider each cell in a lead acid automotive battery is nothing more than a battery in and of itself. That is connected in series with one another and housed together in a plastic case to make up the whole battery as we know of today, which creates the possibility of charging it while discharging it to a load source.

However; some in field of automotive battery technology, they'll overlook the fact when the battery is charged as a whole. We're simultaneously charging and discharging one battery while simultaneously charging and discharging another because of how they're interconnected with one another, which makes it possible to charge it while discharging it to a load source as well.

It seems like there's a disconnection between how some in the field of automotive battery technology design the battery and how, some think its cycling processes will be carried out within the realm of physics as well. Wherein; facts versus perceptions, it comes very murky when it comes to a conversation about charging a lead acid automotive starting battery while discharging it.

I'm not just talking about a lead acid automotive starting battery in particular; but, other types of secondary cell batteries as well. Having energy to go in our batteries, the same way it has to come out. I guess it'll be impossible to charge them while discharging them due to the laws of physics because like charges will repel one another and follow the path of least resistance as well.

However; the assumption that, we can't charge a lead acid automotive starting battery while discharging it due to the laws of physics. It'll not hold true if energy doesn't have to go in the battery the same way it has to come out. Likewise; neither the laws of entropy, inertia nor the laws of conservation of energy will prevent it from being charged while being discharged as well.

Those assumptions made by some in the field of automotive battery technology, they're not be consistent with the laws of entropy, inertia or the laws of conservation of energy as well. Given that; the battery, it'll not be considered a closed or perpetual system under those laws of physics if we're adding energy back to it while taking energy out; thus, those laws of physics will not apply.

Most in the field of automotive battery technology fail to realize, a simultaneous cycling process of a battery means energy will be added back to it while energy is taken out as well. Therefore; we can't use, either the laws of entropy, inertia or the laws of conservation of energy to discredit a simultaneous cycling of the battery.

Then the assumption in and of itself, it'll contradict those laws of physics if we're adding energy back to the battery while taking it out as well. Likewise; the skeptics, they can't use the laws of electrochemistry to discredit a simultaneous cycling of the battery. Given that, we could already add or subtract energy from the same chemical composition of it as well.

Therefore; the only thing, we've to worry about is the electromagnetic laws of physics. Since electrons, they've to go in the battery the same way they've to come out in order to charge its plates of for them to be discharged. Then like charges, they'll prevent it from be charged while being discharged because they repel one another and follow the path of least resistance as well.

Not only will the electromagnetic laws of physics, they'll be an issue if electrons have to go in a battery the same way they've to come out; but, the laws of mechanics will be an issue as well; thus, facts versus perceptions. They become murkier if electrons have to go in the same way they've to come out; given that, either one could prevent the battery from be charged while being discharged as well.

I found a host of things, other than the laws of physics will prevent a secondary cell battery from being charged while being discharged. It has little to do with the laws of physics in and of themselves; but, it has more to do with the laws of mechanics. This is where facts versus perceptions, it gets very murky for some folks because it becomes confusing about which one to blame.

However; focusing on, the laws of mechanics instead of the laws of physics. We could clear up some of this murkiest because we'll be focusing on the right science. See chapter 15. Most people become hyperopic to the fact that, electrons have to go in a battery the same way they've to come out in order to charge its plates or for them to be discharged as well.

On the other hand; most people become hyperopic to the fact that, the design of a lead acid automotive battery. Is the deciding factor in determining how energy will enter or exit it during its cycling processes, not the laws of physics? If we know anything about the mechanics behind its cycling processes, then we'll know this is true as well. See chapter 3 and 4.

Although; those in the field of automotive battery technology, they're the architect behind the design of the battery. However; there's a disconnection between, how they design it and how they think its cycling processes will be carried out as well. It seems like they haven't used their knowledge about its design to know its cycling processes is a product of its design as well.

On the other hand; those in the field of automotive battery technology, they can't foresee the need to change the design of our deep cycle batteries; so, energy could enter and exit them at the same time. Wherein; it seems like, it'll be another option to increase their efficiency and overall capacity without finding a perfect chemical composition for them.

Given that; we could replenish, our secondary cell batteries' chemical energy with electrical energy as well. Thus; if we could add energy back to them while taking it out, then we could increase the overall travel ranges of our vehicles on battery power as well. Then we don't have to worry about finding a perfect chemical composition to a longer-lasting battery as well.

If we're talking about facts versus perceptions, then why do we believe we need to find a perfect chemical composition for our secondary cell batteries when we could replenish their chemical energy with electrical energy as well? If you think about, the whole premise. It doesn't make any sense we need to find a perfect chemical composition for them as well.

On the other hand; if energy, it doesn't have to go in our batteries the same way it has to come out. Then we wouldn't have to worry about the electromagnetic laws of physics or the laws of mechanics as well. Thus; facts versus our perceptions, they become less murky about what we could or couldn't do with our batteries if energy could enter and exit them at the same time as well.

However; on the other hand, the mechanics behind the charging cycle of a lead acid automotive starting battery. It'll clear up some of the murkiness that, it can't be charged while being discharged to a load source because of the laws of physics. Given that; the battery plates in one cell, they're simultaneously charged and discharged to its plates in the next cell as well.

I found the battery plates in each cell are connected in series-parallel with one another by their straps, tab connectors and the electrolyte solution in each cell. And each cell, it's connected in series with its terminal connections as well. In which; it'll allow, its plates in each cell to store energy while supplying it to its plates in the next cell as well.

Although; the battery plates in each cell, they're separated by a non-conducting material called plastic as well. And yet, we're still able to charge its plates in each cell by way of an electrical bus without having to stop charging them in one cell in order to charge them in the next. Wherein; it seems strange, we've to stop its discharging cycle in order to start its charging cycle.

When it comes to facts versus perceptions, it doesn't make any sense because we don't have to stop charging the battery plates in one cell in order to charge them in the next; although, they're separated by a non-conducting material called plastic. In essence, it's like charging the battery plates while discharging them to a load source as well.

This where facts versus perceptions become less murky because we could charge the battery plates in each cell without having to, stop charging them in one cell in order to charge them in the next cell. All because electrons, they could flow in one direction from its plates in one cell to its plates in the next because of the right mechanics, are in play. See chapter 15.

Based on, the mechanics behind the charging cycle of a lead acid automotive starting battery. A person could conclude that the laws of physics aren't the underlining cause why the battery can't be charged while being discharged to a load source. It seems like the battery isn't built to the right specifications for charging while discharging to a load source as well.

All because we've to charge the battery on the same terminal connection that, it's discharged on. Once we've establish some simple facts about, why it can't be charged while being discharged to a load source. Other than blaming it on the laws of physics in and of themselves, then facts become clearer than perceptions as well.

We'll find it has more to do with the laws of mechanics than the laws of physics why a secondary cell battery can't be charged while being discharged to a load source; such as, an electric motor or a cooling fan. Our perceptions about having to, stop one process in order to start the other because of laws of physics is an illusion created by the design of the battery.

This illusion was created more than a century ago by the first lead acid automotive battery or the first secondary cell battery. Wherein; it stemmed from the fact that, electrons had to go in the battery the same way they had to come out in order to cycle it. Since we never changed the design of the battery, our perceptions never changed as well; so, the illusion continues.

Likewise; the assumption, it'll be no benefits in adding energy back to a secondary cell battery while taking it out due to either the laws of entropy, inertia or the laws of conservation of energy as well. It's an illusion created by the lack of information about the process because it has never been done before; so, the assumption stemmed for speculations about the process.

Here's why; since energy, it has to go in the battery the same way it has to come out. Then we can't add it back to the battery while taking it out; so, the idea it'll be no benefits in adding it back to the battery while taking it out because of the laws of entropy, inertia or the laws of conservation of energy is nothing more than speculations because it can't be done.

Fasts versus perceptions or speculations is a different story because if energy. It could flow in one direction from a battery plates in one cell to its plates in the next. Then there's a benefit in it because we don't have to stop charging its plates in one cell in order to charge them in the next; thus, they could store energy while supplying it to the plates in the next cell as well.

For more than a century those in the field of automotive battery technology, they've designed our secondary cell batteries, so, energy has to go in them the same way it has to come out. During all that time, it seems like no one venture outside the status quo to see if it was necessary to stop their discharging cycles in order to start their charging cycles as well.

On the backs of Gaston Plante', who made the first lead acid automotive battery or the first secondary cell battery in 1859. Others in the field of automotive battery technology, they took that information at face value. And then, package it without think is it necessary to stop its discharging cycle in order to start its charging cycle because of the laws of physics?

I think it's a good question to ask since we've to charge the battery on the same terminal connections that, it was discharged on. Wherein; it seems strange, we think it's due to the laws of physics because we've to stop one process in order to start the other. When it's obvious because we've to charge it on the same terminal connections that, it was discharged on.

However; the million dollar question is that, do we've to stop one cycling process of a battery in order to start the other because of the laws of physics? I found the answer is no because the laws of physics, they've little to do with it. Only to the extent that, energy has to go in the battery the same way it has to come out and then they'll be a problem.

Given that; like charges, they'll repel one another and follow the path of least resistance as well. If we expand on the fact that, energy has to go in the battery the same way it has to come out. Then our focus shouldn't be on the laws of physics; but, it should be on the laws of mechanics because they determine how, energy will enter or exit the battery and not the laws of physics.

Focusing the laws of mechanics, they'll help us separate facts from perceptions. Since a lead acid automotive starting battery's exterior structure, it gives a false impression we've to stop its discharging cycle in order to start its charging cycle because of the laws of physics. If we look at the mechanics behind its charging cycle, it'll prove otherwise.

I found it's not necessary to stop the discharging cycle of the battery in order to start its charging cycle; as long as, electrons could flow in one direction from its plates to a load source. Then the laws of physics, they shouldn't an issue because they're not an issue, when electrons have to flow in one direction from its plates in one cell to its plates in the next cell as well.

Although; the battery plates in each cell, they're separated by a non-conducting material called plastic. And yet, we could charge them in one cell while charging them in another without the laws of physics being an issue. When it comes to facts versus perceptions, then our perceptions is flawed due to the exterior design of the battery because it creates this false perception.

Then the concept of charging a lead acid automotive starting battery while discharging it to a load source, it'll challenge our perceptions about the process itself because we can't see any other way to carry out its cycling processes; but, to stop its discharging cycle in order to start its charging cycle because of the laws physics; in which, it's a false perception as well.

When we're talking about facts versus perceptions, our perceptions will lose because facts tell us the only reason. We've to stop the discharging cycle of the battery in order to start its charging cycle because of its exterior design. Since we don't have to, stop charging its plates in one cell in order to charge them in the next; although, they're separated by a non-conducting material.

Since we've to charge the battery on the same terminal connection that, it was discharged on. Then the process itself prevents it from being charged while being discharged because electrons have to go in it the same way they've to come out in order to charge its plates or for them to be discharged as well; thus, it's clear not due to the laws of physics as well.

Here's why; the interior design of a lead acid automotive starting battery, it shows it's not due to laws of physics why it can't be charged while being discharged to a load source. They've little to do with why we've to stop its discharging cycle in order to start its charging cycle because they allow its plates in one cell to be charged and discharged to its plates in the next cell as well.

Although; the battery plates in each cell, they're separated by a non-conducting material called plastic. Therefore; its exterior design, it creates a false impression we've to stop its discharging cycle in order to start its charging cycle because of the laws of physics. Given that; electrons, they've to go in it the same way they've to come out as well.

If we focus on electrons being simultaneously added and subtracted from the battery plates in one cell to its plates in the next cell. Then it'll confirm, it's not due to the laws of physics why we've to stop its discharging cycle in order to start its charging cycle. However; it's due to, how we go about adding or subtracting electrons from its plates.

Facts versus perceptions could become less murky if we take an in depth analysis of the mechanical structures of our secondary cell batteries. To see how, they'll play a role in their cycling processes. Then we'll find, not only do we need to upgrade their chemical compositions from time to time; but, we need to upgrade their mechanical structures as well.

To facilitate, the usage they're needed for in the modern world of automotive technology. For more than a century, we've been designing our secondary cell batteries the same old way. That is we've to charge them on the same terminal connections that, they were discharged on. If we think battery technology, it only comes in the form of a chemical solution.

Then we're sadly mistaking because it could come in the form of a mechanical solution; given that, we could replenish our secondary cell batteries chemical energy with electrical energy as well. Therefore; back to facts versus perceptions because it's not necessary to find, a perfect chemical composition to a longer-lasting battery as well.

Thus; we could use, a charging system to increase the efficiency and overall capacity our secondary cell batteries. After all, we could replenish their chemical energy with electrical energy as well. We fail to realize, we just can't focus on finding a perfect chemical composition; but, we've to utilize our battery technology already on the shelf in the right manner as well.

If we don't take an in depth analysis, behind the mechanical structure of our secondary cell batteries and see how, they'll play a role in their cycling processes. Then we'll never be able to separate facts from perceptions about what we could or couldn't do with our batteries because of the laws of physics, it'll always be a blur between facts versus perceptions as well.

Likewise; if we don't figure out facts from perceptions, then we'll never advance our battery technology without finding a perfect chemical composition to a longer-lasting battery as well. Since our secondary cell batteries' chemical compositions, they're two directional. Then it's a great opportunity for us to advance our battery technology with a mechanical solution as well.

However; if we continue to design our batteries, so, we've charge them on the same terminal connections they were discharged on. It limits our opportunity to increase their efficiency and overall capacity with a mechanical solution because energy, it has to go in them the same way it has to come out in order to cycle them as well.

I find we can't advance our battery technology with a mechanical solution because we've a mechanical issue, when it comes to our batteries energy input and output process. Let's focusing on the exterior design of a modern day cell phone battery for a moment, it'll help us separate facts from perceptions about what we could or couldn't do with our batteries as well.

The innovated design of a modern day cell phone's battery will allow us to charge it without missing a call because it has more than one negative and one positive terminal connection.

Although; its respective terminal connections, they're interconnected by the same strap; but, they allow electrons to flow to or from the same strap as well.

Therefore; we don't have to stop using, the cell phone in order to charge its battery; unlike, the early days when cell phones were first introduced to the general public. We were either using the cell phone or charging its battery; but, we couldn't do both. However; new cell phone battery technology, it allows us to receive or make calls while the phone is in its charging mode.

However; it'll be difficult to charge, the cell phone's battery while using it because it might consume electrons faster than its charging apparatus. It could create a surplus of electrons at the phone's voltage and current regulatory system to reverse the current flow back into its battery. Given that; its respective terminal connections, are interconnected by the same strap.

If we take an in depth analysis of the mechanical structure of today cell phones' batteries, then we'll find it has little to do with the laws of physics why it'll be difficult to charge them while using them; although, they've multiple terminal connections as well. However; if we focus on facts versus perceptions, then we'll find it has more to do with mechanics than anything else.

In other words; we'll find the wrong mechanics, are in play to charge the batteries while using them. While acknowledging the fact that, it's an upside for the batteries' respective terminals sharing the same straps because it allows us to receive or make calls when the cell phones are in their charging modes as well.

Likewise; the batteries' respective terminals sharing the same straps, it allows the cell phones voltage and current regulatory systems to be connected to one respective terminal to power the phones. While the other respective terminals, they'll allow thermistors to be connected to the batteries for overcharge protection for them as well.

Thus; electrons, they could simultaneously flow from the cell phones charging apparatuses to their batteries and from them to their thermistors for overcharge protection. As a result; electrons, they don't have to flow from the same terminal to flow to the cell phones' voltage and current regulatory systems or flow to their thermistors as well.

Therefore; the design concept of the modern day cell phones' batteries allow us to receive or make calls while the batteries, are in their charging modes; unlike, the old cell phone battery technology when the phone was in its charging mode. We couldn't receive or make calls because electrons, they had to flow from the same terminal to power the phone or charge its battery.

In essence; our electric vehicles, they work the same way. We either charging their batteries or using them; but, we can't do both because of the contemporary design of their batteries. Since we've to charge them on the same terminal connections that, they were discharged on because energy has to go in them the same way it has to come out as well.

However; our perceptions, they tell us it's due to the laws of physics why we've to stop the discharging cycles of our batteries in order to start their charging cycles because facts versus perceptions seem murky. Given the exterior design of our batteries, they give us a false perception it's due to the laws of physics why we've to stop one process in order to start the other as well.

But; in reality, it's due to the laws of mechanics why we've to stop one process in order to start the other because energy, it has to go in our batteries the same way it has to come out in order to charge their plates or for them to be discharged to a load source as well. The reason is never going to change why we've to stop one process in order to start the other; unless, we change it.

On the other hand; a lead acid automotive starting battery chemical composition, it doesn't generate its own electrons because it's just a storage device for electrons. Therefore; it'll need an internal electromotive force to, subtract electrons from it and an external electromotive force to add electrons back to it when it comes to its cycling processes.

What I learned was that, it's a matter of which electromotive force will have the highest voltage potential between the charging voltage and the battery voltage. It'll determine which direction electrons will flow between the negative and the positive electrode of the battery. Since it only has two electrodes to, reverse the current flow between during its cycling process.

Wherein; it reinforces our perceptions, it's due to the laws of physics why we've to stop one cycling process in order to start the other. However; taking an in depth analysis of the mechanical structure of the battery, then we'll find it's due to its mechanical structure why we've to stop one cycling process in order to start the other and not due to the laws of physics.

Basically; it's due to the battery having, only one negative and one positive electrode to reverse the current flow between during its cycling process. Then facts versus perceptions become clearer, it's due to its mechanical structure why we've to stop one cycling process in order to start the other because electrons have to go in the same way they've to come out as well.

On the other hand; our perceptions, they've long told us that the charge and the discharging cycle of a lead acid automotive starting battery. They're two different processes; therefore, they can't exist at the same time. It's nothing more than an illusion created by the design of the battery because it has only two electrodes to reverse the current flow between.

No matter how we slice it; line upon line and precept upon precept, it always comes down to the design of the battery why we've to stop its discharging cycle in order to start its charging cycle. Let's see what facts versus perceptions will tell us if the battery has more than two electrodes to reverse the current flow between during its cycling process as well.

Facts will tell us that, the charge and discharging cycle of the battery. They could exist at the same time because electrons don't have to travel the same path in order to enter or to exit it; thus, we don't have to stop its discharging cycle in order to start its charging cycle. This is where facts, they're separated from perceptions about what we could or couldn't do with it as well.

Since the battery, it'll need an internal electromotive force to subtract electrons from it and an external electromotive force to add electrons back to it. Given which electromotive force will have the highest voltage potential will prevail over the other. Then it makes, it possible for the battery charge and discharge cycle to exist at the same time as well.

For example; if a battery, it has more than two electrodes to reverse the current flow between. Then electrons, they don't have to travel the same path in order to enter or to exit the battery. Therefore; the charging action of the battery, it'll not affect its discharging action because it'll have more than two electrodes to reverse the current flow between.

However; which electromotive force has the highest voltage potential between the charging voltage and the battery voltage, it'll prevail over the other. Therefore; we'll not have, two electromotive forces competing against one another; thus, it'll be no conflict between the charge and discharging cycle of the battery if we attempt to charge it while discharging it.

Dismissing the assumption that, the battery charge and discharging cycle. They can't exist at the same time because of the laws of physics as well. It's nothing more a false perception created by the battery having, only one electrode for current to enter or to exit it. Separating facts from perceptions, then it's clear it's a matter of mechanics and not a matter of physics.

When some skeptics make their arguments, it's due to the laws of physics why a battery charge and discharging cycle can't exist at the same time. Their arguments become flawed because they're not acknowledging current has to go in the same way it has to come out. Then facts versus perceptions become murky because it's a matter of mechanics and not a matter of physics.

Let me reiterate what's an electromotive force is and it might help separate facts from perceptions about what we could or couldn't do with our secondary cell batteries as well. An electromotive force or voltage, it moves electrons; in which, current is the combination of voltage and electrons moving in one direction around a conducting source.

There're five different types of electromotive forces; heat, pressure, light, chemical and magnetic will move electrons. However; we usually, use an electromagnetic force to charge a lead acid automotive starting battery and a chemical electromotive force to discharge it. Since both electromotive forces, they could achieve the same goal of moving electrons as well.

Then the only difference between the charge and discharging cycle of the battery, it's the direction that electrons are flowing between its negative and positive electrodes. Since we've to add electrons to it before we could subtract electrons from it, given its chemical composition doesn't generate its own electrons as well.

Therefore; it makes, it possible to add and subtract electrons from the battery at the same time. Given that; adding, electrons to back it will not stop the process of subtracting electrons from it because it's nothing more than a storage device for electrons. This is where facts, are separated from perceptions if we attempt to charge it while discharging it as well.

Remember if a battery plates in each cell, they're connected in series-parallel with one another by their straps, tab connectors and the electrolyte solution in each cell. Thus; electrons could flow from a strap to a tab connector on a plate and then, flow across that plate through electrolyte solution to an adjacent plate without flowing to and from the same tab connector.

So; the battery plates in each cell, they're simultaneously charged and discharged amongst themselves in each cell. If the battery cell, they're connected in series with its terminal connections. Then its plates in one cell, they're simultaneously charged and discharged to its plates in the next cell because of how they'll be interconnected with another.

Wherein; during the charging cycle of the battery, a charge and discharging cycle of its plates will exist at the same time. Since adding, electrons to back it will not stop the process of subtracting electrons from it because it's nothing more than a storage device for electrons. Facts versus perceptions become clear because it's about, how we add or subtract electrons from it.

If a battery, it had more than two electrodes to reverse the current flow between. Then it'll be possible to charge its plates while discharging them to a load source as well. Given that; current, it doesn't have to flow to the battery plates. The same way it has to flow from them in order to charge them or for them to be discharged to the load source as well.

Here's the thing when it comes to facts versus perceptions, it always come down to a person experience or perception. For instance; some in the field of electrochemistry, they think it'll be inconceivable or farfetched to charge a lead acid automotive starting battery while discharging to a load source because of how they view the cycling processes of the battery.

Some in the field of electrochemistry think in order to charge a battery, the polarity between its negative and positive electrode must to be in reverse opposite of the polarity between them during its discharging cycle. Thus; they think, its charge and discharging cycle can't exist at the same time based on their perceptions.

If we separate facts from perceptions; given that, the battery has only two electrodes to reverse the current flow between. And current, it could only enter or exit one of those electrodes in order to carry out the cycling processes of the battery as well. Then it would be impossible to charge the battery while discharging it because of the laws of physics.

Here's why; the polarity between the two electrodes of the battery, they must be in reverse opposite of one another in order for current to enter or to exit the battery. In other words; current, it has to go in the same way it has to come out; given that, it could only enter or exit one of the electrodes; thus, the battery can't be charged while being discharged as well.

If those in the field of electrochemistry, they took an in depth analysis of the mechanical structure of a lead acid automotive starting battery. Then they'll see why its charge and discharging cycle, they can't exist at the same time. Given that; the battery, it has only one opening for current to enter or to exit it as well.

Thus; which voltage potential has the highest potential between the charging voltage and the battery voltage, it'll prevail over the other. Wherein; it'll determine which direction, current will flow between the negative and the positive electrode within the battery. Therefore; current, either flowing in or out of the battery; but, it can't do both because of the design of the battery.

However; if the battery, it had more than two electrodes to reverse the polarity between. Then the assumption made by some in the field of electrochemistry, its charge and discharging cycle couldn't exist at the same time becomes flawed. This is where facts are separated from perceptions because those electrochemists, their argument wouldn't hold true.

However; the electrochemists' argument, it would prove my theory. It would be possible to charge the battery while discharging it if it had more than two electrodes to reverse the current flow between. Given that; if the polarity was reversed between two of its electrodes, then it would be reversed between the other electrodes as well.

Then facts versus perceptions become clearer when it comes to charging the battery while discharging it. For example; based on the conventional current flow concept, current has to enter or exit the positive electrode of a battery in order to cycle it. Therefore; if a battery, it had two positive and one negative electrode to reverse the current flow between.

Then the charging voltage could force electrons through one positive electrode and out the other while voltage is forcing electrons toward the negative electrode. Thus; the argument made by the electrochemists, it'll become flawed because it'll possible to charge the battery while discharging it; given that, it'll have more than two electrodes to reverse the current between.

It'll have little to do with the laws of physics why a secondary cell battery can't be charged while being discharged; but, it'll have more to do with the contemporary design of the battery. Thus; facts versus perceptions, they become less murky once we understand the design of the battery is the deciding factor in determining how its cycling processes will be carried.

On the other hand; the assumptions that, it'll be no benefits in charging a lead acid automotive starting battery while discharging it due to the laws of entropy, inertia or the laws of conservation of energy. Wherein; it'll be a zero gain because energy lost to heat, it'll occur in both of its conversion processes. It'll be hard to believe if we're adding energy back to it while taking energy out as well.

If we separate facts from perceptions, we can't use those laws of physics to discredit a simultaneous cycling process of the battery. Since the laws of entropy, they refer to energy being turned into heat by a mechanical action. The laws of inertia or the laws of conservation of energy refer to energy running down at some point within a system if energy isn't added back.

The third law of thermodynamics states when energy is lost to heat within a closed system. Then energy can't be used to do any more work than it already has done; that is, created heat. Thus; energy, it has to be added back to the system with an outside power source to do more work; therefore, if we're adding energy back to a battery while taking it out.

Then the battery will not be considered a closed system under those laws of physics. There's a disconnection between how we think and how those laws of physics will actually play a role in a simultaneous cycling process of a secondary cell battery. If we focus on facts versus perceptions, then facts say we're focusing on the wrong science when it comes to the process.

My research on the mechanics behind, a run capacitor and a power transformer energy input and output process. It helped me to separate facts from perceptions when it comes to the energy input and output process of a lead acid automotive starting battery. We're not bound or restricted by the laws of physics; but, only restricted by the contemporary design of the battery.

In the next chapter; I'm going to show, we're only restricted by the contemporary design of our secondary cell batteries when it comes to their energy input and output processes. We could charge them while discharging them if energy could enter and exit them at the same time, like it does our capacitors and transformers as well.

CHAPTER 22

REVERSE ENGINEERING OF AN AUTO BATTERY

WE COULD CONJURE UP all manner of laws of physics to justify why a secondary cell battery can't be charged while being discharged. However; no matter how we slice it; line upon line and precept upon precept. It all boils down to the contemporary design of the battery why it can't be charged while being discharged to a load source.

In figure T-3; at the top of the diagram, it shows the design concept of a lead acid automotive starting battery; in which, it has only one tab connector per plate. Wherein; it's another reason why it can't be charged while being discharged, other than blaming it on the mechanics of the voltage and current alone or the laws of physics as well.

Figure T-3

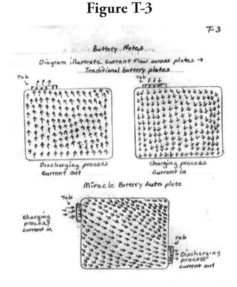

(How electrons will flow across a battery plates)

In figure T-3; I'm using two plates to show, how electrons will flow to and from a lead acid automotive starting battery plates because they only have one tab connector each. I'm hoping it'll shed some additional light on why I was able charge my prototype of the battery while discharging it to tire inflation device. See chapter 19 and 20.

Understanding, how a lead acid automotive starting battery plates are designed. Then you might see, the difference between having a charge and a load source on the same terminal connection of it. Compare to, it having separate terminals for charging and discharging in the same cell and those terminals having their own straps and tab connectors leading to the same plates as well.

Remember the contemporary design of a lead acid automotive starting battery with top and side posts terminal connections. I'm going to use its design to explain why it can't be charged while being discharged because it's not built to the right specification for charging while discharging; although, it has two positive and two negative terminals as well.

Here's my point; the battery, its respective terminal connections share the same strap and tab connectors leading to their respective plates. In which; the strap, it joins the two terminals together in parallel. Not only does it join them together in parallel; but, it makes them equal to one terminal connection for electrons to enter or to exit the battery as well.

Given that, electrons, they've to travel from the same tab connectors in order to enter or exit the battery. It's one more reason why it can't be charged while being discharged to a load source as well. So; it's important to understand, the integrate details about its mechanical structure and how it'll play a role in its cycling processes as well.

If we focus on, the integrate details about the mechanical structure of the battery. Then we'll have no misconceptions about why it can't be charged while being discharged to a load source. We'll know it's due to the contemporary design of the battery; although, it has two positive terminals in one cell and two negative terminals in another cell as well.

It's not just one thing will prevent the battery from being charged while being discharged, it could be a host of things coming together as well. We just can't focus on the mechanics of the voltage and current alone or the laws of physics; although, they could be a part of it. The design of the battery is the main reason why it can't be charged while being discharged.

If we grasp, the mechanical make up of a lead acid automotive starting battery with top and side posts terminal connections. We'll know the wrong mechanics are in play to charge it while discharging it. Wherein; it has little to do with the mechanics of the voltage and current alone or the laws of physics why it can't be charged while being discharged as well.

However; it's due to fact that, the battery respective terminals only equal to one terminal connection. Given that; electrons, they could only travel to and from its respective terminals to their respective plates from the same tab connectors on them. Thus; the two respective terminals, they'll be connected in parallel with one another by their respective plates.

Remember a parallel circuit connection made between two load sources, it'll allow both load sources to receive the same amount of voltage flowing from a power source. But; current flowing to them, it'll be based on their individual resistance. Thus; each load source, it'll be a branch within the circuit in their relationship with the power source. See chapter 5 and 12.

It holds true for a lead acid automotive starting battery plates since they behave like load sources consuming and storing energy during their charging cycle. Thus; the tab connector on each plate, it connects it to its respective strap. In which; the tab connector, it's the branch between the strap and the plate for electrons to travel to and from it as well.

However; electrons, they could flow from one plate to the next by way of the electrolyte solution in each cell without flowing to and from the same tab connector on the same plate; thus, allowing electrons to flow in one direction from one plate to the next. Thus; electrons, they could added and subtracted from one plate, so, they could be added to another plate as well.

Also; straps connecting the plates in one cell together in series with the plates in the next cell, they'll allow electrons to flow from the plates in one cell to the plates in the next cell; although, they're separated by a non-conducting material called plastic. As a result; we don't have to stop charging, the plates in one cell in order to charge them in the next cell as well.

I found it had little to do with the mechanics of the voltage and current alone or the laws of physics. Why we don't have to stop charging, the battery plates in one cell in order to charge them in the next cell; although, they're separated by a non-conducting material called plastic. It was due to straps connecting the plates in one cell together in series with the plates in the next cell.

I found electrons could flow in one direction from the battery plates in one cell to its plates in the next cell. It had little to do with the tab connectors on each plate; but, it had more to do with the electrolyte solution in each cell. It allowed electrons to flow from one plate to another without flowing to and from the same tab connector on the same plate as well.

In which; electrons, they could flow from a strap to a tab connector on one plate. And then, flow across that plate through the electrolyte solution to an adjacent plate. Without flowing to and from, the same tab connector on the same plate. Thus; one plate, it's simultaneously charged and discharged to another plate by way of the electrolyte solution.

Since the battery cells, they're connected in series with one another. Then its plates in one cell, they're simultaneously charged and discharged to its plates in the next cell as well. Given the mechanical make up of a lead acid automotive starting battery, we don't have to stop charging its plates in one cell in order to charge them in the next cell as well.

It's not about the mechanics of the voltage and current alone or the laws of physics. Why we don't have to stop charging, the battery plates in one cell in order to charge them in the next

REVERSE ENGINEERING OF THE DEEP-CYCLE AUTOMOTIVE BATTERY

cell. We achieve this goal by connecting the battery plates in series-parallel with one another in each cell and connecting each cell in series with its terminal connections as well.

Understanding, how the battery plates in each cell is charged. Without having to, stop charging them in one cell in order to charge them in the next cell. Then we don't need to focus on the laws of physics to figure out, why it can't be charged while being discharged; although, it has two positive terminals in one cell and two negative terminal in another cell as well.

Given that; we'll know why, the battery can't be charged while being discharged because of its design. Understanding, the mechanics behind its charging cycle as well. Then we'll know, we could use the same technique; that is, used to simultaneously charge and discharge its plates amongst themselves in each cell.

We could use that technique to simultaneously charge and discharge the battery plates to a load source if we've the right mechanics in play as well. I know it's hard for some folks to believe, it's all about mechanics why we don't have to stop charging its plates in one cell in order to charge them in the next cell; although, they're separated by a non-conducting material.

In figure T-3, it shows a modified version of a lead acid automotive starting battery plate at the bottom of the diagram with two tab connectors instead of one. Changing the design of the battery plates, then we wouldn't need to stop its discharging cycle in order to start its charging cycle because the right mechanics will be in play.

Here's why; if the battery plates had two tab connectors instead of one, then we could simultaneously add and subtract electrons from them without the need of the electrolyte solution. Then we could simultaneously charge and discharge them to a load source; thus; we don't have to stop the discharging cycle of the battery in order to start its charging cycle as well.

However; all the battery plates in each cell, they wouldn't need to have two tab connectors each in order to charge it while discharging it to a load source. If the battery had separate terminals for charging while discharging in the same cell and those terminals, they had their own straps and tab connectors leading to the same plates as well.

Then only the plates joining the charge and the discharging terminal together, they would need two tab connectors each. Wherein; I found that, the battery would only need three terminals instead of four to charge it while discharging it to a load source. Since electrons, they don't need to enter and exit every cell in the battery in order to charge it while discharging it the load source

Given that; electrons, they only need to simultaneously enter and exit one cell in the battery in order to charge it while discharging it to the load source. Providing that; the battery cells, they're connected in series with its terminal connections as well. Although; a capacitor terminals, they've their own plates and they're separated by a non-conducting material.

However; a battery terminals for charge and discharging, they wouldn't need their own plates; but, they would need their own straps and tab connectors leading to the same plates. Given that; we would be using, direct current instead of alternating current to charge the battery while discharging it to a load source.

Since the charge and discharging terminal of the battery, they'll be interconnected by the same plates. Then direct current could flow in one direction from the battery plates in one cell to a load source. In which; the plates joining the charge and the discharging terminal together, they'll create a medium between a charge and a load source in their relationship with the battery.

Those plates joining the charge and discharging terminal together, they'll act like an iron core being a medium between a primary and secondary winding inside a power transformer. The number of turns the wiring, the secondary winding has wrapped around the iron core. It'll determine the amount of current will flow to a load source that is connected to the secondary winding.

I found the connections made between the plates connecting the charge and the discharging terminal together will determine how current. It'll flow to and from those plates; in which, it'll be similar to current flowing from a primary winding to an iron core and then to a secondary winding inside a power transformer as well.

Remember if a battery cells, they're connected in series with its terminal connections. Then the charging voltage, it'll be evenly divided amongst each cell; but, the charging current will be the same throughout each cell within the battery during its charging cycle. The opposite will happen during its discharging cycle.

Here's why; the total voltage potential of the battery, it'll equal to the voltage potential of all of its plates in each cell. But; its current potential, it'll only equal to the current potential of its plates in one cell during its discharging cycle. How its plates and cells are interconnected with its terminals. It'll determine how voltage and current will flow to and from them as well.

Now if a battery cells, they're connected in parallel with its terminals connections. Then the charging voltage flowing through each cell, it'll be the same as the voltage flowing from the charging source. But; the charging current, it'll be divided amongst the battery plates in each cell based on their individual resistance.

Likewise; the opposite, it'll happen during its discharging cycle. Its voltage potential will only equal to the voltage potential of its plates in one cell. However; its current potential, it'll equal to the sum of all of its plates in each cell. How voltage and current will flow in and out of the battery, it'll be based on how its plates and cells are interconnected with its terminals as well.

Focusing on, how a battery plates and cells are interconnected with its terminal connections. It'll tell us how voltage and current will enter or exit it during its cycling processes. It has little to do with the mechanics of the voltage and current alone or the laws of physics; but, it has more to do with the laws of motion and the effect of forces on bodies. See chapter 5.

Understanding, the design of a lead acid automotive starting battery is the deciding factor in determining how voltage and current will enter or exit it. I started comparing it to the energy transfer process of a run capacitor and a power transformer. Then I was sure, I could charge the battery while discharging it with only three terminals instead of four terminals.

If the terminals were in the same cell and they had, their own straps and tab connectors leading to the same plates in that as well. See figure 3B. I realized if there's a charge on one terminal and a load source on the other. Then the charging electrons, they've to flow into one cell in the battery before flowing to the load source.

Thus; the surplus of electrons, they would be created in the cell with the charge and discharging terminal in it. Wherein; the charging current, it'll be reversed back toward the other cells in the battery and toward the load source at the same time. It'll be the reverse engineering of the charging cycle of the battery when it's a charge and a load source on it at the same time.

Since electrons, they no longer have to flow toward the load source before they're reversed back into the battery to charge it; as a result, we're charging it while discharging it to the load source. Wherein; I find that, it's nothing more than the extension of the battery's charging cycle. Since electrons, they've to flow into it before flowing to the load source.

In figure 3B, it shows an unusual design of a deep cycle (lead acid) automotive battery with only three terminals. Two of them, they're in the same cell and they've their own straps and tab connectors leading to the same plates. The diagram is based on the electron current flow concept where current, it has to enter or exit the negative terminal of the battery.

Figure 3B

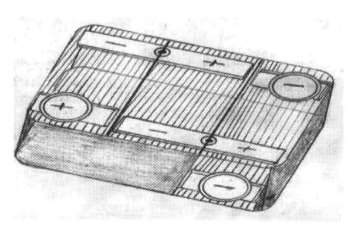

(Reverse Engineering of an Auto Battery)

With this design concept of the battery, I found it's possible to charge it while discharging it to a load source. Here's why; the battery plates in each cell, they're connected in series-parallel with one another by their straps, tab connectors and the electrolyte solution in each cell. And each cell, it's connected in series with its terminal connections as well.

Given how the battery is already constructed, all I had to do was add another terminal to a cell with a terminal already in it. And those terminals, they need their own straps and tab connectors as well. I didn't have to make major changes to the battery in order to charge it while discharging it to a load source.

What I learned during my research is that, the battery charging cycle is the reverse process of its discharging cycle; in other words, current has to go in the battery the same way it has to come out as well. While compiling and comparing data about, how voltage and current will flow to and from the battery plates when they're connected in <u>parallel</u> with one another.

I found there was another meaning why the battery charging cycle is the reverse process of its discharging cycle. Other than current having to go in it the same way current has to come out. Wherein; new meaning, it helped me to understand it was possible to charge it while discharging it with only three terminal connections as well.

Here's why; a lead acid automotive starting battery negative and positive terminal connections, they're connected in <u>parallel</u> with their respective plates in each cell. Then voltage and current will enter the battery. It'll be in reverse of how voltage and current will exit it as well. This is why we need know, voltage will flow through the battery different than current will.

We must separate voltage from current if we're going to understand why it'll be possible to charge a battery while discharging it to a load source. Remember current is a combination of voltage and electrons; but, voltage will behave differently than current will. Since current, it'll consist of electrons and voltage doesn't because it's an electromotive force. See chapter 12.

Understanding, how the mechanics of voltage and current will behave flow in and out of the battery based on its design. Remember its plates in each cell, are connected in series-parallel with one another by their straps, tab connectors and the electrolyte solution in each cell. In short; all of its negative plates in each cell, they're connected in parallel by a single strap.

Likewise; all of the battery positive plates in each cell, they're connected in parallel by a single strap as well. However; its negative and positive plates in each cell, they're connected in series with one another by way of the electrolyte solution in each cell. Then how voltage and current will flow to and from each plate, it'll be based on the laws governing series-parallel circuitry.

This is where facts, they're separated from perceptions when it comes to voltage and current flowing to and from the battery plates. Since its terminal connection, they'll be connected in parallel with its plates by a strap and tab connectors. Then voltage and current will flow from a strap to its plates.

I found it'll be in reverse of how voltage and current will flow from the battery plates to a strap as well. In other words; voltage and current, they'll flow from a strap to a plate different than they'll flow from a plate to a strap. This is why we've to understand, there's a difference in how voltage and current will flow through the battery.

Although; current, it's a combination of voltage and electrons; however, voltage will flow through the battery different than current will based on which cycling process is carried out as well. On the other hand; voltage and current, they'll flow through a series, a parallel or a series-parallel circuit connection differently as well.

Likewise; when there's a charge and a load source on a battery at the same time, voltage and current will flow to and from the battery differently based on the type of connections made between it and the load source in their relationship with the charging source. This is why we've to, separate voltage from current as well. See chapter 12 as well.

For instance; based on the electron current flow concept, current has to flow in and out the negative terminal of a battery in order to cycle it. Therefore; charging voltage and current, they've to enter a cell through the negative terminal of the battery. Since its negative terminal, it'll be connected in parallel with its negative plates by way of a strap and tab connectors in that cell.

Then the charging current will be divided amongst the negative plates in that cell based on their individual resistance as it flows across them. However; each negative plate, it'll receive the same amount of voltage flowing from the charging source. Thus; current, it'll flow to the battery plates different than voltage will because the plates will be connected in parallel with the terminal.

There's a twist when voltage and current starts to flow from a battery plates to a strap connecting them together in parallel. When current, it starts to flow from a battery plates to a strap connecting them together in parallel. Then the sum of the current flowing across each plate will combine at the strap connecting the plates together in parallel.

Although; current, it was divided amongst the battery plates based on their individual resistance when it was flowing from the strap connecting the plates together in parallel. When voltage starts flows from a battery plates to a strap connecting them together in parallel. Then the sum of the voltage flowing to a strap, it'll be the same as the voltage flowing across one plate.

Although; voltage, it'll be the same flowing across each plate when flowing from a strap to a battery plates; but, voltage will not combined at a strap when it's flowing from the plates to a strap connecting them together in parallel like current does. This is another reason why, a battery charging cycle is the reverse process of its discharging cycle as well.

Not only does current has to go in the battery the same way current has to come out in order to cycle it. Since its terminal connections, they'll be connected in parallel with their respective plates by a strap and tab connectors. Then current will flow into the battery in reverse of it flowing out of the battery as well.

In other words; how, voltage and current will flow <u>to</u> a battery plates from a strap connecting them together in parallel. It'll be in reverse of how voltage and current will flow to a strap <u>from</u> a battery plates connecting them together in parallel. This is why it'll be possible to charge a battery while discharging it to a load source with only three terminal connections as well

Providing that; the battery, it has separate terminals for charging and discharging in the same cell. And they had, their own straps and tab connectors leading to the same plates in that cell as well. Then the charging terminal, it'll be the reverse opposite of the discharging terminal because they'll have their straps and tab connectors leading to the same plates as well.

Therefore; how, voltage and current will flow <u>to</u> the plates connecting the charge and the discharging terminal together in parallel. It'll be <u>in reverse</u> and <u>independent </u>of voltage and current flowing <u>from</u> those plates as well. In other words; how, voltage and current will flow to those plates. It'll not affect how voltage and current will flow from those plates as well.

Here's why; although, the charge and the discharging terminal. They'll be in parallel with one another by way of the same plates; but, the parallel connection made between them. It'll not act like a parallel connection that exists, when there's a charge and a load source on the same terminal of the battery; but, it'll be a twist because of the different circumstances involved.

Given that; the charging current, it has to go into the battery before flowing to the load source; thus, it'll flow into the battery based on its resistance because of the laws of parallel circuitry. Since current, it has to flow directly into the battery from the charging source. So; the resistance at the load source, it'll not dictate the amount of current flowing into the battery.

However; current, it'll be flowing from the battery to the load source based on its resistance; in other words, based on the laws governing parallel circuitry as well. Given that; current, it has to flow into the battery first and then flow to the load source based on its resistance while current is flowing through the other cell in the battery based on the laws of series circuitry.

Given that; current, it only has to simultaneously enter and exit one cell in the battery in order to power the load source while the battery is charging as well. It'll be like a capacitor being charged on one terminal while being discharged on another as well. Designing, the battery with separate terminals for charging while discharging in the same cell is the key.

Designing, the battery with separate terminals for charging while discharging in the same cell is the key. Here's why; the plates joining, the charge and discharging terminal together in parallel. They'll receive current based on the laws governing parallel circuitry because they'll be connected in parallel with the charging source by way of the charging terminal.

However; the other plates in the battery, they'll receive current based on the laws governing series circuitry because they'll be connected in series with the plates joining the charge and discharging terminal together in parallel. On the other hand; current, it'll flow to the load source based on the laws governing parallel circuitry.

Since the discharging terminal, it'll be connected in parallel with the same plates as the charging terminal. Then the load source, it'll receive the same amount of voltage flowing from the charging source before a voltage drop will occur within the battery. Given that; the charging terminal, it'll connected in parallel with the same plates as the discharging terminal as well.

Now the other cells in the battery, they'll receive a voltage from the charging source because they're connected in series with the plates connecting, the charge and the discharging terminal together in parallel as well; thus, the charge and the discharging terminal will create a medium between the charge and the load source in their relationship with the battery.

It'll be similar to an iron core being a medium between a primary and a secondary winding inside a power transformer. Wherein; the iron core, it allows current to flow from the primary to the secondary winding based on how they're interconnected with the iron core as well. In view of my findings, we could create a hybrid battery for our hybrid vehicles as well.

Then we could achieve the goal of adding energy to back our batteries while on the go. We wouldn't need to stop their discharging cycles in order to start their charging cycles. Our batteries having separate terminals for charging and discharging, it'll change the relationship between them and our vehicles' voltage and current regulatory systems as well.

Remember traditionally when there's a charge and a load source on the battery at the same time. Then charging current, it has to flow toward a vehicle's voltage and current regulatory system before it's be reversed back into the battery to charge it; however, it'll depend on the amount of the resistance at the voltage and current regulatory system as well.

In which; the resistance at the voltage and current regulatory system, it'll dictate the amount of current flowing into the battery to charge it. If the charging current, it has to flow into one cell in the battery before flowing to the voltage and current regulatory system. It'll work a little different than current flowing toward the voltage and current regulatory system first.

Let me explain what I'm alluding to because it's not the same as having a charge and a load source on the same terminal connection of a battery. For instance; if it was only a charge on the battery, then current has to enter into one cell in it before current could flow to the other cells in it during its charging cycle.

Then current, it'll flow to the other cells in the battery based on how they're interconnected with the cell with the terminal connection. If the other cells in the battery, they're connected in series with the cell with terminal connection in it. Then the charging voltage, it'll be evenly divided amongst those plates in the other cells.

However; the charging current, it'll be the same throughout each cell within the battery. This is where it's important to understand, the difference between voltage and current because they'll behave differently flowing through different types of circuit connections; such as, series, parallel and series-parallel circuit connections as well.

Remember if a battery cells, they're connected in parallel with its terminal connection. Then the charging voltage will be the same across each cell; but, current will be divided amongst each cell based on their individual resistance. This tells us that, a parallel circuit connection made between a battery and a load source in their relationship with a charging source.

It'll be insufficient to charge the battery while discharging it to the load source. Given that; the charging current, it'll flow between the battery and the load source based on their individual resistance. In which; the battery resistance to current, it's different than the load source resistance to current.

Wherein; the battery resistance to current, it'll reflect current away from it; however, the load source resistance to current. It'll be based on the number of turns the wiring has inside of it to conduct current. See chapter 12. What it really boils down to, it's the connections made between a battery and a load source in their relationship with a charging source.

Here's why; remember if we add another terminal to a cell with a terminal connection already in it. Then the charging voltage and current, they still have to flow into that cell before flowing to the other cells in the battery as well. Let's assume the battery will have a charge on one terminal and a load source on the other terminal connection as well.

Wherein; the other cells in the battery, they'll be connected in series with the cell with the two terminal connections in it. Thus; one has to assume, the charging voltage will still be evenly divided amongst the battery plates in each cell; likewise, the charging current will be the same throughout each in the battery as well.

If the battery plates and cells, they're connected in series with its terminal connections. Then; adding, another terminal to the cell with a terminal already in it does change a thing concerning how voltage and current will flow through the battery during its charging cycle. The only difference is a load source will have access to the voltage and current flowing through the battery.

Then voltage and current will flow to the load source based on how, it's connected to the cell with the two terminal connections in it. Since the charging voltage and current, they've to flow into one cell in the battery before flowing to the load source. Then how the two terminals are interconnected by the same plates, it'll make that determination as well.

Then voltage flowing to the load source, it'll be the same as the voltage flowing from the charging source. Given that; when voltage, it reaches the strap of the other terminal connection.

It'll be the same as the voltage flowing across one plate from the charging source; in which, it'll be the same as the voltage flowing from, the charging source as well.

Since the two terminals, they're connected in parallel with one another by way of the same plates. However; when current, it reaches the strap of the other terminal connection as well. Then current flowing across each plate, it'll be combined at the strap connecting the plates together in parallel with the terminal connection.

Thus; the load source, it'll receive the same amount of current flowing across each plate. Since it'll be connected in parallel with each plate by a strap and each plate will be connected in parallel with the charging source as well. Here's the deal; current, it'll flow directly to the load source based on its resistance and the voltage applied by the power source.

Therefore; if current, it flows directly to the load source from the plates connecting the two terminals together in parallel. Then the amount of current flowing to the load source, it'll be based on its resistance and not the resistance between it and the battery in their relationship with the charging source.

Thus; matter and energy, they'll behave differently under different circumstances because they're what we find them to be under different circumstances. Remember if a battery plates in each cell, they're connected in series-parallel with one another. Then the laws of series-parallel circuitry will govern how electrons will flow to and from them.

If electrons, they've to flow into one cell in a battery before flowing to a load source. Compared to flowing toward the load source first and then, reversed back into the battery to charge it. Then matter and energy, they'll behave differently because of the different circumstances involved as well.

We can't assume a battery can't be charged while being discharged to a load source based on the contemporary design of the battery; in which, our traditional wisdom about its cycling processes will be based on its contemporary design as well. When the charging current, it has to flow to a load source before it's reversed back into the battery to charge it.

If the charging current has to flow into one cell in the battery before flowing to the load source, then we can't expect the same results; as though, the charge and the load source were on the same terminal connection of the battery. Given that; matter and energy, they'll behave differently under different circumstances as well.

Remember if a battery cells, they're connected in series with its terminal connections. Then the charging voltage, it'll be evenly divided amongst the battery cells; but, current will be the same throughout each cell in it during its charging cycle. During its discharging cycle, its voltage potential will be the sum of all of its cells; but, its current potential will only equal to one cell in it.

If the charging voltage and current, they'll have to flow into one cell in the battery before flowing to a load source. Then it'll be a different outcome, other than the outcome when current has to flow toward the load source before current is reversed back into the battery to charge it.

I found if a battery has separate terminals for charging and discharging in the same cell and those terminals, they had their own straps and tab connectors leading to the same plates in that cell as well. Then a charge and a load source will be in <u>series-parallel</u> with one another in their relationship with the battery.

In the next chapter; let me explain, why a series-parallel circuit connection made between a charge and a load source in their relationship with a battery. It'll allow us to use reverse engineering on charging cycle of the battery; so, it could be charged while being discharged to the load source as well.

CHAPTER 23

THE LAWS OF SERIES-PARALLEL CIRCUITRY

I FOUND IT'S A big difference between a battery and a load source being in parallel with one another in their relationship with a charging source. Compared to, a charge and a load source being in series-parallel with one another in their relationship with a battery. One will allow the mechanics of the current to dictate which path, it'll travel; but, the other will don't.

All this time, it has been in plain sight how to increase the efficiency and overall capacity of our secondary cell batteries without finding a perfect chemical composition for them. It's our ability to replenish their chemical energy with electrical energy. We just have to figure out how to add energy back to them while taking it out. We do it with our capacitors why not our batteries.

My training in the field of heating, cooling and refrigeration, it helped me to understand. The design of a battery has more influence on its cycling processes than the mechanics of the current alone. I learned it's the type of connections made between load sources in their relationship with a power source. It'll determine how voltage and current will flow between them.

I found the same thing applies to our secondary cell batteries because it's the type of connections made between them and a load source in their relationship with a charging source. It'll determine how voltage and current will flow between them as well. In which; it's the key in adding, energy back to our secondary cell batteries while taking it out as well.

While venturing into the field of heating, cooling and refrigeration, I learned the benefits in connecting our load sources together; such as, electric motors, fans and relay switches in series-parallel with one another. Not only will electrons flow to them based on how they're interconnected with a power source; but, how they're interconnected with one another as well.

In chapter 5, we talked about connecting three load sources together in series with a power source. They'll receive the same amount of current flowing from the power source; but, it'll be a voltage drop across each load source. If each load source required the same amount of voltage to operate, they'll not operate efficiently because of the connection made between them.

If we connected all three load sources together in parallel with the same power source, they all will receive the same amount of voltage; but, the amount of current flowing to them will be based on their individual resistance. If the power source couldn't produce enough current for all three load sources to operate, then they wouldn't operate efficiently.

Now if two of three load sources required the same amount of voltage to operate and one required less voltage than the other two; but, it required the same amount of current as the other two. Understanding the energy requirements of all three load sources, then we could figure out how to make them operate efficiently with the same power source as well.

For example; since two of the load sources, they required the same amount of voltage to operate. Then those two, they could be connected in parallel with one another. Since one of the load sources, it required less voltage than the other two. Then it could be connected in series with one of the load sources; that is, connected in parallel with another load source.

Since two of the load sources will be connected in parallel with one another, they'll receive the same amount of voltage; but, current flowing to them will be based on their individual resistance. The load source that is connected in series with another load source, it'll receive a voltage drop because voltage has to pass through another load source in order to flow to it.

However; the load source that is connected in series with another load source will receive the same amount of current as the load source, it's connected in series with. Understanding the energy requirements of all three load sources, then we could connect them together in a manner; so, they could all operate efficiently with the same power source as well.

Likewise; if we wanted to charge, a secondary cell battery while discharging it. Then we must know what's required to carry out such a process as well. The process might require changing, the design of the battery as well as having the right connections made between it and a load source in their relationship with a charging source as well.

Since our batteries' plates, they behave like load sources consuming energy during their charging cycle. Then the same laws apply to our electrical devices; such as, electric motors, fans or compressors will apply to our batteries' plates as well. It's not hard to imagine they could be charged while being discharged because we could add or subtract energy from them as well.

All we've to do is figure out the right connections needed between a battery and a load source in their relationship with a charging source. To charge the battery while, using it to power the load source; such as, an electric motor, a fan or a compressor. It's not about the laws of physics; such as, electrochemistry, entropy, inertia or the laws of conservation of energy as well.

It's about the laws of motion and the effect of forces on bodies or the laws of mechanics when it comes to charging a battery while using it to power a load source; such as, an electric motor, a fan or a compressor as well. When it comes to adding and subtracting electrons from a secondary cell battery, it's about the laws of motion and the effect of forces on bodies.

Remember a lead acid automotive starting battery plates in each cell, they're connected in series-parallel with one another. By their straps, tab connectors and the electrolyte solution in each cell and each cell, it's connected in series with its terminal connections as well. Given how the battery is constructed, it determines how electrons will be added or subtracted from it.

Here's why; during the charging cycle of the battery, fixed electrons on one plate will reflect free electrons coming from a charging source toward an adjacent plate by way of the electrolyte solution in that cell. Thus; electrons, they could flow to and from the same plate without flowing to and from the same tab connector on the same plate.

Therefore; electrons, they could flow across one plate through the electrolyte solution to an adjacent plate flowing in one direction. As a result; electrons, they could be simultaneously added and subtracted from one plate while they're simultaneously added and subtracted from another without flowing to and from the same tab connector on the same plate.

Since the battery cells, they're connected in series with its terminal connections. Then its plates in one cell, they're simultaneously charged and discharged to the plates in the next cell as well. We're using the laws of series-parallel circuitry in conjunction with the laws of series circuitry to simultaneously add and subtract electrons from the battery plates.

Although; the battery plates in each cell, they're separated by a non-conducting material called plastic. And yet, we're still able to charge its plates in each cell without having to stop charging them in one cell in order to charge them in the next cell. So; it has little to do with the laws of electrochemistry, entropy, inertia or the laws of conservation of energy as well.

It has more to do with the laws of motion and the effect of forces on bodies when it comes to adding and subtracting electrons from the battery plates. Here's why; the battery plates in each cell, they're simultaneously charged and discharged amongst themselves by using mechanics to replenish their chemical energy with electrical energy during their charging cycle.

The question is that, can we use mechanics to simultaneously charge and discharge the battery plates to a load source as well? I find the answer is yes if we create a series-parallel circuit connection between a charge and a load source in their relationship with the battery. Then it's possible to replenish its chemical energy with electrical energy while using it as well.

Here's why; if the battery, it had separate terminals for charging and discharging in the same cell. And those terminals, they had their own straps and tab connectors leading to the same plates in that cell. This design concept of the battery, it'll create a series-parallel circuit connection between a charge and a load source in their relationship with it.

Remember in order to have, a series-parallel circuit connection there must be three or more load sources interconnected by the same power source. At least two of the load sources, they must be connected in parallel. One or more load sources, they must be connected in series with one of the load sources; that is, connected in parallel with another load source as well.

Since the charge and the discharging terminal of the battery, they'll be interconnected by the same plates and they could consume or store energy during their charging cycle. Then those plates will be connected in parallel with a load source by way of the discharging terminal. Thus; we've two or more load sources, connected in parallel with one another as well.

However; the plates in the other cells in the battery, they'll be load sources connected in series with the plates connecting the charge and the discharging terminal together in parallel. If we're going to charge the battery while discharging it to a load source, then we'll have load sources connected in series-parallel with one another in their relationship with a charging source.

Since the battery plates, they'll behave like load sources consuming energy during their charging cycle. Then the same laws that apply to our load sources; such as, electric motors, fans or compressors will apply to the battery plates as well. Thus; it's possible to charge, its plates while using them to power a load source; such as, an electric motor, a fans or a compressor as well.

The straps connecting the plates together in parallel with the charge and discharging terminal, they'll create a series-parallel circuit connection between a charge and a load source in their relationship with the battery; thus, allowing it to be charged while being used to power a load source; such as, an electric motor, a fans or a compressor as well.

If a battery has separate terminals for charging while discharging, then its charging terminal will be in parallel with the charging source; but, the discharging terminal of the battery. It'll be in parallel with the load source while the other plates in the battery will be in series with the plates connecting the charge and the discharging terminal together in parallel.

We'll have a series-parallel connection between the charge and the load source in their relationship with the battery. Thus; voltage and current, they'll flow to and from the straps connecting the plates together in parallel with the charge and discharging terminal based on the laws of parallel circuitry; but, flowing to the other plates in battery based on the laws of series circuitry.

In other words; voltage and current, they'll be flowing from the charging source to the battery and from the battery to the load source based on the laws governing parallel circuitry. While voltage and current, they'll be flowing to the other plates in the battery based on the laws governing series circuitry; as a result, charging the battery while using it to power the load source as well.

While creating, a more efficient cycling process for the battery to increase its efficiency and overall capacity as well. However; some in the field of automotive battery technology, they claim it's impossible due to the laws of physics. I find their claim doesn't add up with the mechanics behind the charging cycle of a lead acid automotive starting battery as well.

We could use the same technique that is used to simultaneously charge and discharge the battery plates in one cell to its plates in the next cell. Without having to, stop charging them in one cell in order to charge them in the next. We could use it to simultaneously charge and discharge them to a load source within the realm of physic as well.

After analyzing, how electrons will flow in and out of a capacitor or a power transformer based on their designs. I found it was a matter of mechanics and not a matter of physics why we couldn't simultaneously add and subtract electrons from our secondary cell batteries. I realized it was a mechanical solution to increase the efficiency and overall capacity of our batteries as well.

If we create, a series-parallel circuit connection between our power and load sources in their relationship with our batteries. Then we could add energy back to them to increase, their efficiency and overall capacity while using them to power our load source as well. Then we wouldn't need to find a perfect chemical composition to a longer-lasting battery as well.

In order to, achieve the goal of creating a series-parallel circuit connection between a charge and a load source in their relationship with a group of batteries. All we've to do is design one battery in a group of batteries with separate terminals for charging and discharging; thus, it would allow us to charge a group of batteries connected in series with it as well. See figure 5.

In figure 5; it shows, a series-parallel connection between a charge and a load source in their relationship with two batteries. Using, the electron flow concept to explain how to charge a battery while discharging it to a load source. I got carried away with my drawings; so, I ended up with two batteries instead of one with separate terminals for charging while discharging in figure 5 and 6 as well.

Figure 5

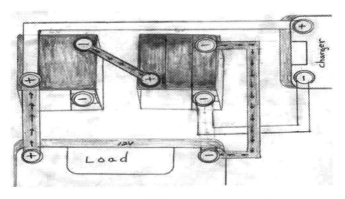

(A series-parallel circuit connection)

Only one battery in a group of batteries will need to have separate terminals for charging and discharging. In order to charge, a group of batteries connected in series with one another. Thus; we don't need to change, the design of every battery needed to power a load source. If the other batteries' cells, they're connected in series with their terminals connections as well.

Remember if a battery cells, they're connected in series with its terminals. Then only the current potential of one cell in it will be used to power a load source. This apply to a group of batteries if their cells, are connected in series with their terminals as well. Only the current potential of one cell in one battery in a group of batteries will be used to power a load source as well.

Therefore; if a battery plates in each cell, they're connected in series-parallel with one another and each cell is connected in series with its terminal connections as well. Wherein; it makes, it possible to simultaneously charge and discharge the battery plates in one cell to a load source. While charging, the plates in the other cells in the battery as well.

Likewise; it makes, it possible to simultaneously charge and discharge the plates in one cell in one battery in a group of batteries to power a load source. While charging, the plates in the other cells in the batteries as well. In figure 6; it shows, we could use an electromotive force created by a charging source to simultaneously charge and discharge a group of batteries to a load source.

What makes this possible is that, a series-parallel circuit connection exists between a charge and a load source in their relationship with the batteries. Given that; one of the battery, it'll have separate terminal for charging and discharging in the same cell. And those terminals, they'll have their own straps and tab connectors leading to the same plates in that cell as well.

Figure 6

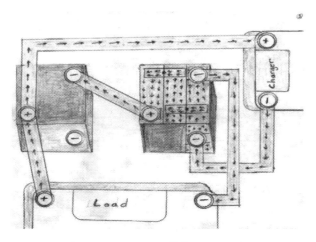

(Using Series-Parallel Circuitry)

Using the laws of series-parallel circuitry, we could substitute the batteries' voltage with the charging voltage. To simultaneously charge and discharge, the plates in one cell in one battery to a load source. While reversing, the current flow back through the other cells in the batteries to charge their plates as well.

What makes, it possible to reverse the current flow back through the other cells in the batteries to charge them. They'll not be using their voltage potential to supply electrons to the load source. Thus; the plates in one cell in one battery could be simultaneously charge and discharged to the load source while reversing, the current flow back through the other cells in the batteries as well.

The pictorial diagram in figure 5, it shows what it means to use the batteries' voltage to discharge them to a load source while they're connected in series-parallel with one another. In figure 6, it shows what it means to use a charging source to discharge two batteries to a load source while they're in series-parallel with one another as well.

I realize not only the design of a battery will determine how voltage and current will enter or exit it; however, the connections made between it and a load source in their relationship with a charging source will as well. In which; a host of things have to, come together in order to charge a battery while discharging it to a load source as well.

Wherein; it seems like, we haven't become aware of those hosts of things because we still don't know what's required to charge a battery while discharging it to a load source or we think it's impossible. Given that; we blame it on the laws of physics why, we can't charge the battery while discharging to the load source as well.

Let's forget about the laws of physics for a moment and just focus on the laws of mechanics. In chapter 22, it showed an unconventional design of a lead acid automotive starting battery plate in figure T-3. To show why, it'll be possible to simultaneously added and subtracted electrons from its plate if it had two tab connectors instead of one.

Imagine if the battery had separate terminals for charging and discharging with their own straps and tab connectors leading to the same plates as well. Then the charging voltage and current will flow <u>to</u> those plates from the charge terminal in reverse of voltage and current flowing from those plates to the discharging terminal as well.

Likewise; how, the charging voltage and current will flow <u>to</u> those plates from the charge terminal. It'll be independent of how voltage and current will flow <u>from </u>those plates to the discharging terminal as well. Since the charge and discharging terminals, they'll be connected in parallel with one another by way of the same plates.

Then the amount of voltage flowing from the charging terminal <u>to</u> those plates, it'll be the same as the voltage flowing from those plates to the discharging terminal. In other words; it'll be no voltage drop before voltage, it reaches a load source that is connected to the discharging terminal since it'll be connected in parallel with the charging terminal by the same plates as well.

Thus; the amount of current flowing across each plate, it'll be combined at the strap connecting the plates together in parallel with the discharging terminal; thus, the sum of the current flowing to the load source from the discharging terminal. It'll equal to the sum of the current flowing across each plate; that is, connected in parallel with the discharging terminal as well.

Understanding, how voltage and current will flow in and out of a battery with its cells connected in parallel with its terminal connections. Then we'll know the significant of having separate terminals for charging and discharging in the same cell and those terminals having their own straps and tab connectors leading to the same plates in that cell as well.

Given that; the charge and the discharging terminal, they'll be in reverse and independent of one another as well. The plates connecting the charge and the discharging terminal together in parallel, they'll produce the same results as a battery with its cells connected in parallel with its terminal connections going through its normal charge or discharging cycle.

Here's why; understanding, the rate at which current will flow to a load source will be based on two factors; the resistance of the load source and the voltage applied by a power source. Then the amount of current flowing from the discharging terminal to the load source, it'll be based on its resistance and the voltage applied by the charging source.

Since the charging terminal, it'll be connected in parallel with the discharging terminal by way of the same plates. When the charging current enters that cell, it'll energize those plates and ionize the electrolyte solution in that cell as well. Thus; transferring current to, the adjacent plates in that cell and then to the plates in the next cell by way of an electrical bus as well.

The plates in each cell will receive the same amount of current flowing from charging source because they're connected in series with the plates connecting the charge and discharging terminal together in parallel as well. Since the charge and the discharging terminal, they'll be connected in parallel with one another by the same plates as well.

Then a load source that is connected to the discharging terminal, it'll receive current based on its resistance and the voltage applied by the charging source because the load source is connected in parallel with the charging source as well. Thus; the load source, it'll receive the same amount of voltage flowing from the charging source by way of the discharging terminal as well.

However; the other plates in the battery, they'll receive a voltage drop from the charging source because they'll be connected in series with the plates connecting the charge and the discharging terminal together in parallel as well. Other than that, the first cell in the battery will produce the same results as a battery with its cells connected in parallel with its terminals.

On the other hand; during the simultaneous cycling process of the battery, the plates in the other cells in it will go through a normal charging cycle. Like a battery with its cells connected in series with its terminals. Only the plates connecting the charge and the discharging terminal together in parallel will go through a simultaneous cycling process.

Since electrons, they could only flow from the charging source to those plates connecting the charge and discharging terminal together in parallel and from those plates to a load source at the same time. Knowing that, electrons can't simultaneously enter and exit the other cells in the battery during a simultaneous cycling process of it.

Given that; once electrons, they leave the plates connecting the charge and the discharging terminal together in parallel. Then electrons, they could only flow to and from the other plates in the battery flowing in one direction at a time because they're connected in series with the plates connecting the charge and the discharging terminal together in parallel.

Understanding the mechanics behind the cycling processes of a lead acid automotive starting battery, we'll know it has little to do with the mechanics of the voltage and current alone or the laws of physics as well. It'll have more to do with the laws of motion and the effect of forces on bodies or the laws of mechanics when it comes to its cycling processes.

Let me explain this in a different way; so, you could understand what I'm alluding to. If you take note of the cycling processes of a battery with its cells connected in parallel with its terminal connections, then you might understand what I'm alluding to. During the charging cycle of the battery, the charging voltage will be the same across each plate in each cell.

However; the charging current, it'll flow across each plate in each cell based on its individual resistance. On the other hand; during the discharging cycle of the battery, its voltage potential will only equal to the voltage potential of its plates in one cell; but, its current potential will equal to all of its plates in each cell during its discharging cycle.

Likewise; if you take note of the cycling processes of a battery with, its cells connected in <u>series</u> with its terminal connections. Then you'll find during its charging cycle, the charging voltage will be evenly divided amongst its plates in each cell; but, the charging current will be the same across each plate in each cell during its charging cycle.

However; during the discharging cycle of the battery, its voltage potential will equal to the voltage potential of all of its plates in each cell; but, its current potential will only equal to its plates in one cell during its discharging cycle. If we take note of the cycling processes of both batteries based on, how their cells are interconnected with their terminal connections.

Then we'll realize voltage and current will flow in and out of the two batteries based on, the laws of motion and the effect of forces on bodies or the laws of mechanics. In other words; it'll be based on, the design of the batteries; in which, their designs will determine how voltage and current will flow through them and not the mechanics of the voltage and current alone.

This is what I've been alluding to throughout my book because the design of a battery is the deciding factor in determining, how its cycling processes will be carried out. Then logic tells us if we could combine some aspects of a battery with its cells connected in series and a battery with its cells connected in parallel with its terminal connections as well.

Combining the two different aspects of the batteries together, then we could use their advantages and disadvantages to our advantage. We could imitate the cycling processes of a battery with its cells connected in parallel with its terminal connections in one cell of a battery with its cells connected in series with its terminal connections as well.

By adding separate terminals for charging and discharging with their own straps and tab connectors leading to the same plates in one cell in a battery with its cells connected in series with its terminals. Then that cell, it'll produce the same results as a battery with its cells connected in parallel with its terminals going through its normal charge or discharging cycle as well.

In figure 4B, I'm going to show you what I'm alluding to if combine some aspects of a battery with its cells connected in series and a battery with its cells connected in parallel with its terminal connections. Then we could imitate the cycling processes of a battery with its cells connected in parallel with its terminals in one cell of a battery with its cells connected in series with its terminals.

Thus; it'll make, it possible to simultaneously charge and discharge the battery to a load source. Without being impeded, by the mechanical structure of the battery or the laws of physics as well.

Figure 4B

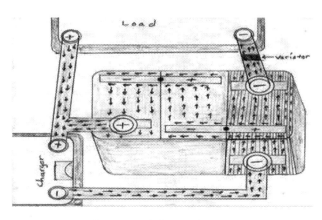

(Charging while Discharging)

Using the electron current flow concept, I'm going to use the negative terminal connection of a secondary cell battery to describe how electrons will flow through it during a simultaneous cycling process of it. In figure 4B, the arrows will show the direction the electrons will be flowing through the battery during a simultaneous cycling process of it.

Wherein; electrons, they'll be flowing from the charging source through the charging terminal of the battery. And then, flow across the negative plates connected in parallel the charging terminal by way of a strap. Thus; energizing, the negative plates and ionize the electrolyte solution in that cell before electrons could flow through the discharging terminal.

As a result; energizing, the adjacent positive plates in the cell with the charge and the discharging terminal in it as well. Therefore; transferring, electrons to the negative plates in the next cell by way of an electrical bus. At the same time, electrons will be flowing toward a load source that is connected to the discharging terminal of the battery as well.

Since the charge and the discharging terminal, they'll be interconnected by the same plates. Then the load source will receive the same amount of voltage flowing from the charging source before a voltage drop will occur within the battery. Current flowing to the load source will be the sum of the current flowing across each plate connected in parallel with the discharging terminal.

In the mean time, the other plates in the battery will receive a voltage drop from the charging source; given that, they'll be connected in series with the plates connecting the charge and the discharging terminal together in parallel as well. We'll only be using the charging voltage to simultaneously charge and discharge the battery the load source, shown in figure 4B.

Since like charges will repel one another and follow the path of least resistance, then we could use those aspects of the laws of physics to simultaneously charge and discharge the battery to a load source as well. Given that; electrons, they could enter and exit the battery at the same time; however, it's more to it than electrons entering and exiting it at the same time as well.

In the lower left hand corner of figure 4B, it shows the positive terminal connection of a battery. In which; it shows, how the charging voltage will return back to its source from a load source and from the battery at the same time. It's one of the hosts of things has to be in play in order to charge a battery while discharging it to a load source; but, we overlook this as well.

It'll be similar to voltage returning back to its source from a power transformer. Remember voltage will return back to its source from a primary winding while voltage is returning back to its source from a load source; that is, connected to a secondary winding inside a power transformer by way of an electrical bus as well.

Untraditional design of a secondary cell battery with separate terminals for charging and discharging in the same cell and those terminals, they'll have their own straps and tab connectors leading to the same plates in that cell. It'll allow the charging voltage to return back to its source from the battery and from a load source at the same time as well.

In which; it's one of the key factors in charging, a battery while discharging it to a load source as well. It would be obvious if we focus on the mechanics taking place within a power transformer during its energy transfer process. See chapter 17. In which; in figure 4B, it shows how this process is similar to the energy input and output of a power transformer as well.

Here's why; the charging voltage, it'll be able to return back to its source from the battery and from the load source at the same time because on the inside of the battery. The charging source, it'll be connected in series with the positive terminal of the battery within it; but, on the outside of it the charging source will be connected in parallel with its positive terminal.

Wherein; the charging voltage, it'll be able to force electrons to the load source by way of the discharging terminal. As a result; the charging voltage, it'll return back to its source from the load source; as though, the battery was going through its <u>normal</u> discharging cycle when it's just a load source on it.

However; on the inside of the battery, the charging terminal will allow the charging voltage to force electrons through each cell in it to charge the plates in each cell. As a result; the charging voltage, it'll return back to its source by way of the positive terminal connection of the battery; as though, it's going through its <u>normal</u> charging cycle when it's just a charge on it as well.

Thereby having separate terminals for charging and discharging with their own straps and tab connectors leading to the same plates in one cell in a battery with its cells connected in series with its terminals. Then it'll make it possible to simultaneously charge and discharge it to a load source without being impeded by the mechanics of the voltage and current as well.

In order to understand, how this is possible. We've to understand voltage and current will flow through a series, a parallel or a series-parallel circuit connection differently. Then we'll know the mechanics of the voltage and current alone, they'll not determine how they'll flow between a battery and a load source in their relationship with a charging source.

However; we'll know that, the connections made between the battery and the load source in their relationship with the charging source will. We'll know this if we separate voltage from current because they behave differently under different circumstances. Thus; we'll know that, we just can't focus on the mechanics of the voltage and current alone as well.

Given there's a host of things will come into play to prevent a battery from being charged while being discharged, other than the mechanics of the voltage and current alone as well. Since voltage, it'll behave differently than current will flowing through a series, a parallel or a series-parallel circuit connection because they're not the same.

In which; voltage, it'll react differently to resistance than current will because voltage doesn't consists of electrons like current does. Let's forget about the mechanics of the voltage and current for a moment. Let us focus on the design of the battery in figure 4B for a moment given we don't have to focus on any else, other than its design.

Since the battery charge and discharging terminal, they'll be in the same cell and they'll have their straps and tab connectors leading to the same plates in that cell as well. Then the design concept of the battery alone, it'll allow us to charge it while discharging it to a load source because of the mechanics of voltage and current alone.

In which; current, it consists of voltage and electrons; thus, voltage will force electrons through the body of the battery based on its design. This is where the laws of mechanics, they'll come into play because it's not based on the mechanics of the voltage and current alone because they don't determine how they'll enter or exit the battery.

Since electrons, they don't have to travel to and from the same terminal in order to enter or to exit the battery in figure 4B. Then electrons could simultaneously enter and exit it because it has a charge and a discharging terminal in the same cell. And those terminals, they've their own straps and tab connectors leading to the same plates in that cell as well.

In which; it's another key in understanding why, it'll be possible to charge the battery while discharging it to a load source. Given that; voltage and current, they can't flow through every cell in the battery and then flow to a load source. If the battery cells, they're connected in series or in parallel with its terminal connections as well.

Wherein; the load source, it might not get the proper amount of voltage needed to supply electrons to it if the battery cells are connected in series with its terminal connections. Or the battery might not get the proper amount of current needed to charge its plates if its cells are connected in parallel with its terminal connections as well.

If voltage and current, they've to flow through every cell in the battery before flowing to the load source. Then we'll not be able to charge the battery while discharging it to a load source because it's simple mechanics 101. If voltage and current is flowing through a series or a parallel circuit connection, we'll not get the same results because they'll behave differently.

If a battery has separate terminals for charging and discharging in the same cell and those terminals, they've their own straps and tab connectors leading to the same plates in that cell as I've been alluding to. Then the charging voltage and current has to flow to and from those same plates connecting the charge and the discharging terminal together in parallel.

Then voltage and current will flow <u>to</u> those plates from the charging terminal in reverse of voltage and current flowing <u>from</u> those plates to the discharging terminal. Likewise; voltage and current will flow to those plates <u>from</u> the charging terminal, it'll be independent of voltage and current flowing from those plates <u>to</u> the discharging terminal as well.

In other words; voltage and current, they'll flow <u>to</u> and <u>from</u> those plates differently because of the parallel circuit connection made between the plates and the terminal connections. Remember if a battery plates, they're connected in series-parallel with one another by their straps, tab connectors and the electrolyte solution in each cell.

Then electrons could flow from a strap to a tab connector on one plate. And then, flow across that plate through the electrolyte solution to an adjacent plate without flowing to and from the same tab connector on the same plate. Thus; the battery plates in one cell, they're simultaneously charged and discharged amongst themselves as well.

The same thing will be happening with the battery in figure 4B with the separate terminals for charging while discharging in the same cell. Then the only different between the battery in figure 4B and a battery with its cells connected in series with its terminals. The battery in figure 4B, it has separate terminals for charging and discharging in the same cell.

Wherein; a load source, it'll have access to the process that is taking place within the battery during its charging cycle that is the difference. Mechanically; the plates connecting the charge and discharging terminal together, they could be simultaneously charged and discharged to a load source without being impeded by the mechanics of the voltage and current.

All this talk about, we can't simultaneously charge and discharge a lead acid automotive starting battery to a load source because of the laws of physics. It's nothing more than an illusion created by the exterior design of the battery because it goes through simultaneous cycling process each time, it's charged because of its interior design as well.

In the next chapter; let's see why, I'm not alone to assume it's conceivable to simultaneously charge and discharge our secondary cell batteries if we change their contemporary designs; so, energy could enter and exit them at the same time as well.

CHAPTER 24

IS IT CONCEIVABLE OR FARFETCHED?

WHEN I PUBLISHED MY first book "Miracle Auto Battery", I didn't take the time to explain why it was conceivable to charge a lead acid automotive starting battery while discharging it and it'll be a benefit in it as well. I left room for some of my readers to criticize my finding and think I didn't know what I was talking about and my assumptions didn't add up with what they know about the battery as well.

After publishing my first book in 2011, I found others in Australia who believes as I did. If we change the contemporary design of our secondary cell batteries, so, electrons could enter and exit them at the same time. Then it would be possible to charge them while discharging them to a load source as well. This information, it verified my research and experiments as well.

However; it spoiled, my chances of getting a patent for a battery with four terminals for charging while discharging. The idea of having a mechanical rather than a chemical solution to increase the efficiency and the overall capacity of a battery to increase the overall travel range of a vehicle on battery power without finding a perfect chemical for its batteries is unheard of.

I find those types of ideas, are continuously swept under the rag by those in the oil industry and those who are in cahoots with them as well. However; by chance, I stumble upon the Australians concept for charging a secondary cell battery while discharging when I was searching the U.S. Patent website after I was told by it was a battery already out there with four terminals for charging while discharging since 2007. See Figure B-10.

Figure B-10

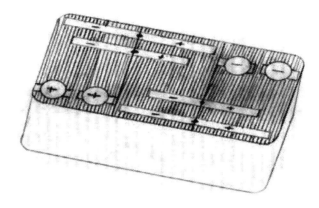

(Australians battery for charging while discharging)

In figure B-10, it shows a drawing of the Australians battery for charging while discharging. I found on the U.S. Patent website and the international publication number is WO 2007/042892-A1. So; my readers, they could see the idea of charging a battery while discharging it is neither inconceivable nor farfetched as some folks believes.

However; I found charging, a battery while discharging it could be accomplished by changing the design of the battery; so, electrons could enter and exit it at the same time. On a short note; I was told that, the inventor of the Australians battery with four terminals for charging while discharging is Human, Jan, Petrus of South Africa; but, I've not verified this yet.

The Australians approach to charging a battery while discharging it is more radical than my approach at first after I figured out that, a battery needs only three terminals for charging while discharging. Months after publishing my first book, I stumbled upon this new concept for charging a secondary cell battery while discharging it to a load source. See chapter 22.

I found if we created a series-parallel circuit connection between a charge and a load source in their relationship with a battery. Then we could charge it while discharging it with only three terminals instead of four. Let's analyze the Australians' battery for a moment because not only do they believe, it'll need four terminals to carry out a simultaneous process.

However; Australians, they believe each terminal will need its own set of plates. If you notice in figure B-10, the Australians' battery has four terminals and they've their own set of plates, straps and tab connectors as well. Also; it's two sets of negative and positive plates in each cell with, their own straps and tab connectors as well. See figure B-10.

Basically; the Australians, they've is two batteries in one. Upon analyzing it, I found the cell <u>without</u> a terminal connection in it. Wherein; electrons, they could only flow in or out of that cell flowing in one direction at a time. Since the battery cells, they're connected in series with its terminal connections as well.

Thus; electrons could only flow toward, a positive terminal or flow from a negative terminal flowing in one direction at a time within the cell <u>without</u> a terminal connection in it. The cell <u>without</u> a terminal connection; although, it has two sets of negative and positive plates in it and they've their own straps and tab connectors as well.

However; it'll not contribute to, a simultaneous cycling process of the battery. Remember the positive and the negative plates in each cell in a lead acid automotive starting battery. They're setup in sequence because it'll be a positive and then, a negative and then a positive plate setup in this sequence. Until; each cell, it has an equal number of positive and negative plates in it.

Thus; electrons, they've to flow across one positive plate and then flow through the electrolyte solution to an adjacent negative plate or vice versa in order to enter or exit the cell <u>without</u> a terminal connections in it. Therefore; electrons, they could only flow in or out of that cell flowing in one direction at a time because of the design of the battery.

Wherein; electrons, they can't flow through the cell <u>without</u> a terminal connection in it while flowing from a positive terminal toward a negative terminal. While electrons, are flowing in the opposite direction through that cell from the other positive terminal toward the other negative terminal. If we believe otherwise, then we're sadly mistaking.

It's impossible for like charges to pass one another while travelling on the same path and flowing in the opposite direction of one another because they'll repel one another in the opposite direction. It would be a waste of money and time to design a battery like the Australians did because it'll not contribute to a simultaneous cycling process of it.

I'm not saying the Australians' battery can't be charged while being discharged. What I'm saying is that, it's not necessary to design a battery like the Australians have. Although; the battery, it does have two terminals in the same cell and they've their own straps and tab connectors. In which; I found that, it's the foundation for charging a battery while discharging it.

However; based on, the conventional current flow concept. We could only add or subtract electrons from the cell with the two positive terminal connections in it; given that, it has two openings for electrons to enter or exit it at the same time. The cell <u>without</u> a terminal in it may have two sets of negative and positive plates and they may have their own straps and tab connectors as well.

Nevertheless; we could only add or subtract electrons from that cell; but, we can't do both. Maybe; the Australians had an illusion that, electrons were flowing from every cell in the battery to a load source. Since the battery has two negative terminals in one cell and two positive terminals in another cell, and two different sets of plates in each cell as well.

In which; the Australians through, they were simultaneously charging and discharging every cell in the battery; but, they were sadly mistaking because electrons can't simultaneously enter and exit the cell <u>without</u> a terminal connection in it; although, it has two different sets of negative and positive plates in it as well.

Since the battery negative and positive plates in each cell, they'll be set up in sequence because it'll be a positive and then, a negative plate setup in this sequence and electrons have to pass through the electrolyte solution as well. Thus; electrons, they wouldn't be able to simultaneously enter and exit a cell <u>without</u> a terminal connection in it because of the design of the battery.

Given that; electrons, they've to flow across one plate and through the electrolyte solution to flow to an adjacent plate because of how the plates, are setup in sequence in each cell as well. So; whichever, electromotive force has the highest voltage potential between the charging voltage and the battery voltage. It'll prevail over the other as well.

Therefore; electrons, they can't simultaneously enter and exit the cell <u>without</u> a terminal connection in it; although, it has two sets of positive and negative plates in it. Thus; it's a waste of money and time to, design a battery with two sets of negative and positive plates in each cell with their own straps and tab connectors because it'll be no benefits in it as well.

On the other hand; designing, a battery with four terminals and each terminal has its own set of plates is a waste of money and time as well. Here's why; the Australians, they could have achieved the same goal if they interconnected the terminals by their respective plates. Then the terminals would only need their own straps and tab connectors leading to the same plates.

Providing that; the battery cells, they're connected in series with its terminal connections. Then the battery, it wouldn't need four terminals for charging while discharging; but, it would only need three. It would be the most cost effective way to design it because we would only need to simultaneously charge and discharge the plates joining the two terminals together.

It seems like the Australians didn't figure this out before they applied for their patent; likewise, it seen like they didn't do the proper analysis of their battery to see if it was necessary for it to have four terminals for charging while discharging or each terminal needed its own set of plates, or each cell needed two sets of negative and positive plates in it with their own straps and tab connectors as well.

I guess Monday morning quarterbacking goes a long way after the game is over? If we don't do the proper analysis on our secondary cell batteries to determine, what's required to charge them while discharging them? Then we'll never know, the most cost effective way to design them; so, they could be charged while being discharged as well.

If we analyze the mechanical make-up of a lead acid automotive starting battery with top and side posts terminal connections, then we'll know why it can't be charged while being discharged; although, it has two positive terminals in one cell and two negative terminals in another cell as well. We still have to stop one cycling process in order to start the other as well.

Here's why; remember the battery's top posts terminal connections, are only connected to their respective plates by a strap and tab connectors; but, its side posts terminal connections aren't connected to their respective plates by a strap and tab connectors; however, they're only connected to their respective top posts terminal connections by a strap.

Wherein; this type of design concept for the battery, it'll not allow it to be charged while being discharged to a load source regardless of it having two positive terminals in one cell and two negative terminals in another cell as well. Although; the battery plates in each cell, they're simultaneously charged and discharged amongst themselves and to its plates in the next cell as well.

Upon my analysis of the battery, I found its cycling processes will be a product of its design. In which; it'll determine, how its cycling processes will be carried out; that is, one process at a time or both processes simultaneously. In other words; we determine, how its cycling processes will be carried out. Not the mechanics of the voltage and current alone or the laws of physics as well.

Although; each cell in the battery, they're separated by a non-conducting material called plastic. But; we're still able to charge, all of its plates in each cell during their charging cycle. Without having to, stop charging them in one cell in order to charge them in the next since they're connected in series-parallel with one another in each cell and each cell is connected in series with its terminals as well.

I found the interior structure of the battery shed light on how to charge it while discharging it to a load source as well. Given how the battery plates in each cell, they're interconnected with one another in their relationship with its terminal connections as well. Once I did the proper analysis on, the mechanics behind the charging cycle of the battery.

I found it had little to do with the mechanics of the voltage and current alone or the laws of physics. Why the battery, it couldn't be charged while being discharged to a load source; but, it had more to do with the wrong mechanics in play. Given the side posts terminal connections of the battery didn't have their own straps and tab connectors leading to their respective plates.

Once I understood that, I found it was conceivable to charge the battery while discharging it to a load source as well. Providing the battery's top and side posts terminal connections, they had their own straps and tab connectors leading to their respective plates as well. I also realized it would only need three terminals to charge it while discharging it to a load source as well.

However; the Australians' battery, it made me realize I wasn't alone in my quest to change a battery while discharging it to a load source as well. Designing a battery with separate terminals for charging while discharging, it makes it possible to charge it at will. As Richard A. Perez stated in his book; Complete Battery Book, written in 1985.

I can't say for sure, the idea of charging an automotive battery at will was brewing in his head at the time or whether he was making a suggestion to illustrate a point about a more efficient way to cycle our automotive batteries as well. Maybe; the idea of charging the battery at will was overshadowed by the fact we've to charge it on the same terminal that, it was discharged on as well.

After reading books written by those in the field of automotive battery technology about, a lead acid automotive starting battery's cycling processes. I found it was scientifically possible to charge it at will if we design it for that purpose; in which, it'll give it a more efficient cycling process as Richard A. Perez theorized more than four decades ago.

Being able to charge a secondary cell battery at will, it'll be conceivable that we could increase its efficiency and overall capacity with an intermittent charge on it. Given that; it'll decrease, its internal and electrical resistance while replenishing its chemical energy as well; thus, increasing the overall travel range of a vehicle on battery power as well.

Charging a secondary cell battery at will is neither inconceivable nor farfetched; but, it could be accomplished by changing its contemporary design; so, electrons could enter and exit it at the same time. It's a simple concept because we could already add or subtract electrons from the same chemical composition of the battery plates as well.

All we've to do is eliminate the process of charging the battery on the same terminal connection that, it was discharged on. In which; it makes, it conceivable to charge it at will. I'm not talking about major changes to the battery; but, just add an additional terminal connection to the cell with a terminal connection already in it which isn't a major change as well.

If we make the right mechanical changes to the battery, so that, electrons could enter and exit it at the same time. Then we don't have to stop its discharging cycle in order to start its charging cycle because we could charge it at will. We could achieve this goal if the battery had two terminals in the same cell for charging and discharging, which is simple as well.

Those terminals having their own straps and tab connectors leading to the same plates, then it would allow us to overcome the challenges that the mechanics of the current poses when there's a charge and a load source on the battery at the same time. Given that; current, it wouldn't have to go in the battery the same way it has to come out as well.

Making simple changes to the battery and then charging it at will, it would not defy the laws of physics. Given that; Gaston Plante', he proved this more than a century ago when he made the first lead acid automotive battery. By using mechanics, he was able to charge the battery plates in each cell without having to stop charging them in one cell in order to charge them in the next.

Maybe; we weren't paying attention then and we're not paying attention now because Gaston Plante', he only connected the battery plates together in series-parallel in each cell and each cell is connected in series with its terminal connections as well. It's simple mechanics to use, so, we don't have to stop charging its plates in one cell in order to charge them in the next.

We're not talking about physics here; but, we're talking about using mechanics. So; we don't have to, stop charging the battery plates in one cell in order to charge them in the next without the laws of physics being an issue. Maybe; back then and now, we haven't really taking an in depth analysis of the mechanics behind the cycling processes of the battery.

To see if, it's necessary to stop the discharging cycle of the battery in order to start its charging cycle because of the laws of physics. I found once Gaston Plante' achieved the goal of adding or subtracting electrons from the same chemical composition of the battery plates. Then it was no longer about the laws of electrochemistry; but, it became more about the laws of mechanics.

Are focusing on the wrong science when it comes to adding or subtracting electrons from the battery? I say the answer is yes because remember if a battery cells, are connected in series with its terminal connections. Then we could receive the voltage potential of all of its plates in each cell; but, only the current potential of its plates in one cell during its discharging cycle.

On the other hand; during the charging cycle of the battery, voltage will be evenly divided amongst its plates in each cell; but; current will be the same across each plate in each cell. If we take notes on, how a battery is designed. Then we'll find its cycling processes will be carried out by the laws of motion and the effect of forces on bodies as well.

Wherein; the laws of electrochemistry, entropy, inertia or the laws of conservation of energy, they'll have little to do with the cycling processes of a battery; given that, they don't determine how electrons are added or subtracted from the battery plates. However; the laws of mechanics, they determine how electrons are added or subtracted from the battery plates.

A person whom I encountered on a business trip asked, what's the purpose of charging our batteries while using them to power our vehicles? When we could, use the energy right away to power our vehicles instead of charging their batteries. At first; I was amazed, by the question because that person worked in the field of automotive battery technology as well.

Nevertheless; I replied, if we could add electrical energy back to our batteries at will. Then we could increase their efficiencies and overall capacities while using them as well. Therefore; increasing, the overall travel range of a vehicle on battery power as well. Without finding that, perfect chemical composition for a longer-lasting battery as well.

All these things, they go hand and hand; given that, adding electrical energy back to our secondary cell batteries restores their chemical energy as well. On the other hand; we could use, an on board charging system to charge a vehicle batteries. While using them if we could add, energy back to them while taking it out as well.

Wherein; the vehicle batteries, they could help power its electric motor in conjunction with its charging system to propel it during a simultaneous cycling process of its batteries. While diminishing, their internal and electrical resistances while on the go; in which, it'll be less energy lost to heat from their internal and electrical resistance as well.

Finding a perfect chemical composition to a longer-lasting battery, it's not the only solution to increase the overall travel range of a vehicle on battery power since a battery chemical energy could be replenished with electrical energy as well. We can't see the benefits in charging it at will since we can't see beyond finding a perfect chemical composition for it as well.

Overlooking, the fact our secondary cell batteries' chemical energy could be replenished with electrical energy. In which; it doesn't make, any sense to search for a perfect chemical composition for them as well. Wherein; this mistake, it causes us to rely more heavily on our gasoline engines than our electric motors to propel our basic form of transportation as well.

It shouldn't be inconceivable or farfetched to think it will be beneficial to add energy back to our secondary cell batteries while taking it out; given that, we could replenish their chemical energy with electrical energy as well. Then it should be obvious why we should add energy back to them while taking it out as well.

Likewise; it should be obvious why, we should change the contemporary design of our batteries; so, energy could enter and exit them at the same time as well. Then the idea of adding energy back to them while taking it out, it shouldn't seem so inconceivable or farfetched as well. If it does, then we can't see beyond energy going in the same way it has to come out as well.

Designing our secondary cell batteries with separate terminals for charging and discharging, it makes it conceivable to add energy back to them while taking it out as well. Given that; we wouldn't need to stop, their discharging cycles in order to start their charging cycles as well. Then it's obvious why we should change their contemporary designs as well.

From a mechanical perspective, it makes sense why our batteries should have separate terminals for charging while discharging; so, we could charge them on one terminal while discharging them on another terminal. Then we could get around the problems the mechanics of the current poses when there's a charge and a load source on our batteries at the same time as well.

The idea of charging our secondary cell batteries while discharging them isn't out of the realm of physic as some may believe. A matter of fact it has little to do with the laws of physics; but, it has more to do with the laws of motion and the effect of forces on bodies proven by the mechanics behind the charging cycle of a lead acid automotive starting battery.

Wherein; charging our batteries while discharging them, it'll be nothing more than a mechanical process because we could already add or subtract electrons from the same chemical composition of their plates. Therefore; it's a matter of how, we go about adding or subtracting electrons from their plates; but, some think it's a matter of physics wherein it's not.

Here's why; we could charge, a lead acid automotive starting battery plates in each cell without having to stop charging them in one cell in order to charge them in the next by using mechanics. Although; the battery plates in each cell, they're separated by a non-conducting material called plastic. Basically; electrons, they're taken from one cell to be added to another.

Remember the battery plates in each cell, are connected in series-parallel with one another by their straps, tab connectors and the electrolyte solution in each cell. Thus; electrons, they could flow from a strap to a tab connector on one plate. And then, flow across that plate through the electrolyte solution to an adjacent plate without flowing to and from the same tab connector.

Then the battery plates in each cell, they're simultaneously charged and discharged amongst themselves and to its plates in the next cell; given that, they're connected in series with one another by way of an electrical bus. So; it seems like, it's neither inconceivable nor farfetched to charge a secondary cell battery while discharge it to a load source as well.

The reason we could assume this is because electrons, they're taken from the battery plates in one cell to be added to its plates in the next cell; although, they're separated by a non-conducting material called plastic. Then what's the difference between taken electrons from one cell, so, it could power a load source that is connected it by way of an electrical bus as well.

If we focus on the laws of mechanics and not the laws of physics, then we'll be focusing on the right science. For instance; a lead acid automotive starting battery amp draw, it only comes from a cell with a terminal connection in it. Since the other cells in the battery, they'll be connected in series with the cell with the terminal connection in it to increase its voltage potential.

Since the battery amp potential only comes from a cell with a terminal connection in it, then it's conceivable to charge it while discharging it if it had two terminals in the same cell with a charge on one terminal and a load source on the other. Thus; it's conceivable, we could increase its efficiency and the overall capacity with the help of the charging source as well.

Here's why; if the charging voltage, it has to flow into the cell with the two terminal connections in it before flowing to a load source. Then the charging voltage, it could increase the amp potential of the cell with the two terminals in it. Given that; its voltage potential times amp potential, it equals total amp capacity for the battery as well. See chapter 26.

I've compiled and compared data about, the mechanics behind the charging cycle of a lead acid automotive starting battery to come to a conclusion. It's scientifically possible to charge the battery while discharging it to a load source. It'll have little to do with the laws of physics; but, it has more to do with the laws of mechanics and this not an assumption.

Here's why; because it's based on, how electrons are added or subtracted from a lead acid automotive starting battery plates during their cycling processes. Thus; if we allow the charging electrons to flow into one cell in the battery before flowing to a load source, then the charging source. It becomes the power source for the battery and it becomes the power for the load source.

One thing I've been alluding to in my book is that, we've to look at a lead acid automotive starting battery from a different perspective because it's not new science. It's old science when it comes to charging the battery while discharging it because it happens every time, we charge it. We just don't realize it yet because of our different perspectives.

Here's why; we don't have to, stop charging the battery plates in one cell in order to charge them in the next cell; although, they're separated by a non-conducting material. In which; this tells us that, we don't have to stop its discharging cycle in order to start its charging cycle as well. We just don't know it yet because we haven't realized it yet as well.

Charging a lead acid automotive starting battery while discharging it to a load source, it's not about new science; but, it's about old science as well. Given that; the plates in each cell in the battery, they consume and store energy during their charging cycle. Then it'll be no different than them consuming and energy while supplying it to a load source as well.

Understanding the science behind the charging cycle of the battery, then it shouldn't be unattainable to charge it while discharging it to a load source if we've the right mechanics in play. Then we wouldn't need to stop the discharging cycle of the battery in order to start its charging cycle since we don't need to stop charging its plates in one cell in order to charge them in the next as well.

However; if we don't see the need to charge our batteries while discharging them, then we won't see the need to change their designs as well. We know it's possible to replenish our batteries' chemical energy with electrical energy as well. But; some say that, it'll be no benefits in it if we add energy back to them while taking it out due to the laws of physics as well.

Wherein; it doesn't make any sense to use either the laws of entropy, inertia or the laws of conservation of energy to discredit a simultaneous cycling process of our batteries. Since energy, it's lost to heat in both conversion processes of our batteries as well. Although; energy lost to heat, it occurs during their charging cycles. And yet, we're still able to charge them.

It doesn't make any sense to use the laws of entropy, inertia or the laws of conservation of energy to discredit a simultaneous cycling process of our batteries; given that, we're charging them as well. The difference between a simultaneous cycling process and stopping one process in order to start the other, energy doesn't have to go in the same way it has to come out.

We've to ask ourselves if energy, it doesn't have to go in the same way it has to come out. Is it conceivable or farfetched to charge our batteries while discharging them without being impeded by the laws of entropy, inertia or the laws of conservation of energy as well? The answer is yes because if we're adding energy back to our batteries while taking it out as well.

Then the laws of entropy, inertia or the laws of conservation of energy become irrelevant because they'll not apply. It's just a misconception because they're applied incorrectly due to us focusing on the wrong science as well. Given that; those laws of physics, they don't determine if electrons going to be added or subtracted from our batteries as well.

There're other extenuating circumstances will prevent us from charging our batteries while discharging them, other than those laws of physics; such as, our batteries having one opening for energy to enter or to exit them or their resistance to current versus our load sources resistance to current, or the type of connections made between them as well.

Those types of extenuating circumstances will make us think, it's due to either the laws of entropy, inertia or the laws of conservation of energy why we can't charge our batteries while discharging them as well. See chapter 5 and 12. This is where facts versus perceptions come into play, about what we could or couldn't with our batteries because of the laws of physics as well.

It's not complicated why we blame it on the laws of physics why we can't charge our batteries while discharging them; wherein, it's due to the lack of information about the mechanics behind their cycling processes. It seems like we haven't acquired enough information to figure out what's required to charge them while discharging them as well.

Wherein; the contemporary design of our batteries, they cause us to believe it's due to the laws of physics they can't be charge while being discharged them because they're not designed for that purpose. They're designed for us to stop their discharging cycles before we could start their charging cycles. It's a matter of mechanics and not a matter of physics.

Those in the field of automotive technology, they figured out. How to charge an automotive starting battery while it has a load source on it; although, it has only one opening for electrons to enter or to exit it and it'll be in parallel with the load source in their relationship with the charging source as well. It wasn't rocket science; but, it was simple mechanics.

Those in the field of automotive technology, they design an electronic alternator regulator control system; so, it'll allow an alternator charging system to create a surplus of electrons at a vehicle voltage and current regulatory system. In order to, reverse the current flow back into its battery to charge it; given that, they would be in parallel with one another.

If a charging system could produce more electrons than needed, then any additional electrons flowing toward the vehicle's voltage and current regulatory system. It would be reversed back into its battery to charge it because they're connected in parallel with one another; in which, current could only flow to the voltage and current regulatory system or to the battery.

Let's be realistic because creating a surplus of electrons at a vehicle voltage and current regulatory system in order to reverse the current flow back into its battery to charge it. Wherein;

it just might work fine for a gasoline power automobile, it probably wouldn't work for an all-electric or a hybrid vehicle because of the different circumstance involved.

Since a charging system, it probably couldn't keep up with the demand of a vehicle's electric motor. Given the wind resistance and the type of terrain the vehicle might encounter; in which, its electric motor might consume electrons faster than its charging system. It could create a surplus of electrons at the vehicle's voltage and current regulatory system.

In order to, reverse the current flow back into the vehicle's batteries to charge them. This type of scenario will only occur if the vehicle's batteries, they're connected in parallel with its voltage and current regulatory systems in their relationship with its charging system. It would make it hard for a charging system to power an electric motor and charge batteries at the same time.

Even if the charging system, it could produce twice the energy needed to power the electric motor and charge the batteries at the same time. Given the parallel connection will be made between them. The charging electrons could flow directly to the electric motor to power it and not flow into the batteries at all because of how electrons are added or subtracted from them.

In which; it makes, it inconceivable or farfetched to charge the batteries while powering the electric motor. Although; the charging system, it could produce twice the energy needed to power the electric motor and charge the batteries at the same time as well. However; it's not inconceivable or farfetched to, manipulate the direction of the current.

By creating a surplus of electrons at a vehicle's voltage and current regulatory system in order to reverse the current flow back into its battery to charge it; although, it'll be connected in parallel with the vehicle's voltage and current regulatory system in their relationship with its charging system. Those in the field of automotive technology, they took advantage of the situation.

Given that; one aspect of the mechanics of the current is that, it'll follow the path of least resistance. Thus; it makes, it possible to charge an automotive starting battery while it has a load source on it; although, it has only one opening for current to enter or exit it and it'll be in parallel with the load source in their relationship with the charging system as well.

Likewise; it shouldn't be inconceivable or farfetched to manipulate, the mechanics of the current with the design of a battery in order to charge it while discharging it to a load source as well. Wherein; it seems like, it has little to do with the laws of physics; but, it has more to do with the laws of motion and the effect of forces on bodies as well.

In which; it seems like, we haven't acquired enough information to figure out what's required to charge a battery while discharging it to a load source without being impeded by the laws of physics as well. Once we figure it out, then the only thing left to consider is where the energy input going to come from to charge our batteries on the go? In the next chapter, let's explore those possibilities as well.

CHAPTER 25

ENERGY INPUT FOR AN AUTO BATTERY

IF WE UNDERSTOOD, THE value of charging our secondary cell batteries while discharging them. Then we might want to change their energy input processes; so, energy could enter and exit them at the same time. Then we could minimize the effects their discharging cycles will have on their efficiency and overall capacity and the overall travel range of a vehicle on battery power as well.

Here's the deal; if energy isn't added back to our batteries right away, then they become less and less efficient to power our vehicles because their internal and electrical resistance. Given that; their chemical compositions, they work like a double edged sword because energy is released from them as well.

However; our batteries' chemical compositions, they'll limit the amount of energy could be released from them as well. Since the chemical residue left on their plates from their chemical discharging cycles, it'll limit the amount of energy could be released from them as well. Remember the internal electrical resistance of a lead acid automotive starting battery.

It's calculated by squaring the current in amperes and multiplying it by the internal resistance of the battery; in which, its electrical resistance is compounded by its internal resistance as well. The factors affecting the efficiency and overall capacity of the battery, they're all interrelated as well. See chapter 4.

Therefore; if we change one factor, then we change them all. Thus; if we could add energy back to a secondary cell battery while taking it out, then it'll limit the amount of the chemical residue accumulating on its plates; thus, keeping their resistance as low as possible. If we've an intermittent charge on them, during their discharging cycles as well.

With the intermittent charge on the battery, then we could minimize its internal and electrical resistance affecting its efficiency and overall capacity as well. Remember adding electrical energy back to a secondary cell battery, it'll diminish the chemical residue that accumulated on its plates during their discharging cycles.

Therefore; it's another reason why, we should charge our batteries while discharging them; so, we could minimize the effects their discharging cycles will have on their efficiency and overall capacity as well. It'll be another option to increase the overall travel range of our vehicles on battery power without finding a perfect chemical composition for their batteries as well.

Since adding electrical energy back to our secondary cell batteries, it reduces their internal and electrical resistance and restores their chemical energy as well. If we focus on these facts, then it seems there will be a benefit in adding electrical energy back to our batteries while taking it out; providing that, we change their energy input process as well.

Then we could do something about the inherent energy lost to heat affecting our batteries' efficiency and overall capacity providing we had an intermittent charge on them during their discharging cycles. And then, we could increase their efficiency and overall capacity by maintaining a constant state of charge within our batteries while on the go as well.

Thus; increasing, the overall travel ranges of our vehicles on battery power by using the same techniques. That is used to simultaneously charge and discharge, a lead acid automotive starting battery plates amongst themselves in each cell and to its plates in the next cell. Without having to, charging them in one cell in order to charge them in the next cell as well. See chapter 22.

All we've to do is, create a series-parallel circuit connection between our vehicles' charging systems and their voltage and current regulatory systems in their relationship with their batteries. Then we don't have to stop their discharging cycles in order to their charge cycles; thus, we could charge them while discharging them as well. See chapter 23.

A series-parallel circuit connection made between our vehicles' charging systems and their voltage and current regulatory systems in their relationship with their batteries. It'll allow us to charge their batteries while using them to power their electric motors to propel them as well. Providing that; we change the way, we carry out their batteries energy input process as well.

Here's why; if we design our batteries, so, electrons could enter them on one terminal and exit them on another terminal connection at the same time. Then electrons, they've to flow into one cell in our batteries before flowing to a vehicle's voltage and current regulatory system. Thus; the amount of resistance at the vehicle's voltage and current regulatory system, it wouldn't have any bearing on the amount of electrons flowing into its batteries as well.

Therefore; the amount of wind resistance and the type of terrain that, the vehicle might encounter. It'll have no bearing on the efficiency and overall capacity of its batteries because we could add energy back to them while taking it out as well. Since the charging electrons, they've to flow into one cell in the batteries before flowing to the vehicle's voltage and current regulatory system.

Therefore; the charging system, it doesn't have to create a surplus of electrons at the vehicle's voltage and current regulatory system. Given that; the surplus of electrons, they'll be created in one cell in the vehicle batteries before flowing to its voltage and current regulatory system.

Thus; the charging system, it doesn't need to produce twice the energy needed to power the vehicle's electric motor and charge its batteries. Since its batteries, they'll not be in parallel with its voltage and current regulatory system in their relationship with its charging system as well.

Then it'll seem conceivable for the vehicle's charging system to, charge its batteries while they're powering its electric motor to propel it because we change the energy input process of its batteries. Then the question is that, where's the energy input going to come from to charge its batteries while on the go?

Although; we've some electric vehicles, they could travel over three hundred miles on a single battery charge. However; the question still remain, where's their energy input going to come from when it's time to charge their batteries while on the go. Find a charging station may or may not exist in most cities and if so, they're few and far in between.

Then the only other alternative for an all-electric vehicle when travelling far from home, it's not to travel far from home. In which; it defeats the purpose of buying an all-electric vehicle to save on gas when travelling far from home as well. If we're planning on, travelling far from home and don't want to worry about finding a charging station while on the go.

Thus; we might want to buy, a hybrid vehicle because we could charge its batteries while on the go; given that, it has an on board charging system. On the other hand; charging, its batteries on the go. It'll allow its combustion engine to consume more fuel than normal to propel it and charge its batteries at the same time as well.

Therefore; we'll be relying more heavily on, its combustion engine than its electric motor to propel it when travelling far from home. Thus; we'll be better off buying, a gasoline powered automobile if we're going to travel far from home. Then we don't have to worry about finding a charging station when travelling far from home as well.

A gasoline powered automobile will use less fuel than a hybrid vehicle will when travelling far from home. Given that; its combustion engine, it doesn't have to simultaneously turn a drive train and a motor generator system in order to propel the vehicle and charge its battery at the same time.

Remember a hybrid vehicle has two propulsion systems; a combustion engine and an electric motor to propel it. The problem with the vehicle is that, it has to rely more heavily on its combustion engine than its electric motor to propel it. Given that; we can't use, its electric motor to propel it while charging its batteries on the go.

The vehicle's batteries, they'll be connected in parallel with its voltage and current regulatory system in their relationship with its motor generator system. Since its batteries have to be charged on the same terminal that, they were discharged on. Then one thing leads to another because of how, the vehicle's batteries energy input process is carried out.

Remember the vehicle's motor generator system might not be able to keep up with the demands of its electric motor and charge its batteries at the same time. Given its electric motor might consume electrons faster than its motor generator system. It could create a surplus of electrons at the vehicle's voltage and current regulatory system to reverse the current flow back into its batteries to charge them because of wind resistance and the type of terrain it might encounter.

Remember this type of scenario will only occur if a parallel circuit connection exists between a vehicle's batteries and its voltage and current regulatory system in their relationship with its motor generator system. Then the parallel circuit connection between them, it'll prevent the vehicle from using its electric motor to propel it while charging its batteries. See chapter 24.

Thus; the vehicle's gasoline engine, it has to simultaneously turn its drive train and motor generator system in order to propel it and charge its batteries at the same time. Wherein; this process, it decreases the overall travel range of the vehicle on a single gallon of gasoline. Given the extra fuel its engine has to consume to turn its drive train and motor generator system at the same time while on the go.

If we focusing on the engineering concept of a hybrid vehicle, then we'll realize we're not utilizing its technology in the right manner to decrease its fuel consumption; although, it has two propulsion systems. But; the way we use them, it defeats the purpose of it having two propulsion systems when it comes to decreasing our fuel consumption as we hope.

Using reverse engineering on the hybrid vehicle; that is, using its gasoline engine to temporarily turn its motor generator system; so, it could temporarily charge its batteries while they're powering its electric motor to propel it. Then we'll be using our automotive technology in the right manner to decrease our fuel consumption. Given that; we could rely more heavily on, the electric motor than the gasoline engine to propel our vehicles.

Since our vehicles, they'll not need two propulsion systems to propel them; but, they'll only need one. Wherein; their gasoline engines, they'll only turn their motor generator systems when their batteries are low on charge and not turn their drive trains as well. In which; the vehicles motor generator systems, they'll temporarily charge their batteries while they're temporarily powering their electric motors to propel them.

Thus; using, less fuel to propel the vehicles and charge their batteries at the same time. I know some skeptics say this will not work because of the load put on the vehicles batteries from their electric motors. Therefore; increasing, the load on their charging systems and their combustion engines as well. Then their engines will consume more fuel to rotate the charge systems in order to charge the batteries.

As a result; skeptics say that, this type of idea will not decrease our fuel consumption if a motor generator system has to charge a vehicle's batteries while they're powering its electric motor to propel it. I'll explain in more detail in chapter 26, why this type of process might work if we're charging our batteries while using them to power our electric motors as well.

I found utilizing our automotive technology in the right manner. We could decrease our fuel consumption. Given that; we could rely more heavily on, the electric motor than the gasoline engine to propel our vehicles. If we don't use their engines to turn their drive trains and motor generator systems at the same time. Also; we could use natural gas or propane to power their combustion engines, so, we don't have to emit poisonous gases into atmosphere on a daily basis as well.

The only way we could rely more heavily on our vehicles' electric motors than their combustion engines to propel them. We've to utilize our automotive battery technology in the right manner as well. The contemporary designs of our secondary cell batteries today. They're good examples of us, not utilizing our battery technology in the right manner. Since we can't use, our hybrid vehicles' electric motors to propel them while charging their batteries.

Given that; our vehicles' batteries, they only have one opening for energy to enter or to exit them; therefore; we can't add energy back to them while taking it out. This type of design concept, it causes us to rely more heavily on a combustion engine than an electric motor to propel our vehicles. Since their batteries have to be charged on the same terminal connections that, they were discharged on.

Then one thing, it leads to another because we're not utilizing our battery technology in the right manner. Charging our batteries on the same terminals that, they were discharged on. It makes it impossible to use a vehicle's motor generator system to charge its batteries while they're powering its electric motor. All because energy has to go in our batteries the same way it has to come out because we're not utilizing, our battery technology in the right manner.

If we change the energy input and output process of our batteries, so, we don't have to charge them on the same terminal connection they were discharged on. Then energy could enter them on one terminal and exit them on another terminal connection. Thus; we don't have to, stop their discharging cycles in order to start their charging cycles; therefore, utilizing our battery technology in the right manner as well.

It doesn't make any sense to design our secondary cell batteries; so, energy has to go in them the same way it has to come out; given that, they're chargeable batteries. In other words; if we could charge and discharge the same battery, then why charge it on the same terminal connection it was discharged on. It doesn't make any sense because it becomes inconvenient and time consuming to carry out its cycling processes; given that, we've to stop one process in order to start the other.

Remember when charging a lead acid automotive starting battery, we don't have to stop charging its plates in one cell in order to charge them in the next cell; although, they're separate by a non-conducting material called plastic. The reason for this because of how its plates and cells, are interconnected with its terminal connections; thus, energy could flow in one direction from its plates one cell to its plates in the next cell as well.

Logic tells us if energy doesn't have to go in our batteries the same way it has to come out. Then we don't have to stop their discharging cycles in order to start their charging cycles as well. Thus; one thing, it leads to another because we wouldn't need to stop using our vehicles in their electric modes in order to charge their batteries while on the go. If we didn't have to charge them on the same terminal connections that, they were discharged on as we do today.

Designing our batteries with separate terminals for charging and discharging, then the familiar problem of having to stop their discharging cycles in order to start their charging cycles will no longer exist. In which; it'll eliminate, the unforeseen problem when it comes to our hybrid vehicles. Given that; we can't use their electric motors to propel them while charging their batteries because of their contemporary designs, which affect their energy input as well.

Since energy, it has to go in our batteries the same way it has to come out. Then our batteries will be connected in parallel with our vehicles electric motors in their relationship with their motor generator systems as well. Thus; we can't charge, our vehicles' batteries while using them to power the vehicles' electric motors because of the connections made between them.

Wherein; it has little to do with the laws of physics, why we can't charge our vehicles' batteries while using them to power the vehicles' electric motors; but, it has more to do with the energy input of our batteries. If they had separate terminals for charging while discharging, then we could change their energy input and output process as well.

Thus; we could increase, the overall travel range of our vehicles on battery power by keeping an intermittent charge on their batteries; therefore, keeping their internal resistances as low as possible as well. In which; it'll minimize, their internal electrical resistance while minimizing the amount of energy lost to heat as well. Charging them while discharging them, it's like having a trickle down charge on them; but, we'll be subtracting energy from them as well.

Then it'll take less time and energy to keep our batteries charged because the trickle down charge, it won't allow them to become highly discharged while subtracting energy from them; thus, shortening their charging duration. In which; it'll be a much faster way of charging, our batteries as well. Some have developed charging systems will charge our batteries within two hours; however, two hours is a long time to wait if you're on the go as well.

Having an intermittent charge on our batteries while on the go, it seems like a much quicker way to charge them while on the go because we don't have to stop using them in order to charge them. The only way we could achieve this goal our batteries have to have separate terminals for charging and discharging in the same cell. And those terminals, they've to have their own straps and tab connectors leading to the same plates in that cell as well.

Wherein; this new design concept of our batteries, it'll allow us to add energy back to them while on the go. Without having to, stop their discharging cycles in order to start their charging cycles. Using our battery technology in the right manner, then we could use reverse engineering on our hybrid vehicles as well.

Given that; the vehicles, they wouldn't need two propulsion systems to propel them and charge their batteries at the same time. Only using, their combustion engines to temporary turn their charging systems to charge their batteries while they're powering their electric motors to propel them as well. We don't have to use gasoline to power their engines; but, we could use other sources of energy; such as, natural gas or propane having less impact on our environment as well.

On the other hand; we could use, solar energy to charge our vehicles' batteries while they're powering their electric motors while on the go as well. If energy, it could enter our batteries on one terminal and exit them on another. Then the amount of resistance at the vehicle's voltage and current regulatory system, it'll no longer dictate the amount of electrons flowing into its batteries.

Since electrons, they've to flow into one cell in the batteries before flowing to the voltage and current regulatory system. Then the path of least resistance for the electrons, it's to flow into one cell in the batteries before flowing to the voltage and current regulatory system. Thus; a charging source, it doesn't need to create a surplus of electrons at the voltage and current regulatory system to reverse the current flow back into the batteries to charge them.

Likewise; we don't have to worry about the amount of wind resistance and the type of terrain, a vehicle might encounter draining its batteries. It'll become irrelevant if electrons could flow into one cell in its batteries before flowing to its voltage and current regulatory system. Utilizing our battery technology in the right manner, then we could eliminate the familiar and unforeseen problems that exist with our hybrid vehicles today.

If electrons, they could flow into one cell in a vehicle's batteries before flowing to its voltage and current regulatory system. Then its motor generator system, it wouldn't need to create a surplus of electrons at the vehicle's voltage and current regulatory system to reverse the current back into its batteries to charge them; thus, we could use much smaller charging systems for our hybrid vehicles as well.

We could use something similar to an electronic alternator regulator control system to monitor the voltage or the current level within our batteries and then, charge them when needed as well. Since an alternator charging system, it could produce a large amount of current. It'll be an idea charging system when it comes to charging our batteries on the go if they had separate terminals for charging while discharging as well.

Having separate terminals for charging while discharging in the same cell and those terminals, they'll have their own strap and tab connectors leading to the same plates in that cell. Then we could change the energy input process of our batteries because electron, they no longer have to flow toward a load source first before flowing into the batteries when there's a charge and a load source on them at the same time.

Then it makes, it plausible to use an electronic alternator regulator control system to monitor the voltage or the current level within our batteries as well. In 1860; the continuous current Dc generator system, it was created by Antonio Pacinotti based on the Wikipedia. Also; it stated that, "Only as cars became more sophisticated and employed more electrical devices, then the need for an onboard power generation become necessary."

"Dc generators at the time, they were simple to manufacture in spite of their operating deficiencies. After World War II, it was the beginning of the end for Dc generators in automobiles. U.S military services, they demanded alternators on their vehicles because of their reliability and increased power to size ratio."

"Once it was an economical way to regulate an alternator three phase Ac output into Dc by silicon rectifiers. Then the path was clear for alternators use in domestic cars as well." See figure 7.

Figure 7

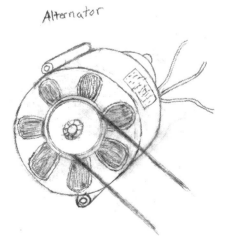

(An alternator charging system)

An alternator charging system, it uses transistors, rectifiers and diodes to turn alternating current into direct current. It could be an idea charging system for our deep cycle automotive batteries because it could produce large amounts of current. Using transistors, rectifiers and diodes, we could determine the energy input for our batteries with an alternator charging system.

However; it'll be a give and take situation, when it comes to voltage versus current output. For example, if we increase, its voltage output from 12 volts to 120 volts. Then it'll decrease its current output from 300 amps to 30 amps; but, its total energy output will never change regardless of the ratio of voltage to current. It'll be an idea charging system to charge our deep cycle automotive batteries while on the go as well.

Remember as long as a vehicle's engine is running, then an electronic alternator regulator control system. It'll monitor the voltage or the current level within the vehicle's battery to insure that, electrons are added back when needed. In which; this type of set-up, it'll allow the vehicle to have a constant source of electrical energy while its engine is running.

We could have a similar set-up with our deep-cycle automotive batteries to insure that, electrons will be added back to them when needed while on the go as well. Wherein; our deep-cycle automotive batteries, they'll remain the primary source of power for the vehicle. Given that; we could still use them to, power the vehicle while charging them providing they've separate terminals for charging and discharging as well. See chapter 26.

Some people say that, we can't use an electronic alternator regulator control system to charge a deep cycle (lead acid) automotive battery because when an alternator charging system throws its full load across the battery, it'll cook it. If we compare, its similarities with a lead acid automotive starting battery mechanical make up. Then it'll seem plausible that, we could use an alternator charging system to charge the deep cycle (lead acid) automotive battery as well.

Since the deep cycle (lead acid) automotive battery and a lead acid automotive starting battery, they're practically the same except the deep-cycle battery plates. They're much thicker and have more surface space than a lead acid automotive starting battery plates. What most people don't know is that, an alternator charging system will cook a lead acid automotive starting battery if it's almost empty or almost full when the alternator charging system throws its full load across it as well.

It's not that a deep cycle (lead acid) automotive battery can't be charged with an alternator charging system as most people suggests. What I found is that, it boils down to the state of the battery when an alternator charging system throws its full load across it. In other words; it depends on, the state of the battery before the alternator charging system starts or stops adding electrons to it; in which, it'll determine if it'll be cooked it or not.

What I discovered was that, the whole purpose of an electronic alternator regulator control system is to add electrons back to a battery when needed without cooking it. Since an electronic alternator regulator control system, it monitors the voltage or the current level within a battery in order to add or stop adding electrons back to the battery. Then it seems like, it'll be an idea charging system for our deep cycle (lead acid) automotive batteries as well.

Here's why; if the battery, it has separate terminals for charging and discharging in the same cell. And those terminals, they'll have their own straps and tab connectors leading to the same plates in that cell as well. Then we could use an electronic alternator regulator control system to monitor the battery voltage or current level; therefore, it'll never be almost empty or almost full when the alternator charging system throws its full load across it.

Also; what I've been alluding to is that, the cell with the charge and the discharging terminal in it will aid a vehicle's charging system. To help power, its electric motor to propel it during the charging cycle of its batteries as well. Given that; the vehicle's charging system, its voltage and current regulatory system will be in series-parallel with one another in their relationship with its batteries; thus, they'll no longer be in parallel as they're traditionally.

Based on, my research on a lead acid automotive starting battery. It seems like those in the field of automotive battery technology, they're hyperopic because they can't foresee energy having to go in our batteries the same way it has to come out. It'll limit their potential more so than their chemical compositions does. Also; energy having to go in the same way it has to come out, it'll limit our batteries energy input and output process as well.

Stopping one cycling process in order to start the other, it'll create an insufficient cycling process for our batteries because it causes us to rely more heavily on the gasoline engine than the electric motor to propel our basic form of transportation. Are those in the field of automotive battery technology, they're myopic as well as they're hyperopic as well? Since they can't see far enough to determine, new battery technology could come in form of changing the design of the battery as well.

Wherein; our automotive batteries, they must be able to facilitate the need they're needed for in the modern world of automotive technology. We can't continue to rely on their contemporary designs if they're going to facilitate the need, we need them for. In other words; we can't continue to charge our batteries on same terminal connections that, they were discharged on if they're going to advance our automotive technology beyond the gasoline engine.

We've to have come up with new battery technology for our all-electric and hybrid vehicles if we're going to decrease our fuel consumption from gasoline from crude oil. We've to change the energy input and output process of our automotive batteries. We can't continue to allow energy to go in the same way that, it has to come out if they're going to facilitate, the need they're needed for in the modern world of automotive technology as well.

For more than a century, we've been searching for that perfect chemical composition to a longer-lasting battery. So; the electric vehicle, it could become our primary source of transportation; but, we always have came up short. Now it's time to look for a mechanical rather than a chemical solution to increase the overall travel range of our vehicles on battery power. Since we could, replenish our batteries' chemical energy with electrical energy as well.

Based on, my research. We already have a mechanical solution to increase the efficiency and overall capacity of our batteries; that is, their charging cycles. Allowing electrons to, flow into one cell in our batteries before flowing to a vehicle's voltage and current regulatory system. It would increase our batteries' efficiency and overall capacity because adding electrons back to them decrease their internal and electrical resistance as well.

Thus; increasing, the overall travel ranges of the vehicle on battery power because we're adding energy back to its batteries while taking it out as well; therefore, they remain its primary source of power since the surplus of electrons will be created in one cell in them before flowing to a voltage and current regulatory system. Thus; the charging current, it'll be reversed back toward them and through the other cells in its batteries as well.

Thus; charging, the vehicle's batteries while discharging them because how we carry out the energy input process of its batteries; wherein, it'll allow us to equip our vehicles with much smaller combustion engines. Since they'll not need to, simultaneously turn a drive train and a motor generator system. In order to, propel the vehicles and charge their batteries at the same time.

Therefore; we could equip our vehicles with much smaller charging systems, similar to an alternator charging system used in a gasoline powered automobiles as well. Since the charging system, it wouldn't need to create a surplus of electrons at a vehicle's voltage and current regulatory system. In order to, reverse the current flow back into its batteries to charge them.

Since electrons, they've to flow into one cell in the vehicle batteries before flowing to its voltage and current regulatory system. In which; its charging system, it doesn't need to produce twice the energy needed to power its electric motor in order to charging its batteries and power its electric motor at the same time as well.

It seems plausible we could use an onboard charging system similar to an alternator charging system or use a solar charging system to charge our batteries while on the go as well. If we allow the charging electrons to, flow into one cell in our batteries before flowing to the vehicle's voltage and current regulatory system; thus, we could charge its batteries while using them to power its electric motor as well.

Remember if a battery, it has separate terminals for charging and discharging in the same cell. And those terminals, they'll have their own strap and tab connectors leading to the same plates in that cell as well. Then the charge and the discharging terminal, they'll be in reverse and independent of one another as well. See chapter 22 and 23 as well.

Since the charging voltage and current, they've to flow into one cell in the vehicle's batteries before flowing to its voltage and current regulatory system. Then the voltage needed to supply electrons to the vehicle's voltage and current regulatory system. It'll not take away from the voltage needed to supply electrons to its batteries to charge them as well.

Having separate terminals for charging and discharging, it'll allow the vehicle's batteries and its voltage and current regulatory system to receive the same amount of voltage needed to supply electrons to them in order to power it and charge its batteries at the same time as well. What makes it possible is that, the charge and the discharging terminal are interconnected by the same plates.

Since the charging current, it has to flow directly into one cell in the batteries before flowing to the vehicle's voltage and current regulatory system. Then current need to power the vehicle, it'll not take away from the current needed to charge its batteries because they'll receive the same amount of current flowing from the charging source based on their resistance.

Likewise; since current, it has to directly to the vehicle's voltage and current regulatory system from its batteries. Then its voltage and current regulatory system, they'll receive the same amount of current flowing across the plates connecting the charge and the discharging terminal together in parallel; but, based on the resistance at the voltage and current regulatory system as well.

Remember voltage and current will flow from the charging terminal to the plates connecting them together in parallel. It'll be independent of voltage and current flowing from those plates to the discharging terminal. Likewise; voltage and current, they'll flow to the plates connecting them together in parallel with charging terminal. It'll be in reverse of voltage and current flowing from those plates to the discharging terminal as well. See chapter 22 and 23.

Given that; the charge and discharging terminal, they'll have their own straps and tab connectors leading to the same plates in the same cell as well. Remember in chapter 23 if a charge and a discharging terminal, they had their own strap and tab connectors leading to the same plates. Then the amount of voltage flowing from the charging terminal <u>to</u> those plates, it'll be the same as the voltage flowing <u>from</u> those plates to the discharging terminal as well.

Since the discharging terminal, it's connected in parallel with the charging terminal by way of the same plates. Then the amount of current flowing across each plate to the discharging terminal, it'll be combined at the strap connecting them together in parallel. Thus; the sum of the current flowing to the vehicle, it'll equal to the current flowing across each plate connecting them together in parallel with the discharging terminal as well.

Therefore; a vehicle's batteries having separate terminals for charging and discharging, it'll allow us to the charge a vehicle's batteries while using them to power its electric motor as well. See chapter 26. Changing the design of our batteries so they'll have separate terminals for charging and discharging, then they could facilitate. The need they're needed for when it comes to increase the overall travel range of our vehicles on battery power as well.

Remember in chapter 2, I talked about the CSX trains as a good example for decreasing our fuel consumption; given that, they rely more heavily on their electric motors than their diesel engines to propel them. In short; a CSX train, it doesn't use its diesel engines as its primary power source to propel it; but, it uses its diesel engines. To power, its generator systems to charge its batteries and they power its electric motors to propel it.

We could achieve this goal if we change the energy input process of our batteries. Then we could use a similar process as the CSX trains to decrease our vehicles fuel consumptions as well. Given that; we could, rely more heavily on the electric motor than the combustion engine to propel our hybrid vehicles if energy could enter and exit their batteries at the same time.

Then we could use smaller combustion engines because the vehicles engines will no longer need to simultaneously turn their drive trains and motor generator systems. Given that; we could, use their batteries in conjunction with their charging systems to help power their electric motors while charging their batteries at the same time as well.

I know it sounds unbelievable for those in the field of automotive and automotive battery technology to add energy back to our batteries while using them to power our vehicles. However; I found that, it's conceivable and not farfetched if we allow the charging voltage and current to flow into one cell in a battery before flowing to a load source.

In the next chapter, I'm going to explain, why allowing the charging voltage and current to flow into one cell in a battery before flowing to a load source. Not only do we change the energy input process of the battery; but, we change its energy output as well.

CHAPTER 26

ENERGY OUTPUT FROM AN AUTO BATTERY

CHARGING OUR BATTERIES ON the same terminal connections that, they were discharged on. Not only does it limits their energy input process; but, it limits their energy output process as well. Utilizing our battery technology in the right manner, then we could increase the overall travel range of our vehicles on battery power without finding a perfect chemical composition for their batteries well.

For more than a century those in the field of automotive battery technology, they always have been talking about finding a perfect chemical composition to a longer-lasting battery so far they haven't. When it comes to our vehicles, we continue to rely more heavily on the gasoline engine than the electric motor to propel them.

I found we could rely more heavily on the electric motor than the gasoline engine with a mechanical solution. If our automotive batteries had separate terminals for charging and discharging in the same cell and those terminals, they had their own straps and tab connectors leading to the same plates in that cell as well.

Thus; we could increase, the overall travel range of our vehicles on battery power without finding a perfect chemical composition for their batteries as well. I know it sounds inconceivable to most in the field of automotive and automotive battery technology; but, it's conceivable if we've the right mechanics in place.

Remember its voltage potential times current potential equals total power when it comes to electricity. Well! The same thing applies to our secondary cell batteries because its voltage potential time's amp potential, it equals total amp capacity for our batteries. We use the words" total amp capacity" instead of total power for our batteries; but, it means same thing as total power as well.

While carrying out my research on a lead acid automotive starting battery, I found if we allow the charging voltage and current to flow into one cell in the battery before flowing to a load source. Then with the help of a charging source, we could increase the amp potential of the cell the charging voltage and current has to enter in first; thus; increasing the energy output of the battery as well.

The plates connecting the charge and the discharging terminal together in parallel, their total energy output will be increased by the total voltage output of the charging source. If a load source is connected to the discharging terminal, then it'll receive the total amp potential of the plates connecting the charge and the discharging terminal together in parallel since its voltage potential time's amp potential equals total amp capacity.

Let me explain this supposition in a different way and then, you might understand why. It'll be possible to increase the efficiency and overall capacity of a battery with a mechanical rather than a chemical solution. Since adding electrical energy back to a secondary cell battery, it restores its chemical energy as well.

During a lead acid automotive battery discharging cycle, it uses its chemical electromotive force to supply electrons to a load source. Thus; the battery internal and electrical resistance, they continue to increase as it continues to power the load source. In which; the internal and electrical resistance of the battery, they'll limit the amount of energy could be used to power the load source as well.

Therefore; the battery, it becomes less and less efficient as it continues to power the load source as well. Thus; if energy, it could enter and exit the battery at the same time. Then we could solve the problem of its internal and electrical resistance limiting the amount of energy could be released from it as well.

Remember a lead acid automotive starting battery chemical composition works like a double edged sword. Not only does it allow electrons to be released from its plates; but, it'll limit the amount of electrons that could be released from its plates as well. Since its chemical composition will leave a chemical residue on its plates, it'll limit the amount of electrons released by its plates during their discharging cycle as well. See chapter 4.

If we don't add electrons back to the battery right away, not only will its chemical electromotive force become less efficient as it continues to be discharged. However; as more electrons, are released from it. Then less and less electrons could be released from it as well. We could avoid this type of scenario by limiting the build-up of the chemical residue on the battery plates with an intermittent charge on them.

Since adding electrons back to the battery, it'll diminish the chemical residue build-up on its plates and will help restore its chemical electromotive forces as well. In other words; restoring, the battery voltage potential; so, more electrons could be released from its plates as well. Understanding what'll happen on the inside of the battery during its charging cycle and then, we could assume what'll happen during simultaneous cycling process of it as well.

If we add electrons back to the battery while taking them out, then we could assume it'll keep the chemical residue build-up on its plates as low as possible; in which, it'll make them more efficient in releasing electrons while restoring their chemical energy as well. It seems like it'll be a benefit in a simultaneous cycling process of the battery based on the mechanics behind its charging cycle as well.

Remember the battery plates in each cell, are connected in series-parallel with one another by their straps, tab connectors and the electrolyte solution in each cell. And each cell, it's connected in series with its terminal connections as well. Then electrons, they could flow from a strap to a tab connector on one plate. And then, flow across that plate through the electrolyte solution to an adjacent plate.

Then one plate is simultaneously charged and discharged to another plate by way of the electrolyte solution in each cell. Without flowing to and from, the same tab connector on the same plate. Since each cell, it's connected in series with the battery terminal connections. Then its plates in one cell, they're simultaneously charged and discharged to the plates in the next cell by way of an electrical bus as well.

Although; the battery plates in each cell, they're separated by a non-conductive material called plastic. We're still able to charge its plates in each cell, then it seems like it'll be no different than simultaneously charging and discharging its plates to a load source. Given that; electrons are taken from one cell, so, they could be added to another cell without the laws of physics being an issue as well.

Since the battery goes through a simultaneous cycling process each time, it's charged. It seems like it has been a serious miscalculation by some in the field of automotive battery technology to assume. It'll be no benefits in a simultaneous cycling process of it due to the laws of physics. Not only electrons having to go in the battery the same way they've to come out. It limits how electrons are added or subtracted from it as well.

It also limits our options down to finding a perfect chemical composition with a longer-lasting chemical reaction for the battery as well. If we change, how its energy input and output process is carried out. Then we wouldn't need to find a perfect chemical composition for it; given that, adding electrical energy back to it restores its chemical energy as well. It seems like it would be a good idea to have the ability to add electrical energy back to it while taking it out as well.

For one to assume, it'll be no benefits in a simultaneous cycling process of the battery due to either the laws of entropy, inertia or the laws of conservation of energy as well. It seems like they haven't taking an in depth analysis of the mechanics behind its charging cycle. Given that;

those laws of physics, they're not an issue when its plates in one cell are simultaneously charged and discharged to its plates in the next cell as well.

Upon carrying out, a number of experiments on the charging cycle of a lead acid automotive starting battery. I found there will be many benefits in a simultaneous cycling process of it; in which, I've already told you about some. However; it's one in particular, I've only mention it in passing. Wherein; this particular benefit, it'll allow us to increase the energy output of the battery during a simultaneous cycling process of it as well.

If we allow, the charging voltage and current to flow into one cell in a secondary cell battery before flowing to a load source as I've been alluding to. Then with the help of a charging source, we could increase the amp potential of the cell the charging voltage and current has to enter in first as I've mention earlier. Thus; we could increase, the energy output of the battery to help power a load source if the battery has separate terminals for charging and discharging.

Understanding, what'll happen on the inside of a lead acid automotive starting battery during its charging cycle? It'll help in understanding if it had separate terminals for charging and discharging in the same cell. And those terminals, they had their own straps and tab connectors leading to the same plates in that cell as well. Not only could we charge the battery at will; but, we could increase its total energy output with the help of a charging source as well.

Here's why; remember voltage potential time's current potential, it equals total amp capacity for a battery. It'll be similar to the total energy output of a magnetic power source; but, for the magnetic power source. It'll be voltage (emf) times amperage (current) or P=E x I; in which, it means total power for the magnetic power source. If we equate a battery energy output to a magnetic power source energy output, we'll realize it means the same thing.

Remember each cell in a lead acid automotive starting battery is nothing more than 2-volt batteries in and of themselves. That is connected in series with one another and housed together in a plastic case to increase voltage and current potential for its plates in one cell. In which; its amp potential, it'll be based on the size and the number of plates in each cell. See chapter 3 and 5 for more information about connecting a battery cells together.

If each cell in the battery, it has the same number of plates and they're the same size. Then each cell will have the same amp potential; therefore, the amount of voltage applied to a cell will determine its total amp capacity. Given that; it'll be the voltage potential applied to the cell time's its amp potential, which will equal its total amp capacity.

For example; if 2-volts were applied to a cell with an amp potential of 125 amps, then it'll have a total amp capacity of 250 amps; that is, 2-volts time's 125 amps. Likewise; if a total of 6-volts were applied to the same cell, then it'll have a total amp capacity of 750 amps; that is, 6-volts times 125 amps. Thus; connecting a battery cells together in series, it only increases the voltage and amp potential of its plates in one cell.

Since the other cells in the battery, they'll be connected in series with a cell with a terminal connection in it to increase its voltage and amp potential. If a cell had separate terminals for charging and discharging in it and those terminals had their own straps and tab connectors leading to the same plates in that cell as well. Then we could increase, its total amp capacity with the help of a charging source as well

This is why I said in chapter 25, it's wrong for someone to assume. We can't charge a battery while discharging it to a load source because putting a load on the battery. It'll increase the load on a charging source; in which, it'll increase the load on an engine to power the charging source. Thus; the engine, it'll consume more fuel than normal because of the increase load on it as well.

So; the skeptics think, we can't charge our batteries while discharging them to a load source. The reason they think this way because they're thinking about the process based on, the traditional cycling process of our secondary cell batteries and not based on an unconventional cycling process of them. By allowing the charging voltage and current to, flow into one cell in our batteries before flowing to a load source.

Then we could increase the total energy output of the cell, the charging voltage and current has to enter in first; thus, helping to help power a load source without increasing the load on the batteries in order to keep up with the demand of the load source; therefore, without increasing the load on an engine to rotate the charging system. Given the total voltage applied to the cell by the charging source will determine its total amp capacity as well.

Here's why; the total voltage applied by the charging source, it'll be transferred to plates connecting charge and discharging terminal together in parallel. Then it'll be transferred to a load source by way of the discharging terminal. Voltage applied by the charging source, it'll increase the total amp capacity of the plates connecting the charge and discharging terminal together in parallel.

For example; if a charging source, it produces 12-volts and the plates connecting the charge and discharging terminal together in parallel has an amp potential of 125 amps. Then they would have a total amp potential of 1,500 amps; that is, 12-volts times 125 amps. Remember its voltage potential time's amp potential equals total amp capacity.

Therefore; allowing, the charging voltage and current to flow into one cell in a battery before flowing to a load source. We could increase the total energy output of the plates connecting the charge and discharging terminal together in parallel. Then an engine rotating the charging system, it doesn't have to increase its rotation in order to increase the rotation of the charging system to increase its energy output in order to keep up with the demand of the battery.

Given that; the charging voltage and current, they've to flow into one cell in the battery first; thus, increasing its energy output to help power the load source without increasing the fuel consumption of the engine as well. Let me shed more light on my supposition; so, you could understand what I've been alluding to. The amp potential of each cell in a battery will be based on the size and the number of plates in each cell as I mention earlier.

Then the total amp capacity of each cell in the battery, it'll be determine by the amount of voltage applied to it if each cell is connected in <u>series</u> with its terminal connections. Therefore; the total amp capacity of the battery, it'll be based on the voltage potential of all of its plates in each cell times the amp potential of its plates in one cell during its discharging cycle.

On the other hand; let's say that, the battery cells are connected in <u>parallel</u> with its terminal connections instead of series. Then its total amp capacity, it'll be based on the voltage potential of its plates in one cell times the amp potential of all of its plates in each cell during its discharging cycle. Thus; how, a battery cells are interconnected with its terminal connections. It'll determine its total amp capacity.

Therefore; if a charge and a discharging terminal, they're connected in parallel with one another by the same plates. Then the total amp capacity of those plates, they'll equal to their amp capacity times the voltage applied to them by a charging source. In which; it'll increase, the total energy output of those plates connecting the charge and discharging terminal together in parallel as well.

Based on, my supposition. It's conceivable we could increase the total energy output of a cell with a charge and a discharging terminal in it with the help of a charging source. Based on the mechanics behind, how we connected our secondary cell batteries plates and cells together with their terminal connections. So; it's not inconceivable or farfetched, we could increase the total energy output of a secondary cell battery with its mechanical structure as well.

Remember the total amp capacity of a lead acid automotive starting battery is based on, the size of its plates and the number of plates in each cell. And how, those cells are interconnected with its terminal connections as well. The chemical structure of the battery will determine <u>if</u> electrons could be added or subtracted from it; but, its mechanical structure will determine <u>how</u> electrons will be added or subtracted from it. See chapter 3 and 4.

If we think, it's more to it than the battery mechanical structure determining its energy input and output process. Then we're focusing on the wrong science or the wrong mechanics as well. The mechanical structure of a lead acid automotive starting battery gives every indication. It's the deciding factor in determining how its energy input and output process will be carried out; that is, one process at a time or both simultaneously.

On the other hand; if we think the laws of physics will determine, how the battery energy input and output process is carried out. Then we've failed to understand, how its mechanical structure will play a role in its energy input and output process as well. Thus; we can't begin to comprehend, why it can't be charged while being discharged to a load source as well; in other words, we don't understand what's required to carrying out such a process?

If we don't understand what's required to carry out such a process, most likely we'll blame it on the laws of physics. Why the battery, it can't be charged while being discharged to a load source as well. Working in the field of automotive battery technology, then common sense tells us. We can't charge the battery while discharging it to a load source because of the contemporary design of the battery won't allow it to be charged while being discharged.

Given that; electrons, they've to go in the same way they've to come out in order to charge the battery plates or for them to be discharged to a load source as well. Therefore; we just can't conjure up all manner of the laws of physics to justify, why we think the battery can't be charged while being discharged to a load source as well.

We just can't focus on the laws of physics; but, we've to focus on the laws of motion and the effect of forces on bodies as well. To determine, what's required to charge the battery while discharging it to the load source? Since electrons, they've to go in the same way they've to come out in order to charge the battery plates or for them to be discharged to a load source as well.

Therefore; all the blame, it doesn't lie with the laws of physics; but, some of the blame lies with the laws of motion and the effect of forces on bodies as well. We've to examine the mechanical structure of the battery to determine if it's part of the problem. Why it can't be charged while being discharged to a load source, other than blaming on the laws of physics; given that, other extenuating circumstances could be involved as well.

For instance; the connections made between, a battery and a load source in their relationship with a charging source or the battery capable of powering the load source based on its energy requirements? Likewise; is the charging source, capable of charging the battery or powering the load source as well? Those questions, we must ask ourselves in order to understand what's required to charge a battery while discharging it to a load source as well.

We just can't blame it on one science; especially, when we could charge or discharge same battery and we've to charge it on the same terminal connection that, it was discharged on. We really have to ask ourselves is it due to the laws of physics or there're other extenuating circumstances involved that, we haven't realized yet as well. See chapter 10 and 15.

On the other hand; we've to understand that, voltage and current. They'll flow through a series, a parallel or a series-parallel circuitry connection differently as well. Likewise; we've to take into consideration, a battery resistance to current is different than a load source resistance to current. Thus; a parallel circuit connection made between two load sources, it doesn't work the same as a parallel circuit connection made between a load source and a battery as well.

Since current, it'll flow to a battery differently than it'll flow to a load source because of the different circumstances involved. We've to understand those different circumstances involved if we're going to understand why current. It'll behave differently flowing to a battery than it would behave flowing to a load source. Understanding the different circumstances involved, then we could understand what's required to charge a battery while discharging it to a load source as well.

For instance; remember in chapter 12, we talked about a battery resistance to current. It'll act like a non-conducting material trying to reflect current away from it. But; a load source resistance to current, it'll be based on the number of turns that wiring has inside of it to conduct current. Thus; we find that, a parallel circuit connection made between a battery and a load source. It doesn't work the same as a parallel circuit connection made between two load sources.

When we're having a conversation about charging, a battery while discharging it to a load source those are some of the things we've to consider. However; it's not just one thing, we've to consider; but, it's a host of things we've to consider. And then, bring them together in order to understand what's required to charge a battery while discharging it to a load source as well. If not, then we might end up focusing on the wrong science as well?

For example; let's say that, we've an 18-volt, 40 amp charging system and a 12-volt lead acid automotive starting battery with an amp capacity of 120 amps. A load source requiring 12-volts and 10-amps to operate, then the question is that. How do we bring all this together, so, we could charge the battery while discharging it to the load source? First of all, we've to change the design of the battery; so, electrons could enter and exit it at the same time as well.

Once we've changed the design of the battery, so, electrons could enter and exit it at the same time. Then we've met the requirement to charge it while discharging it to the load source as well. Here's why; in order to power the load source, the battery has to be at least 12-volts to supply electrons to the load source. However; in order for the charging system to power the load source, it must produce at least 12 volts and 10 amps as well.

On the other hand; in order to charge the battery, the charging system has to produce more than 12-volts in order to reverse the current flow back into the battery to charge it. Therefore; based on, the energy requirements of the load source. The capability of the charging system and the battery, then it seems like we've everything in place to charge the battery while discharging it to the load source as well.

After putting everything in place, then we could anticipate what will happen. Based on, the energy requirements of the load source and the capability of the charging system and the battery. Here's why; the battery, it could receive up to 120 amps per cell to charge the plates in them because it's a six cell battery. Thus; eighteen volts divided by six cells, it equals to 3 volts per cell since the battery cells will be connected in series with one another by an electrical bus.

Then 120 amps will be flowing through each cell; that is, 3-volts time's 40 amps flowing through each cell coming from the charging source. If energy could enter and exit the battery at the same time, then it could be charged while being discharged to the load source. Since the load source, it requires only 120 amps to operate; that is, 12-volts times 10 amps; but, 720 amps will be flowing between them; that is, 18-volts time's 40 amps coming from the charging source.

It'll be more than enough to charge the battery and power the load source at the same time. Some people might disagree with me because they'll be focusing on the traditional cycling process of the battery. Remember when there's a charge and a load source on a battery at the same time, they'll be on the same terminal connection of it; thus, it'll be impossible to charge it while discharging it to the load source as well.

However; I'm talking about, an unconventional cycling process of the battery where the charging voltage and current has to flow into one cell in it before flowing to the load source. My supposition is based on the battery having separate terminals for charging and discharging in the same cell. And those terminals, they'll have their own straps and tab connectors leading to the same plates in that cell as well.

Therefore; having, an 18-volt, 40 amp charging system on one terminal and a load source on the other terminal connection requiring only 120 amps to operate. Then it becomes possible to charge the battery while using it to power the load source. The reasons most people are skeptical because they're looking at it from the traditional cycling process of the battery when it has a charge and a load source on the same terminal connection of it.

Wherein; the charging current, it has to flow toward the load source before its reversed back into the battery to charge it. With an unconventional design of the battery, then the load source will have access to the current potential flowing across the plates connecting the charge and discharging terminal together in parallel; in which, this makes it possible to charge the battery while using it to power the load source as well.

With this unconventional design of the battery, the amps needed to power the load source. It'll not take away from the amps needed to charge the battery; given that, the charging current has to flow into one cell in the battery before flowing to the load source. Thus; the other plates in the battery, they'll receive the same amount of current flowing from the charging source because they're connected in series with the charging terminal.

I'm going to use the electron current flow concept to show how this concept will work. When charging, a battery while using it to power a load source. In figure T-5, it shows an inside view of a pictorial diagram of a deep cycle (lead acid) automotive battery with separate terminals for charging while discharging in the same cell. And those terminals, they've their own straps and tab connectors leading to the same plates as well. See chapter 23 and 24.

Figure T-5

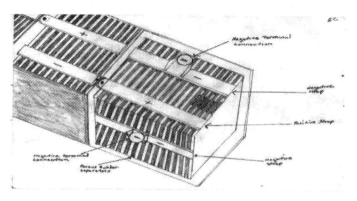

(A cell with separate terminals for charging and discharging)

Let me elaborate a little farther on the pictorial diagram of the battery in figure T-5. So; you could see, what I'm alluding to when I'm talking about charging the battery while using it to power the load source. Since the battery, it'll have separate terminals for charging and discharging in the same cell. And those terminals, they'll have their own straps and tab connectors leading to the same plates in that cell as well.

Remember the battery charge and discharging terminal, they'll be connected in parallel with one another by the same plates. Thus; the battery and the load source, they'll receive the same of voltage flowing from a charging source. Then the voltage needed to supply electrons to the load source, it'll not take away voltage needed to supply electrons to the battery. Given that; the charge and the discharging terminal, they're connected in parallel with one another.

Since the charging current has to flow into one cell in the battery before flowing to the load source, then current needed to power the load source. It'll not take away current needed to charge the battery. Since the plates connecting the charge and the discharging terminal together in parallel, they'll be connected in series with the other plates in the battery; thus, they'll receive the same amount of current flowing from the charging source as well.

With this unconventional design of the battery, then we could increase the amp potential of the plates connecting the charge and the discharging terminal together in parallel. Thus; with the help of the charging voltage, then we could power the load source and charge the battery plates at the same time. Since current, it could enter the battery on one terminal from the charging source and exit the battery on another terminal to the load source as well.

Once the charging voltage and current, they enter the cell with the charge and the discharging terminal in it. Then the laws of parallel circuitry, they begin at the strap connecting the plates together in parallel with the charging terminal. Then voltage and current will flow <u>from</u> the strap connecting the plates together. It'll be in reverse of voltage and current flow <u>to</u> the strap connecting the plates together in parallel with the discharging terminal.

Remember when the charging current flows from a strap connecting, the plates together in parallel with the charging terminal. It'll be divided amongst those plates based on their individual resistance. However; voltage flowing across the plates, it'll be the same as the voltage flowing from the charging source. The opposite happens when voltage and current flows from the plates to the strap connecting them together in parallel with the discharging terminal.

Wherein; current, it'll combine at the strap connecting the plates together in parallel with the discharging terminal. However; voltage, it'll <u>not</u> combine at the strap; but, it'll be the same as the voltage flowing across one plate. Thus; voltage and current, they'll flow into the battery and toward the load source based on the laws governing parallel circuitry. Since the charge and the discharging terminal, they'll be connected in parallel by the same plates as well.

Therefore; voltage and current, they will flow directly to the battery from the charging source; likewise, they'll flow directly from the battery to the load source. Thus; the resistance at the load source, it doesn't dictate the amount of current flowing in the battery as it does when there's a charge and a load source on the same terminal connection of a battery. In which; it's another reason, it makes it impossible to charge a battery while using it to power a load source as well

On the other hand; since the other plates in the battery, they'll be connected in series with the plates connecting the charge and the discharging terminal together in parallel. Then voltage and current will flow to the other plates in the battery based on the laws governing series circuitry. A voltage drop will occur after voltage leaves the plates connecting the charge and the discharging terminal together; thus, it'll be a voltage across the other plates in the battery.

Understanding, voltage and current will flow through the battery based on which cycling process is carried out. If the battery has separate terminals for charging and discharging in the same cell and they've their own strap and tab connectors leading to the same plates in that cell. Then we'll understand how, voltage and current will flow to and from the straps connecting the plates and the charge and the discharging terminal together in parallel will work as well.

With the unconventional design of the battery, we'll know it's possible to increase the energy output of the plates connecting the charge and the discharging terminal together in parallel as well. Wherein; the amount of voltage applied to the plates by the charging source, it'll determine their total amp capacity. Therefore; with the normal voltage output of the charging source, we could increase their total amp capacity as well.

Thus; an engine, it doesn't have to increase its rotation in order to increase the rotation of a charging system to increase its energy output. In order to, keep up with the demand of the battery for it to keep up with the demand of a load source. If the charging voltage and current could flow into one cell in the battery before flowing to the load source, then we could increase the battery energy output without increasing the energy output of the charging system as well.

Remember each cell within a lead acid automotive starting battery is nothing more than a collection of 2-volt batteries connected in series with one another and housed in a plastic case to make-up the whole battery as we know of today. In order to, increase voltage and current potential for a particular application when one battery isn't enough. I find it's no different than a battery having separate terminals for charging and discharging as well.

Remember in chapter 3, I explained how electric vehicle hobbyists had to connect their batteries together in series in order to increase voltage potential if one battery wasn't enough. They had to connect one battery positive terminal to another battery negative terminal connection. Thus; the batteries, they would be connected in series with one another.

On the other hand; in order to increase, voltage potential for a single battery. Those in the field of automotive battery technology, they've to connect the battery negative plates in one cell to its positive plates in the next cell or vice versa throughout each cell within the battery in order to increase its voltage potential for a particular application.

Thus; the battery cells, they would be connected in series with its terminal connections. However; in order to increase, amp potential if one battery amp potential wasn't enough. Then the electric vehicle hobbyists, they've to connect one battery positive terminal to another battery positive terminal. And then, connect their negative terminals together in order to increase amp potential.

Thus; the batteries, they'll be connected in parallel with one another. However; in order to, increase amp potential for a single battery. Those in the field of automotive battery technology, they've to connect all of the battery negative plates together in parallel with its negative terminal and all of its positive plates in parallel with its positive terminal to increase its amp potential.

Thus; the battery cells, they'll be connected in parallel with its terminal connections. If the electric vehicle hobbyists wanted to increase voltage and amp potential if one battery wasn't enough, then they've to connect a certain number of batteries together in series-parallel with one another in order to reach the desired voltage and amps needed.

For instance; if the electric vehicle hobbyists needed, 250 amps and 12-volts to power a vehicle; but, they only had four-six volt batteries and they were only 125 amps each. Then the electric vehicle hobbyists, they've to connect two of the batteries together in parallel. And then, connect the remaining two batteries in series with the other two batteries; that is, connected in parallel with one another to reach the desired voltage and amps needed.

In short; the electric vehicle hobbyists, they've to connect two of the batteries together in parallel to increase current potential to 250 amps and connect two of the batteries together in series with the other two batteries; that is, connected in parallel with one another. In order to, increase voltage potential to 12-volts since they've only four-six volt batteries and they're only 125 amps each.

In essence; the electric vehicle hobbyists, they connected their batteries together in series-parallel on the outside in order to increase their energy output. It's no different than those in the field of automotive battery technology. They connecting, a battery plates and cells together in series-parallel on the inside in order to increase its total energy output as well.

If we summarize, how our batteries are interconnected with one another on the outside or how their plates, cells and terminal connections are interconnected with one another on the inside. We find it's nothing more than a mechanical solution to increase their total energy output. Thinking we can't increase their total energy output with a mechanical solution, then we must be focusing on the wrong science and the wrong mechanics as well.

We usually end up focusing on the size of the batteries' plates and the number of plates in each cell to determine their total energy output. However; we fail to realize, how their plates and cells. They're interconnected with their terminal connections, are the deciding factor in determining their total energy output as well. Thus; we thinking, it's inconceivable or farfetched to increase their total energy output with a mechanical solution as well.

On the other hand; we thinking, it's inconceivable or farfetched to increase the overall travel ranges of our vehicles on battery power with a mechanical solution. Since we think putting a load on our batteries while using them to power our vehicles, it would increase the demand of a charging system; in which, it'll increase the demand of on an engine. Thus, the engine, it would consume more fuel than normal to keep up with the demand of the charging system.

I found this scenario only exist because we've to charge our batteries on the same terminal connection that, they were discharged on. If we didn't, then the scenario wouldn't exist as well. In the next chapter; let me answer, a certain question for some of my readers why those in the field of automotive and automotive battery technology? They haven't thought of a mechanical solution to increase the overall travel range of our vehicles on battery power before?

CHAPTER 27

RELUCTANT TO CHANGE THE STATUS QUO

DEEP CYCLE AUTOMOTIVE BATTERIES, they've been around for more than a century during all that time. Those in the field of automotive battery technology, they've limit the batteries' potential by designing them with only one opening for electrons to enter or to exit them; in which, it was probably okay in the mid 1800s before the advent of the electric car. Once it came on the scene, then it was time to take a different approach concerning its batteries.

Especially; when the electronic alternator regulator control system, it came on the scene in the mid 1900s. You would have thought those in the field of automotive and automotive battery technology, they would have been focusing on. The possibility to charging our deep cycle automotive batteries while using them to power our vehicles as well; given that, we could replenish our batteries' chemical energy with electrical energy as well.

I believe in the mid 1900s, it was probably too late to design our deep cycle automotive batteries; so, they could be charged while being used to power our vehicles. Given that; crude oil, it had became a vital part of our global economy for more than a century. Switching from relying heavily on the gasoline engine to relying more heavily on the electric motor, it could have had an adverse effect on our global economy as well.

This is why I believe those in the field of automotive and automotive battery technology, they were reluctant then and now to implement a plan to charge our batteries while using them to power our vehicles. It would change the status quo and may have an adverse effect on our global economy. If we rely more heavily on the electric motor than the gasoline engine to propel our basic form of transportation as well.

In the year 2010, I talk to a former boss of mine because I had worked in his plant for more than eleven years making automotive parts; plus, he had been in the business for more than fifty years as well. Thus; I thought, he could point me in the right direction when it came to my ideas of charging our secondary cell batteries while using them to power our vehicles. I told him about my ideas and he seen interested at first; but, he was somewhat skeptical as well.

He wanted to talk to some of his friends in the field of automotive battery technology to hear what they had to say. One day, a friend of my former boss sent me an e-mail about my idea. Saying it wasn't viable because the charge and the discharging cycle of a battery can't exist at the same time; given that, they're two different processes. I replied only if energy has to go in the battery the same way it has to come out as well.

I ended my line of communication with that person because it seen like. They were trying to use the contemporary design of a battery to explain why its charge and discharging cycle couldn't exist at the same time. After, my former boss heard the remarks from his friend. He told me he decided to talk too another friend of his because he wanted to have some confirmation either way because he didn't want to lose out or throwing money down a rabbit hole as well.

My former boss, other friend sent me an e-mails asking was I familiar with the laws of conservation of energy; in which, I wasn't familiar with them at that time. The person stated that; even if, energy could enter and exit a battery at the same time. It'll be no benefits in it due to the laws of conservation of energy because energy loss to heat will occur in both of its conversion processes; therefore, no gain because they'll cancel themselves out.

I couldn't respond at first because I wasn't familiar with the concept of the laws of conservation of energy. A week later that the same person e-mailed me and asked, what's the purpose of adding energy back to our batteries while talking it out? When it could be used right away to, power our vehicles. I replied adding energy back to our batteries while taking it out could increase their efficiency and overall capacity as well.

Therefore; increasing, the overall travel range of our vehicles on battery power; given that, adding electrical energy back to our batteries restore their chemical energy as well. I didn't get a reply. However; the truth of the matter, I wasn't ready to compete with my former boss friends about my idea when they started invoking the laws of conservation of energy into the conversation because I wasn't familiar with those laws of physics at the time.

However; I only understood, the mechanical aspects of charging a battery while discharging it; thus, my former boss. He became reluctant to go any further with my idea; in which, I had some doubt about it as well. Since I didn't know, how the laws of conservation of energy would play a role in charging a battery while discharging it; thus, everything went downhill from there.

Two years after I applied for a patent for a battery with four terminals for charging while discharging. The U.S. Patent Office finally contacted me in the early part of 2012 to inform me someone in Australia had already applied for a patent for a battery with four terminals for charging while discharging in the early part of 2007. I couldn't believe it; given that, I had paid a patent attorney to do a patent search in the early part of 2009. See chapter 24.

My patent attorney claimed there were no other patents on record for a battery with four terminals for charging while discharging in 2009. Before I learned about, the Australians' patent in 2012. In the latter part of 2011, I had already discovered a battery could be charged while being discharged with only three terminals as well. While learning, most in the field of automotive battery technology were reluctant to change the status quo as well.

It seems like most in the field of automotive battery technology, they were willing to misinterpret the laws of physics to justify their reluctance to change the contemporary design of our batteries; so, they could be charged while being discharged. I've heard many excuses why it'll be no benefits in it; in which, those excuses sound ridiculous coming from those in the field of automotive battery technology as well.

They should know if a battery plates in each cell, they're connected in series-parallel with one another. And each cell, it's connected in series with its terminal connections as well. Then the battery plates in one cell, they're simultaneously charged and discharged to its plates in the next cell by way of an electrical bus. So; all of its plates in each cell, they could be charged at once during their charging cycle as well.

Therefore; we don't have to stop charging, the battery plates in one cell in order to charge them in the next cell. Although; its plates in each cell, they're separated by non-conducting material called plastic. Then it seems ridiculous to assume, it'll be no benefit in simultaneously charging and discharging its plates to a load source. Since we don't have to, stop charging the battery plates in one cell in order to charge them in the next cell as well.

The only difference between a battery plates in one cell being simultaneously charged and discharged to its plates in the next cell. Compared to them being simultaneously charged and discharged to a load source, it's the direction the electrons are flowing. Wherein; electrons, they've to go in it the same way energy has to come out in order to charge the battery plates or for them to be discharged to a load source as well.

In America some in the field of automotive battery technology, they'll use the laws of physics to discredit a simultaneous cycling process of a lead acid automotive starting battery. Although; they know it goes through a simultaneous cycling process each time, it's charged. Nevertheless; they still say that, it's impossible to simultaneously charge and discharge it to a load source due to the laws of physics as well.

In Australia, some have achieved the goal of charging the battery while discharging it to a load source without the laws of physics being an issue. It seems like in America, some people will conjure up all manner of laws of physics to explain why a battery can't be charged while being discharged to a load source; although, their assumptions. They'll contradict the mechanics behind the cycling processes of the battery as well.

The mechanics behind the charging cycle of a lead acid automotive starting battery proves that. It's not necessary to stop its discharging cycle in order to start its charging cycle because of the laws of physics. If electrons, they doesn't have to go in it the same way they've to come out in order to charge its plates or for them to be discharged to a load source as well. It seems like there's a reluctance to change the status quo when it comes the cycling process of the battery.

It seems like those in the oil industry, the field of automotive and automotive battery technology. They want us to remain heavily dependent on gasoline from crude oil to power our basic form of transportation because it's more money in it. Then the question is that, are most in the field of automotive and the field of automotive battery technology in cahoots with big oil when it comes to our basic form of transportation?

Based on my research, it doesn't make any sense for those in the field of automotive and in the field of automotive battery technology to be talking about. Finding a perfect chemical composition to a longer-lasting battery before the electric vehicle, it could become our primary source of transportation. It seems disingenuous since we could replenish our batteries' chemical energy with electrical energy as well.

Wherein; it seems like, a ruse to keep us dependent on gasoline from crude oil to power our basic form of transportation. On the other hand; the claims, it's due to either the laws of entropy, inertia or the laws of conservation of energy why it'll be no benefits in adding energy back to a secondary cell battery while taking it out. Given that; energy, it'll be lost to heat in both conversion processes of the battery; therefore, no gain.

In which; the claims in and of themselves, they'll contradict the laws of entropy, inertia or the laws of conservation of energy if we're adding energy back to the battery while taking it out as well. We can't use those laws of physics to discredit the process; given that, they're referring to a closed or a perpetual motion system where energy isn't added back to the system with an outside power source.

Thus; one has to conclude that those claims, they're nothing more than disinformation to discredit the process of adding energy back to the battery while taking it out. I found there's no credence to those claims, other than they're erroneous belief handed down from one generation to the next given the mechanics behind the charging cycle of a lead acid automotive starting battery.

Remember any energy lost to heat during the charging cycle of the battery, it'll be made up by the charging source as well. Then one has to conclude that, any energy lost to heat during a simultaneous cycling process of it must be made up by the charging source as well. After all, we're adding energy back to the battery while taking it out as well. Therefore; the claims in and themselves, they're bogus information handed down from one generation to the next.

Here's why; if we attempt to, charge a lead acid automotive starting battery while discharging it. Wherein; it doesn't change the fact that, we're charging it as well. Therefore; any energy lost to heat during the process, it'll be made up by the charging source as well. Energy lost to heat in both conversion processes of the battery will occur; but, it doesn't mean they'll cancel themselves out if we attempt to charge it while discharging it as well.

We just can't focus on the facts that, energy will be lost to heat in both of conversion processes of the battery. In order to get an accurate account of what'll happen during a simultaneous cycling process of it, we've to account for the energy being added back to it if we're charging it while discharging it. Wherein; those false claims, they don't add up with the mechanics behind its charging cycle as well.

Let's focus on the mechanics behind the charging cycle of a lead acid automotive starting battery for a moment. So; you might understand, why it'll be a ruse to suggest it'll be no benefits in charging it while discharging it. In which; it might help you understand, why there's a reluctance to change the status quo when it comes to our basic form of transportation as well.

Although; the resistance of the battery plates, they cause them to consume energy during their charging cycle. See chapter 4. All that energy lost heat, it becomes irrelevant because we're still able to add energy back to them; given that, it'll be made up the charging source as well. It seems like some battery experts, they're overlooking this fact as well.

It shouldn't matter if energy is lost to heat due to the resistance of the battery plates or due to a load source consuming it during a simultaneous cycling process of the battery, it'll be made up by the charging source as well. It seems like some in the field of automotive battery technology, they're focusing on the wrong science as well.

Thinking about the assumption, it'll be no benefits in charging a lead acid automotive starting battery while discharging it because of the laws of entropy, inertia or the laws of conservation of energy as well. It seems like it's nothing more than a ruse because the assumption, it leaves out the fact energy will be added back to the battery while it's taken out as well.

Wherein; it seems like some in the field of automotive battery technology, they're using disinformation as a ruse to hide their reluctance for the status quo to change. That is for us remain heavily dependent on gasoline from crude oil to power our basic form of transportation; given that, there're another options out there as well.

If we're not accounting for the energy being added back to the battery, then we're missing the whole point about charging it while discharging it. If we focus on the mechanics behind the charging cycle of a lead acid automotive starting battery, then we'll understand the point about charging it while discharging it to a load source as well.

Here's why; remember each cell in the battery, it's nothing more than a battery in and of itself. That is connected in series with one another and housed together in a plastic case to make up the whole battery as we know of today. Understanding the mechanical make up of the battery, it's hard not to see the point of charging it while discharging it to a load source as well.

When the battery is charged as a whole, we're simultaneously charging and discharging one battery while simultaneously charging and discharging another in order to charge all the batteries connected in series with one another. Although; energy lost to heat, it'll occur during process; but, we're still able to charge each battery connected in series with one another as well.

Then one has to conclude using the laws of entropy, inertia or the laws of conservation of energy to discredit a simultaneous cycling process of the battery as a whole. It's nothing more than an excuse to hide the reluctance to change the contemporary design of our batteries; so, they could be charged while being discharged to a load source as well.

If we think about it, what's the difference between one battery consuming and storing energy while supplying it to another? Compared to, one battery consuming and storing energy while supplying energy to a load source as well. I find the only difference is mechanics because of how the battery and the load source will be interconnected with one another as well.

If energy has to go in the battery, the same way it has to come out in order to charge the battery plates or for them to be discharged to the load source. Then common sense tells us that, the wrong mechanics. They're in play to charge the battery while discharging it to the load source; thus, using the laws of physics to discredit the process is disingenuous as well.

It begs the question, are we focusing on the wrong science? When we assume, it'll be no benefits in a simultaneous cycling process of the battery because of the laws of entropy, inertia or the laws of conservation of energy? I raise this question because we could charge its plates in each cell without having to stop charging them in one cell in order to charge them in the next.

Although; the battery plates in each cell, they're separated by a non-conducting material called plastic. Wherein; the laws of entropy, inertia or the laws of conservation of energy will not be an issue. When the battery plates in one cell, they're simultaneously charged and discharged to its plates in the next cell as well.

I seems like some in the field of automotive battery technology, they're using, the laws of physics to discredit a simultaneous cycling process of the battery. As another option to increase, the overall travel ranges of a vehicle on battery power. Without finding that, perfect chemical composition to a longer-lasting battery for it as well.

It's nothing more than a campaign of disinformation to hide the reluctance to change the status quo; that is, for us to remain heavily dependent on the gasoline engine than the electric motor to propel our basic form of transportation as well. Given that; the assumption, it doesn't add up with the mechanical make up of the battery as we know of today as well.

If we stop using half of every barrel of oil; that is, pumped out of the earth for gasoline to power basic form of transportation. Then a lot of money could have been made wouldn't be made because we no longer need gasoline from crude oil to power our basic form of transportation as well.

Therefore; those in the oil industry, they'll lose half of their annual sales; in which, it's easy to see their reluctance to change the status quo. On the other hand; imagine, the amount of money those in the automotive industry will lose. If the electric vehicle, it becomes our primary source of transportation as well.

We'll no longer need parts used in the gasoline powered automobile; such as, gas tanks, fuel pumps, fuel injection systems, gasoline engines and other parts that are used for the gasoline powered automobile; but, imagine the number of jobs will be lost in the automotive industry as well.

Not only the jobs lost in the automotive industry will be a major problem; but, other jobs linked to the automotive industry in some form or fashion will be affected as well. What affect one industry, it might affect another because they're all interrelated in some form or fashion when it comes to the gasoline powered automobile as well.

So; it's easy to see, why those in the automotive and oil industry might be reluctant to change the status quo. However; it's not so easy to see, why those in the field of automotive battery technology might be reluctant to change the status quo; that is, for the electric vehicle to become our primary source of transportation as well.

Given that; revolutionizing, the electric vehicle. You would think it'll increase the sales for the deep-cycle automotive batteries; on the other hand, it'll not increase the aftermarket sales for the batteries because of their slow turnover rates; in which, they could be cycle hundreds of times as well.

In contrast to increasing the sales for the deep cycle automotive battery, it'll decrease the sales for the automotive starting batteries for those in the field of automotive battery technology. Economically; it might not be enough incentive for them for the electric vehicle to become our primary source of transportation as well.

After considering everything, it seems like making the electric vehicle our primary source of transportation. It might not be in the best interest of those in the oil industry, the field of automotive and automotive battery technology as well. In which; it might explain, why there's a reluctance to change the status quo by people who fit in those categories as well.

What affects those in the oil industry, it might affect those in the field of automotive and automotive battery technology as well. Given that; they're all interrelated in some form or fashion, when it comes to the gasoline powered automobile; so, they all might be reluctant to change the status quo as well.

Wherein; it begs question, are those in the field of automotive and automotive battery technology in cahoots with big oil? Given a lot of money from crude oil can be made; providing that, gasoline from crude oil remains our primary source of energy to power our basic form of transportation as well.

Do you really think those in the oil industry, the field of automotive and automotive battery technology trying to make the electric vehicle our primary source of transportation? I don't think so because it'll not be in their beat interest! It'll be too much money taking out of the global economy if the electric vehicle becomes our primary source of transportation as well.

It seems like those in the field of automotive and automotive battery technology, are in cahoots with big oil to delay the electric vehicle from becoming our primary source of getting around by using disinformation. About the need to find a perfect chemical composition to a longer-lasting battery before the vehicle, it becomes our primary source of getting around as well.

Likewise; it seems like those in the field of automotive battery technology, they're using disinformation to suggest. It'll be no benefits in adding energy back to a secondary cell battery while taking it out due to the laws of physics. Further supporting the idea, it'll be no benefits in changing the design of the battery; so, energy could enter and exit it at the same time as well.

This continuous campaign of disinformation by those in the oil, automotive and automotive battery technology industries, it keeps us relying more heavily on the gasoline engine than the electric motor; until, they find a more profitable way for the electric vehicle to become our primary source of transportation as well.

If we take an in depth analysis of the claim; that is, we've to find a perfect chemical composition to a longer-lasting battery before the electric vehicle could become our primary source of transportation. It doesn't make any sense because we could replenish a secondary cell battery chemical energy with electrical energy as well.

Likewise; the claim, it'll be no benefits in adding energy back to a secondary cell battery while taking it out due to either the laws of entropy, inertia or the laws of conservation of energy. It doesn't make sense because we could take energy from the battery plates in one cell; so, it could be added to its plates in the next cell without those of physics being an issue as well.

Upon analyzing those claims, I found they're flawed because one can't exist without the other. Given that; we can't add energy back to the battery while taking it out if it has to go in, the same way it has to come out. Thus; those laws of physics, they'll become a factor because we could only add or subtract energy from the battery; but, we can't do both.

Then the battery, it becomes like a closed system because you can't add energy back to it while taking energy out; given that, you're taking it out as well. If we don't change the design of the battery, so, energy could enter and exit it at the same time. Then the battery, it'll be govern by either the laws of entropy, inertia or the laws of conservation of energy as well.

If we change the design of the battery, so, energy could enter and exit it at the same time. Then neither clam will be true if we could add energy back to the battery while taking it out. Given that; neither the laws of entropy, inertia nor the laws of conservation of energy will apply because we'll be adding energy back to the battery while taking it out as well.

After analyzing, the mechanics behind the charging cycle of a lead acid automotive starting battery. It was clear it'll be a benefit in a simultaneous cycling process of it. Given that; its plates in one cell, they're simultaneously charged and discharged to its plates in the next cell. Although; energy, it'll be lost to heat during the electrolysis process of its plates as well.

I found that the battery plates are still charged; wherein; the laws of entropy, inertia or the laws of conservation of energy wasn't an issue during their charging cycle. Although; energy, it was taken from the battery plates in one cell; so, it could be added to its plates in the next cell without having to stop charging them in one cell in order to charge them in the next cell as well.

If we focus on, the mechanics behind the charging cycle of a lead acid automotive starting battery. It seems like the laws of entropy, inertia or the laws of conservation of energy. They'll have little to do with why it'll be no benefits in adding energy back to the battery while taking it out; but, it seems like it has more to do with the laws of mechanics as well.

Then the idea it'll be no benefits in charging the battery while discharging it due to the laws of physics; therefore, it'll be no benefit in changing its design as well. It seems flawed because it denies the facts that. The battery plates in one cell, they're simultaneously charged and discharged to its plates in the next cell without having to stop charging them in either cell as well.

I'm reiterating the facts that neither the laws of entropy, inertia nor the laws of conservation of energy, they'll be an issue during a simultaneous cycling process of the battery. Given that; energy, it's taken from its plates in one cell; so, it could be added to the plates in the next cell. And yet, we're still able to charge its plates in each cell as well.

How could anyone in the field of automotive battery technology come to the conclusion, it'll be no benefits in charging a lead acid automotive starting battery while discharging it to a load source due to the laws of entropy, inertia nor the laws of conservation of energy? If they understand, the mechanics behind the battery's charging cycle as well.

I found any energy lost to heat during the charging cycle of the battery, it'll be made up by a charging source as well. Then the idea it'll be no benefits in charging it while discharging it to a load source because of the laws of entropy, inertia or the laws of conservation of energy. It's nothing more than a ruse to hide the reluctance to change the status quo as well.

Wherein; the ability to add, energy back to the battery while taking it out to power a load source. It'll be up to the ability of the charging source and the connections made between it and the battery in their relationship with the load source. Then blaming it on the laws of physics, it's nothing more than a campaign of disinformation as well.

Wherein; it's intended, keep us relying more heavily on gasoline from crude oil to power our basic form of transportation. Given that; it's a ruse because the only difference between the charging the battery plates in one cell while discharging to its plates in the next cell. Compared to, charging them while discharging them to a load source is the direction energy is flowing as well.

I found energy could flow in one direction from the battery plates in one cell to its plates in the next cell; as a result, we don't have to stop charging its plates in one cell in order to charge them in the next cell. If we want to charge its plates while discharging them to a load source, then we can't due to the wrong mechanics are in play to carry out such a process as well.

If energy could flow in one direction from the battery plates to a load source, then it would be possible to charge them while discharging them to the load source. Given that; we could charge them in one cell without having to stop charging them in the next cell because energy, it could flow in one direction from the plates in one cell to the plates in the next as well.

Wherein; common sense should tell us that, we've to change the way energy has to enter and exit the battery if we're going to charge it while discharging it to a load source. Given that; energy, it has to go in the battery the same way it has to come out in order to charge its plates or for them to be discharged to a load source as well.

The idea it'll be no benefits in changing the design of the battery; so, energy could enter and exit it at the same time. It's nothing more than a campaign of disinformation because if we don't change its designs; so, energy could enter and exit it at the same time. Then we can't charge it while discharging it because we haven't changed its design in order to do so as well.

This is why it's disingenuous to use, the laws of physics to discredit a simultaneous cycling process if we haven't changed its design; so, energy could enter and exit it at the same time because it's no other way to carry out such a process; but, to change the contemporary design of the battery as well.

If it's due to the laws physics why it'll be no benefits in charging the battery while discharging it, then it's nothing we could do about it because it'll be inconsequential to changing its design. However; if it's due to the laws of mechanics, then it's something we could do about charging it while discharging it by changing its design as well.

This is why some in the field of automotive battery technology, they're telling us it's due to the laws physics why the battery can't be charged while being discharged because changing its design becomes inconsequential. There's barely any material written about, how the mechanical structure of a lead acid automotive starting battery will play a role in its cycling processes as well.

Wherein; the material that is available about the battery for the general public, it's mostly dedicated to the chemical reactions. That is take place within the battery during its cycling processes; in which, it keeps the general public in the dark about the battery as well. It seems like there's a lack of information given to the general public about the battery as well.

In which; it keeps, the general public in the dark when it comes to another option to increase the overall range of a vehicle on battery power. Without finding that, perfect chemical composition to a longer-lasting battery for it as well. Wherein; it makes you wonder, what does the general public knows about the mechanics behind our batteries' cycling processes as well.

It seems like most in the field of automotive and automotive battery technology. They don't want to keep the general public in the dark about that type of information. While wanting, the general public to believe there's no other option to increase the overall ranges of our vehicles on battery power; but, to find a perfect chemical composition to a longer-lasting battery as well.

Wherein; it doesn't make sense, given we could replenish our secondary cell batteries' chemical energy with electrical energy as well. On the other hand; those in the field of automotive and automotive battery technology, they want us to believe it's necessary to stop their discharging cycles in order to start their charging cycles because of the laws of physics as well.

In which; it doesn't make sense either, given we don't have to stop charging our batteries' plates in one cell in order to charge them in the next cell. Although; their plates in each cell, they're separated by a non-conducting material plastic called as well. The general public might not understand all the details about the cycling processes of our batteries as well.

However; the general public, they're smart enough to know. If we could replenish our batteries chemical energy with electrical energy, then why should we need to find a perfect chemical composition for them as well? Wherein; it all seems like, a campaign of disinformation to keep us dependent on gasoline from crude oil to power our basic form of transportation as well.

Here's why; I make such a claim, given a lead acid automotive starting battery plates in one cell. They're simultaneously charged and discharged to its plates in the next cell because of how they're interconnected with one another. It leaves a discrepancy between what those in field of automotive battery technology, they say and what we could actually do with the battery as well.

The mechanics behind the charging cycle of a lead acid automotive starting battery. It raises a red flag about the validly of what those in field of automotive battery technology. They're saying about what we could or couldn't do with the battery because of the laws of physics; wherein, it seems like it has been a serious miscalculation made by them as well.

Remember in chapter 3, we examine a lead acid automotive starting battery internal structure to see why it functions the way it does. Its plates in each cell, they're connected in series-parallel with one another by their straps, tab connectors and the electrolyte solution in each cell. And each cell, it's connected in series with its terminal connections as well.

In which; it allows, the charging electrons to flow in one direction from its plates in one cell to its plates in the next cell. Wherein; electrons, they could flow from a strap to a tab connector on one plate. And then, flow across that plate through the electrolyte solution to an adjacent plate. Without flowing to and from, the same tab connector on the same plate as well.

Thus; fixed electrons on one plate, they could reflect free electrons coming from the charging source toward an adjacent plate by way of the electrolyte solution in each cell. Since each cell is connected in series by way of an electrical bus, then fixed electrons on the plates in one cell could reflect free electrons coming from charging source toward the plates in the next cell as well.

Wherein; we don't have to, stop charging the battery plates in one cell in order to charge them in the next cell because of how they're interconnected with its terminal connections. As a result; the battery plates in each cell, they're simultaneously charged and discharged amongst themselves and to the plates in the next cell as well.

So; all the battery plates in each cell, they could be charged at once during their charging cycle. Then talk about we've to stop its discharging cycle in order to start its charging cycle because of the laws of physics. It doesn't add up with the mechanics behind its charging cycle. Are we focusing on, the right science when it comes to its energy input and output process?

Based on, the mechanics behind the charging cycle of the battery. It seems like it has been a serious miscalculation made by some in the field of automotive battery technology. To assume, we've to stop its discharging cycle in order to start its charging cycle because of the laws of physics; maybe, it's just a campaign of disinformation orchestrated by big oil as well.

No matter how we slice it, line upon line or precept up on precept. It'll always come down to the design of the battery when it comes to its energy input and output process. Likewise; energy lost to heat in both of its conversion processes, it has no bearing on whether a simultaneous cycling process of it will be beneficial or not as well.

All the assumptions about what we could or couldn't do with our secondary cell batteries because of the laws of physics. They're nothing more than ruses intended to hide the reluctance to change the status quo. In the next chapter; I'm hoping to shed some light on, why those ruses exist because of the technology already on the self to decrease or eliminate our fuel consumption altogether.

CHAPTER 28

TECHNOLOGY ALREADY ON THE SHELF

IF WE LOOK BACK at the history of the electric vehicle in a nut shell, we could draw a valid conclusion. Why some or most in the field of automotive and automotive battery technology, they're in cahoots with big oil because of the reluctance to change the status quo. It shouldn't be any surprise because they all interrelated in some form or fashion when it comes to our automobiles as well.

Gaston Plante' made the first lead acid automotive battery in 1859 and then came the discovery of crude oil around that same time. In the late 1890s, the first electric car came on the scene. However; in the early 1900s, the first gasoline powered automobile came on the scene. The idea of the electric car becoming our primary source of transportation was swept under the rug.

Until; the mid 1990s, then General Motors started experimenting with the EV1 or the all-electric vehicle. For some reason or another, the EV1 practically disappeared overnight and then came the hybrid vehicles in the late 1990s. They've two propulsion systems; a gasoline engine and an electric motor to propel them as well.

If you notice the engineering concept of the hybrid vehicles, they weren't designed to decrease our fuel consumption; but, to keep us dependent on gasoline from crude oil. Although; the hybrid vehicles, they've two propulsion systems. However; it doesn't necessarily mean, they'll decrease our fuel consumption; but, more likely it'll increase it.

Remember hybrid vehicles, they save on fuel when they're idling or driven around town on battery power. When the time comes to charging their batteries on the go, it'll take away all those savings because their gasoline engines will burn more fuel than normal to turn their drive train and motor generator systems at the same time as well.

We'll find hybrid vehicles are nothing more than a ruse to cover up our fuel consumption; in which, they're just another stall tactic used by big oil and those who are in cahoots with them. To keep us, dependent on gasoline from crude to power our basic form of transportation. In hope, it'll keep the general public from complaining about high gas prices.

However; in the mean time, big oil and those who are in cahoots with them will keep on raking in billions of dollars each year from our fuel consumption. Nothing will change if we buy a hybrid vehicle because we still have to rely more heavily on its gasoline engine than its electric motor to propel it as well.

No doubt in the late 1990s, big oil and those who are in cahoots with them came up with the idea of the hybrid vehicles to get rid of the all-electric vehicle in the mid 1990s. Not only did big oil and those who are in cahoots with them managed to keep us dependent on gasoline from crude oil to power our basic form of transportation as well

However; big oil and those who are in cahoots with them, they managed to increase our fuel consumption as well. Likewise; without a ton of evidence, we could draw a valid conclusion why the all-electric vehicle hasn't become our primary source of transportation for more than a century as well. Given that; a lot of money, it could be made from gasoline for crude oil as well.

Here's the thing those in the oil industry, the field of automotive and automotive battery technology. They're all interrelated in some form or fashion when it comes to the gasoline powered automobile. What effects one, it might affect them all; so, they all might be reluctant to change the status quo as well.

Thinking about the electric vehicle in terms of its job creation, it wouldn't have product the volume of jobs as the gasoline powered automobile has done for more than a century as well. Likewise; switching from, the gasoline powered automobile to the all-electric vehicle. After crude oil, it has become a vital part of our global economy for more than a century as well.

It might have an adverse effect on our global economy for years to come since revolutionizing the electric vehicle. It would eliminate many jobs associated with the production of the gasoline powered automobile and some jobs indirectly related as well. Thus; a lot of people, other than those who work in the oil and the automotive industry might be out of a job as well.

On the other hand; those in the field of automotive battery technology, they keep telling us. They don't have the right battery technology on the shelf to make the electric vehicle our primary source of transportation. It seems like they've been telling us this same old story for more than a century; in which, it begins to sound like a stall tactic as well.

Given that; we could replenish, our secondary cell batteries' chemical energy with electrical energy. Thus; it seems like, we already have the right battery technology on the shelf to make the electric vehicle our primary source of transportation. But; it seems like those in the field of automotive battery technology, are reluctant to utilize it in the right manner as well.

I believe this is the case if you're old enough to remember when the first portable transistor radios came on the scene in the mid 1950s or the first boom boxes came on the scene in the early 1970s. The general public didn't have rechargeable batteries to put in their portable transistor radios in the mid 1950s or their boom boxes in the early 1970s as well.

Although; the first rechargeable battery, it was made almost a hundred years before the first transistor radio and more than a hundred years before the first boom box came on the scene as well. And yet, we didn't have rechargeable batteries to put in our transistor radios in the mid 1950s or our boom boxes in the early 1970s as well.

Do you know what a boom box is? It's a portable stereo system with an AM & FM radio, a cassette player and ten inch speakers. However; once people used up, their non-rechargeable batteries back in the mid 1950s or the early 1970s. Then they had to buy more batteries if they wanted to keep using their transistor radios or their boom boxes during those time periods.

It makes me wonder why the general public didn't have rechargeable batteries to put in their transistor radios in mid 1950s or their boom boxes in the early 1970s? Since technology for rechargeable batteries, it was on the shelf since 1859. Is battery technology more than a century behind other technologies; that is, developed for the general public?

Unless; those in the field of battery technology during the mid 1950s and the early 1970s, they weren't ready for rechargeable batteries to be available for the general public. Given that; they could make, more money by selling non-rechargeable batteries compared to selling rechargeable batteries during those time periods as well.

Wherein; it's easy to explain, why the general public didn't have rechargeable batteries to put in their transistor radios in the mid 1950s or their boom boxes in the early 1970s as well. On the other hand; the first electric car, it was made in the late 1800s. And then, the first gasoline powered automobile in the early 1900s; so, what happen the electric car?

Almost seventy years after the first gasoline powered automobile was made, humankind went to the moon. About fifty years after the first moon landing, we're still no closer in making the electric vehicle our primary source of transportation than it was in the late 1800s. Today; those in the field of automotive battery technology still say that, they don't have right battery technology.

What so ironic about us going to the moon was that, the first vehicle on the moon was an electric car in the early 1970s; in which, it was called the moon Rover. It makes you wonder what's going on with the electric car here on earth today since we had technology back in the early 1970s to build and send one to the moon; but, not to drive one around here on earth.

The technology we've today must be more advanced and sophisticated than it was back in the early 1970s. So; it should raise serious questions about, why the electric car hasn't become our primary source of transportation today. The idea we've to find a perfect chemical composition before it could become our primary source of transportation.

It seems like it's nothing more than a ruse created by those in oil industry and those who are in cahoots with them to keep the general public from putting political pressure on the automotive and automotive battery technology industries. To produce, a battery would allow the electric vehicle to become our primary source of transportation as well.

In which; the ruse, it's designed to keep us dependent on gasoline from crude oil to power our basic form of transportation; given that, we could replenish our secondary cell batteries' chemical energy with electrical energy as well. So, why we need to find a perfect chemical composition to a longer-lasting battery, it doesn't make sense as well.

Wherein; this ruse about finding a perfect chemical composition to a longer-lasting battery, it'll allow big oil and those who are in cahoots with them to keep on raking in billions of dollars each year. In which; big oil agenda for us, it's for us to remain heavily dependent on gasoline from crude oil to power our basic form of transportation as well.

We probably will never get the whole story why the electric vehicle, it hasn't become our primary source of transportation. In the mean time; those in the field of automotive battery technology, they'll keep on misleading us about finding a perfect chemical composition to a longer-lasting battery as well.

If we look at all the circumstantial evidence, we'll find the truth why the electric vehicle hasn't become our primary source of transportation for more than a century. It's not because of the lack of battery technology; but, it's to maintain the status quo. That is for us to remain dependent on gasoline from crude oil because it's more money in it and jobs as well.

If the electric vehicle, it had become our primary source of transportation more than a century ago. It wouldn't have produce the volume of jobs as the gasoline powered automobile has done for more than a century. It's real obvious why gasoline from crude oil should remain our primary source of energy to power our basic form of transportation because it's more jobs in it.

The idea about finding a chemical composition with a longer-lasting chemical reaction, it had to be implemented by big oil and those who are in cahoots with them as a stall tactic to delay the vehicle electric from becoming our primary source of transportation; so that, they could continue to rake in billions of dollars each year from our fuel consumption as well.

For instance; back in the late 1970s or the early 1980s, it was rumored that a person had invented a carburetor would allow an automobile to get a hundred miles on a single gallon of gasoline. Remember a carburetor is a device for an internal combustion engine that mixes liquid fuel and air in the correct proportions that vaporizes them and transfer them to cylinders to turn the engine.

Today; we've fuel injection systems that, mixes liquid fuel and air in the correct proportions and then transfer them to cylinders to turn the engine. However; the rumor was that, big oil bought the rights to the carburetor to make sure it'll never materialize in our gas guzzling automobiles; given that, it'll take money out of their pockets.

It seems like we already have the right technology on the shelf to decrease our fuel consumption; but, it's not available for use. This is why, I believe some or most in the field automotive battery technology. They're in cahoots with big oil because they haven't changed the contemporary design of our batteries; so, energy could enter and exit them at the same time as well.

It seems like there's a joint effort by big oil and those who are in cahoots with them to keep us dependent on gasoline from crude oil to power our basic form of transportation. I'm making this claim because we could charge or discharge the same battery. However; we've to charge it on the same terminal connection that, it was discharged on as well.

Wherein; the design concept of our secondary cell batteries, they're designed for us to stop their discharging cycles in order to start their charging cycles. It has nothing to do with the laws of physics as most in the field of automotive battery technology claims. However; it has more to do with, how they design our secondary cell batteries than anything else.

On the other hand; some in the field of automotive battery technology, they want to claim. It'll be no benefits in charging our batteries while discharging them; even if, energy could enter and exit them at the same time because of the laws of physics. This claim doesn't add up with the mechanics behind the charging cycle of a lead acid automotive starting battery as well.

Given that; the battery plates in one cell, they're simultaneously charged and discharged to its plates in the next cell; although, they're separated by a non-conducting material called plastic. Basically; we're taking energy from the plates in one cell, so, it could be added to the plates in the next cell. And yet, we're still able to charge them in each cell as well.

It seems like we already have the right technology on shelf to charge our secondary cell batteries while using them to power our vehicles. In order to, increase their overall travel ranges on battery power without finding a perfect chemical composition with a longer-lasting chemical reaction for their batteries as well.

Here's why; we could, use the same technique that is used to simultaneously charge and discharge our batteries' plates in one cell to their plates in the next cell; although, they're separated by a non-conducting material called plastic. We could use the technique to charge our batteries while using them to power vehicles as well.

Remember a lead acid automotive starting battery plates in each cell, they're connected in series-parallel with one another by their straps, tab connectors and the electrolyte solution in each cell. Thus; the charging electrons, they could flow from a strap to a tab connector on one plate. And then, flow across that plate through the electrolyte solution to an adjacent plate.

Using the electrolyte solution in each cell, then electrons don't have to flow to and from the same tab connectors on the same plate in order to flow to and from the same plate. Therefore; fixed electrons on one plate, they could reflect free electrons coming from the charging source toward an adjacent plate by way of the electrolyte solution in each cell as well.

As a result; the battery plates in one cell, they're simultaneously charged and discharged amongst themselves in each cell. Since each cell is connected in series with one another by way of an electrical bus, then the plates in one cell. They're simultaneously charged and discharged to its plates in the next cell by way of the electrical bus as well.

Thus; we don't have to stop charging, the battery plates in one cell in order to charge them in the next cell; so, we already have the right technology on the shelf to charge our batteries while using them to power our vehicles to increase their overall travel ranges on battery power as well. We just have to utilize our battery technology in the right manner.

The mechanics behind the charging cycle of a lead acid automotive starting battery tells us that. We don't have to stop its discharge cycle in order to start its charging cycle if we've the right mechanics, are in place. It casts doubt on the claim we don't have the right technology on the shelf to increase the overall travel ranges of our vehicles on battery power as well.

If electrons, they could flow in one direction from our batteries' plates to a load source. Then we could charge their plates while using them to power the load source to increase their efficiency and overall capacity. Without finding that, perfect chemical composition with a longer-lasting chemical reaction because adding electrical energy back to them restores their chemical energy as well.

We already have the right technology on the shelf; given that, we could charge or discharging the same battery and replenish its chemical energy with electrical energy as well. We just haven't utilized our battery technology in the right manner to increase the efficiency and overall capacity of our batteries with a mechanical rather than a chemical solution.

Do you really think, we don't already have the right battery technology on the shelf? For us to rely, more heavily on the electric motor than the gasoline engine to propel our basic source of transportation? Those in the field of automotive battery technology, they want us to believe we don't because they want to keep talking about finding a perfect chemical composition.

Since we could, simultaneously charge and discharge our secondary cell batteries' plates in one cell to their plates in the next cell to replenish their chemical energy with electrical energy during their charging cycles. Then we've the right technology on the shelf to increase the overall travel ranges of our vehicles on battery power as well.

Here's why; the only difference between, simultaneously charging and discharging our batteries' plates to a load source while replenishing their chemical energy is mechanics. Given that; energy, it could flow in one direction from our batteries' plates in one cell to their plates in the next cell in order to charge them in each cell as well.

However; electrons, they can't flow in one direction in order to charge a secondary cell battery plates while discharging them to a load source. Given that; electrons, they've to go in the same way they've to come out in order to charge the battery plates or for them to be discharged to the load source. It's a mechanical issue and not an issue concerning the laws of physics.

Some people want us to believe, it's a matter concerning the laws of physics why we've to stop the discharging cycle of a battery in order to start its charging cycle. It's obvious that, it's a matter of mechanics and not a matter of physics because electrons have go in the battery the same way they've to come out in order to charge its plates or for them to be discharged as well.

For example; analyzing, how a series-parallel circuit connection will between a battery plates in each cell and how a series circuit connection will work between each cell. Then we'll find a series-parallel and a series circuit connection. They'll allow the battery plates to be charged in each cell without having to stop charging them in one cell in order to charge them in the next.

I found it's an ingenious solution to get around the problems that, the mechanics of the current pose. If we're trying to the charging, a battery plates in each cell without having to stop charging them in one cell in order to charge them in the next; although, they're separated by a non-conducting material called plastic. Its old technology; but, it's very affective as well.

It's technology that is already on the shelf to help us understand, how to charge our batteries while using them to power our vehicles. To increase, their overall travel ranges on battery power. Without finding that, perfect chemical composition with a longer-lasting chemical reaction for their batteries while increasing their efficiency and overall capacity as well.

Understanding the old technology for our secondary cell batteries, then we'll realize that. We already have the right technology on the shelf, so that, we don't have to stop their discharging cycles in order to start their charging cycles because of the problems that, the mechanics of the current will pose if we attempt to charge them while discharging them as well.

Carrying out experiments on a lead acid automotive starting battery, I found it's not design to conform to the mechanics of the voltage and current; but, they conform to its design. Remember its plates in each cell, are connected in series-parallel with one another. And each cell, it's connected in series with its terminal connections as well.

During the battery discharging cycle, its voltage potential is the sum of all of its plates in each cell; but, its current potential will only equal to the current potential of its plates in one cell. However; during its charging cycle, the charging voltage will be evenly divided amongst its cells; but, the charging current will be the same throughout each cell in it.

Thus; the mechanics of the voltage and current, they'll conform to the design of the battery. Then why those in the field of automotive battery technology will suggest that, the laws of physics are the deciding factor in determining how our batteries' cycling processes will carry out; that is, one process at a time or both processes simultaneously as well.

I find this suggestion is nothing more than a ruse to hide the battery technology already on the shelf to increase the efficiency and overall capacity of our secondary cell batteries. With a mechanical rather than a chemical solution without finding that, perfect chemical composition with a longer-lasting chemical reaction for them as well.

Remember each cell in a lead acid automotive starting battery is nothing more than a battery in and of itself. That is connected in series with one another and housed together in a plastic case to make up the whole battery as we know of today. When it's charged as a whole, we're charging and discharging one battery while charging and discharging another as well.

Thus; our inability to charge our secondary cell batteries while discharging them to a load source, it has little to do with the laws of physics in and of themselves; but, it has more to do with the process itself or how we go about adding and subtracting electrons from our batteries; in which, it determines how their cycling processes will be carried out as well.

You might think I'm crazy or a conspiracy theorist and you might think, I don't know what I'm talking about as well. Nevertheless; it doesn't make any sense to assume, we can't increase the overall travel ranges of our vehicles on battery power. Without finding that, perfect chemical composition with a longer-lasting chemical reaction for their batteries as well.

Likewise; it doesn't make any sense to assume, we've to find a perfect chemical composition with a longer-lasting chemical reaction before we could depend more heavily on the electric motor than the gasoline engine to propel our vehicles. Given that; we could replenish, our secondary cell batteries' chemical energy with electrical energy as well.

The assumption we need to find a perfect chemical composition with a longer-lasting chemical reaction for our secondary cell battery. Before we could, depend more heavily on the electric motor than the gasoline engine to propel our automobiles. It seems like nothing more than a campaign of disinformation to keep us more dependent on gasoline from crude oil as well.

The ability to replenish our secondary cell batteries' chemical energy with electrical energy, it proves there's a mechanical rather than a chemical solution to increase their efficiency and overall capacity. Without finding that, perfect chemical composition with a longer-lasting chemical reaction for them as well.

If we use the technology already on the shelf for our batteries, then we could increase their efficiency and overall capacity by adding energy back to them while taking it out as well. I'm not talking about anything outside the realm of physic; but, I'm talking about the process use to simultaneously charge and discharge their plates in one cell to them in the next cell as well.

Adopting this process, we don't have to stop our batteries' discharging cycles in order to start their charging cycles. Given that; we'll not be using, their chemical electromotive force to discharge them while charging them. However; we'll be using, an electromotive force created by a charging system to discharge them while charging them as well.

We already have the technology on the shelf to increase the overall travel ranges of our vehicles on battery power. Without finding that, perfect chemical composition with a longer-lasting chemical reaction for their batteries. All we've to do is change their contemporary designs, so, energy could enter and exit them at the same time as well.

We've to overlook the assumptions, it'll be impossible to charge our batteries while discharging them and it'll be no benefits in it; even if, energy could enter and exit them at the same time because of the laws of physics. It's nothing more than a campaign of disinformation to hide the battery technology already on the shelf that could decrease our fuel consumption.

If energy could enter a battery on one terminal and exit it on another, then we wouldn't need to stop its discharging cycle in order to start its charging cycle. The idea it'll be no benefits in it because energy lost to heat occurs in both conversion processes become flawed. Given that; we only using, a charging source to add and subtract energy from the battery as well.

Then any energy lost to heat during the process, it'll be made up by the charging source. Upon carrying out a number of experiments on the charging cycle of a lead acid automotive starting battery, I found the amount of energy lost to heat during its charging cycle. It doesn't determine if it's charging or not, it just determine if it's getting hot or not during its charging cycle.

We've to overlook those assumptions, it'll be impossible to charge our batteries while discharging them and it'll be no benefits in it; even if, energy could enter and exit them at the same time as well. Wherein; these assumptions, they're nothing more than erroneous beliefs handed down from one generation to the next to hide the technology already on the shelf.

Those in the field of automotive battery technology, they've failed to utilize our battery technology already on the shelf in the right manner to increase the overall travel ranges of our vehicles on battery power. Without finding that, perfect chemical composition with a longer-lasting chemical reaction for their batteries because of a reluctance to change the status quo.

It's the same reason gasoline is our primary energy source to power our basic form of transportation because those in the field of automotive battery technology. They've failed to design our deep cycle batteries to facilitate, the need they're needed for when it comes to relying more heavily on the electric motor than the gasoline engine as well.

Like I said before, you might think I'm crazy or a conspiracy theorist. And you might think, I don't know what I'm talking about as well. You don't need rocket science or regular science, you only need common sense to figure out why gasoline from crude oil remain our primary source of energy to power our basic form of transportation for more than a century.

We already have the right technology on the shelf to increase the overall travel ranges of our vehicles on battery power. Without finding that, perfect chemical composition with a longer lasting chemical reaction for their batteries. All we've to do is utilize our battery technology in the right manner to depend more on the electric motor than the gasoline engine.

However; it seems like, it's a deliberate attempt by big oil and those who are in cahoots with them to keep us dependent on gasoline from crude oil to power our basic form of transportation. Have you seen the movie "On deadly ground" with Steven Seagal and Michael Caine? Steven Seagal had to stop Michael Caine's oil company from taking land from the Inuit tribe in Alaska.

Michael Caine's oil company had temporary drilling rights to the land; but, it was for a limited time period. In which; Michael Caine's oil company, they had to prove it was profitable to drill for oil on the land before they could get permanent drilling rights. Steven Seagal worked for Michael Caine's oil company; but, he decided to help the Inuit tribe to keep their land.

However; at the end of the movie, Steven Seagal talked about many different technologies that is already on the shelf to help us decrease or eliminate altogether our dependence on gasoline from crude oil to power our basic form of transportation. He indicated those types of technologies, are kept from the general public because of money to be made from oil.

I recommend you to watch the end of the movie because it'll give some insight into why, gasoline from crude oil remains our primary source of energy to power our basic form of transportation; although, we've alternative sources of energy. It's not being used because of profits to be made from gasoline from crude oil to power our basic form of transportation.

There're many different web sites showing, a variety of energy sources capable of powering our basic form of transportation. One web site in particular is www.panacea-bocaf.org. In which; Ismael Aviso of the Philippines, he has developed a self charging battery system for the electric vehicle called repelling force technology.

Also; the Gyro kinetic engine or the Papp engine, which Bob and Tom Rohner are trying to restore its lost technology developed by Joseph Papp whom Bob and Tom Rohner worked with on the engine. Joseph Papp didn't reveal the combination of the inert gases used to power the Papp engine before his death; but, go to Bob@RGEneryg.com for additional information.

These are some of the technologies; that is, already on the shelf that could be used to decrease or eliminate altogether our gasoline consumption from crude oil. Our secondary cell batteries, they're old technology that is already on the shelf that could decrease our fuel consumption if we utilize them in the right manner as well.

Remember if a secondary cell battery cells, they're connected in series with its terminal connections. Then its total amp capacity during its discharging cycle, it'll only equal to the amp potential of its plates in one cell times the voltage potential of all of its plates in each cell. Now if the battery cells, they're connected in parallel with its terminal connections.

Then its total amp capacity during its discharging cycle, it'll equal to all of its plates in each cell times the voltage potential of its plates in one cell. Thus; the total, energy output of a secondary cell battery has nothing to do with the laws of physics in and of themselves; but; it has more to do with the design of the battery than anything else.

Understanding how a secondary cell battery cells, they're interconnected with its terminal connections tell us. There's a mechanical rather than a chemical solution to increase its efficiency and overall capacity if we could add energy back to it while taking energy out as well. Then it could be a mechanical solution to increase overall travel range of a vehicle on battery power as well.

Given that; voltage and current, they'll flow in and out of a secondary cell battery based on its design. Then its energy input and output process, they're a product of its design. If we make the right changes to it, so, it'll allow its charging cycle to transition into its discharging cycle. Then we could increase its total amp capacity with the help of a charge source as well.

It's nothing more than a mechanical solution to increase the efficiency and overall capacity of the battery; as long as, we could replenish its chemical energy with electrical energy. Then we already have the right battery technology on the shelf to increase the overall travel range of a vehicle on battery power if we use our technology in the right manner as well.

What's strange about the idea of finding a perfect chemical composition to a longer-lasting battery, it's not that we haven't found one yet. It's the fact that we could replenish a secondary cell battery chemical energy with electrical energy. Not only should we be looking for a perfect chemical composition for it; but, we should be looking for a batter way to cycle it as well.

Most people think if it was possible to add energy back to a secondary cell battery while taking it out, then why those in the field of automotive battery technology haven't thought of this before. It's not because they haven't; but, it seems like they're trying to sweep it under the rug. Some battery experts in Australia, they've perused this idea for awhile and developed it as well. See chapter 24.

In America it seems like most battery experts, they're not interested in adding energy back to a secondary cell battery while taking it out; likewise, they can't imagine the benefits in it as well. While some battery experts in Australia, they understand the benefits in adding energy back to a secondary cell battery while taking it out as well.

In America some battery experts, they keep on saying it's impossible to add energy back to a secondary cell battery while taking it out power a load source due to the laws of physics.

FRANK EARL

However; some battery experts in Australia, they've already achieve this goal without the laws of physics being an issue as well.

In America it seems like some if not, most in the field of automotive battery technology. They've aligned themselves with big oil and those who are in cahoots with them to maintain the status quo; that is, for us to remain depend on gasoline from crude oil to power our basic form of transportation as well.

Some folks, they're reluctant to give up crude oil because it has become a vital part of our global economy for more than a century. They think if we switching from relying heavy on the gasoline engine to relying more heavily on the electric motor to propel our basic form of transportation. It would have an adverse effect on our global economy as well.

It might explain why most businesses directly or indirectly related to the production of the gasoline powered automobile. They're not in a big rush for the electric vehicle to become our primary source of transportation since it wouldn't produce the volume of jobs as the gasoline powered automobile has done for more than a century as well.

After it's all said and done, I found it has little to do with the lack of battery technology why the electric vehicle hasn't become our primary source of transportation for more than a century. It has more to do with the reluctance to change the status quo if you need more proof, compare the parts needed to build a gasoline engine to the parts needed to build an electric motor.

One vehicle might create fewer jobs than the other, so, switching from a gasoline powered automobile to an all-electric vehicle might have an adverse effect on our global economy. It might explain why those in field of automotive and automotive battery technology, they haven't been in a big rush to use the technology already on the shelf as well.

Relying heavily on the electric motor than the gasoline engine to propel our basic form of transportation, it might be a risky proposition since the gasoline powered automobile has shaped our global economy for more than a century; thus, we find ourselves in a dilemma for more than a century as well.

FINAL THOUGHTS

When I was a teenager in the early 1970s, I watched futuristic television series; such as, Space 1999, Star-Trek and a few others as well. In some of those futuristic television series, our basic form of transportation hovered above the ground as we travelled; also, we had space stations on the moon and other planets as well.

In the early 1970s, I would have never imagined that. We would still be using gasoline from crude oil to power our basic form of transportation in the twenty-first century as well. Remember all the rocket engines that, NASA used to go to the moon or go into space during the twentieth century. They were one way engines because they couldn't bring us back to earth.

In our science fiction movies, our rocket engines could take us into space and bring us back to earth; although, it was science fiction at it bests. However; in the twentieth-first century, we've figured out how to build rocket engines that could take us into space and bring us back to earth as well. Although; it took, more than fifty years to figure it out; but, we figured it out.

I found there was a desire to build rocket engines that could take us into space and bring us back to earth as well. There's no desire to rid ourselves of gasoline from crude oil to power our basic form of transportation. It's a lot of profit to be made from gasoline from crude oil; so, the gasoline powered automobile lives on in the twenty-first century as well.

I found we haven't utilized our technology in the right manner to decrease or eliminate our gasoline consumption from crude oil; so, we continue to spew poisonous gases into our atmosphere on a daily bases. There's no desire to figure out, how to add energy back to our batteries while taking it out as an option to increase their efficiency and overall capacity as well.

Most in the field of automotive battery technology, they think adding energy back to a battery while taking it out is impossible due to the laws of physics. If anyone thinks otherwise, it's nothing more than wishful thinking; although, we could charge its plates in each cell without having to stop charging them in one cell in order to charge them in the next as well.

Although; adding, electrical energy back to our batteries restores their chemical energy. It's amazing no one wants to expand on this old technology as another option to increase the overall travel range of a vehicle on battery power. Without finding that, perfect chemical composition with a longer-lasting chemical reaction for its batteries as well.

How difficult it would be for those in the field of automotive battery technology to devise away to add, electrical energy back to our batteries while taking it out? Given that; we don't have to stop charging their plates in one cell in order to charge them in the next cell; although, they're separated by a non-conducting material called plastic as well.

Then charging our batteries while using them to power a load source, it shouldn't be any different as well. In the twenty-first century, it has been a big push by those in the field of automotive and automotive battery technology to revolutionize the electric vehicle. But; one question remains, how do we replenish our batteries' chemical energy while on the go?

Find a charging station that may or may not exist in most cities and if so, they're few and far in between. Those who are advocating to get rid of gasoline from crude oil, they might find it's impossible to have their cake and eat it too. Maybe; this is what big oil intended, when the hybrid vehicles came on the scene in the late 1990s to have their cake and eat it too.

So far, big oil and those who are in cahoots with them, their plan hasn't worked because we haven't utilized our battery technology in the right manner to achieve such a goal. However; if energy could enter and exit our batteries at the same time, then we could increase the overall travel ranges of our hybrid vehicles on battery power as well.

Given that; we could rely, more heavily on the vehicles' electric motors than their gasoline engines to propel them; thus, decreasing their fuel consumption while saving our planet in the process as well. New battery technology doesn't always come in the form a chemical composition; but, it could come in the form of changing the design of a battery as well.

SOURCES OF RESEARCH MATERIALS

I've complied and compared data from the following books to come to a logical conclusion. It's scientifically possible to carry out a simultaneous cycling process of our secondary cell batteries. Providing that; we change their contemporary designs, so, energy could enter and exit them at the same time.

Using the knowledge, I acquired in the field of heating, cooling and refrigeration about the mechanics of the voltage and current. Then I applied that knowledge to the cycling processes of a lead acid automotive starting battery to devise away to design it; so, it could be charged while being discharged as well.

The Complete Battery Book by: Richard A. Perez. 1985 Edition
Electronic Circuit Fundamentals by: Walter J. Weir. 1987 Edition
Build Your Own Electric Vehicle by: Seth Leitman and Bob Brant. 1994 Edition
Fundamentals of Direct Current by: Robert E. Armstrong. 1986 Edition
Electricity (for refrigeration and heating and AC) by: Russell E. Smith: 2007 Edition.

Printed in the United States
by Baker & Taylor Publisher Services